Spikes, decisions, and actions

The dynamical foundations of neuroscience

Spikes, decisions, and actions
The dynamical foundations
of neuroscience

HUGH R. WILSON

Visual Sciences Center, University of Chicago

OXFORD

UNIVERSITY PRESS

OXFORD
UNIVERSITY PRESS

Great Clarendon Street, Oxford OX2 6DP
Oxford University Press is a department of the University of Oxford.
It furthers the University's objective of excellence in research, scholarship,
and education by publishing worldwide in

Oxford New York

Auckland Cape Town Dar es Salaam Hong Kong Karachi
Kuala Lumpur Madrid Melbourne Mexico City Nairobi
New Delhi Shanghai Taipei Toronto

With offices in

Argentina Austria Brazil Chile Czech Republic France Greece
Guatemala Hungary Italy Japan Poland Portugal Singapore
South Korea Switzerland Thailand Turkey Ukraine Vietnam

Oxford is a registered trade mark of Oxford University Press
in the UK and in certain other countries

Published in the United States
by Oxford University Press Inc., New York

First published 1999   1004898164

Reprinted 2002, 2003, 2005

British Library Cataloguing in Publication Data
Data available

Library of Congress Cataloging in Publication Data

ISBN 0-19-852431-5 (Hbk)
ISBN 0-19-852430-7 (Pbk)

Printed in Great Britain by
Antony Rowe Ltd., Chippenham, Wiltshire

To Fran, whose loving support
and intellectual companionship
made this book possible

Preface

There is one central reason why all neuroscientists and biopsychologists should be conversant with nonlinear dynamics. Nonlinear dynamics reveals and elucidates a range of phenomena that are simply inconceivable in the more mundane world of linear systems theory. Memory and forgetting, decision making, motor control, action potentials, and perhaps even free will and determinism can no longer be intelligently conceptualized without a basic understanding of nonlinear systems. At the deepest level, therefore, my decision to write this book was predicated on the belief that an understanding of brain function and behavior must be grounded in a conceptual framework informed by nonlinear dynamics.

When I received my Ph.D. in theoretical chemistry in 1969 I was not familiar with many of the nonlinear dynamical phenomena developed in this book, and others such as chaos were just being discovered and understood by mathematicians. Theoretical chemistry and physics of the time did deal with nonlinearities, but the general approach was to hope that they were small and then expand solutions in a Taylor series retaining only the first few terms. When I was given the opportunity to switch my focus to mathematical biology by Stuart Rice, my Ph.D. advisor, and Jack Cowan, who offered me a postdoctoral position in the Department of Theoretical Biology at The University of Chicago, my awareness of nonlinear dynamical phenomena began to expand immensely.

As a result of this new exposure to nonlinear dynamics, I began to use these dynamical concepts in my own research on the visual system, and I designed a graduate level course in 1973 to teach these concepts to others. For the first few years the course was populated by a combination of theoretical biology, physics, and chemistry students. Then the physics and chemistry departments began offering this material to their students, so my course was discontinued. It was due to the recent upsurge of interest in neural modeling, connectionist and otherwise, that I decided to resurrect the course and restructure it to focus exclusively on neuroscience. Just as physicists are aided in learning mathematics by using examples such as oscillating springs and planetary motion, so I believe that neuroscientists can more easily grasp the mathematical concepts if examples involve action potentials, hysteresis in memory, and so on.

Spikes, decisions, and actions is intended for three audiences: advanced undergraduates in neuroscience or physiological psychology, graduate students, and professionals in these areas who wish to develop neural simulations relevant to their own interests. When I speak of neuroscience here, I intend it in the broadest sense that encompasses cellular and molecular neurobiology, systems neurobiology, psychophysics, and other fields where the object of study is the nervous system. As the book is based on a one-semester course that I have taught at the University of Chicago intermittently over the past 25 years, the book can readily be used in similar courses. In addition, the book may be used to supplement standard courses in advanced calculus, differential equations, or computer modeling when an introduction to nonlinear dynamics in biology is desired.

Preface

The mathematical background assumed by the book is a minimum of one to two years of calculus. Ideally, the reader will have been exposed to an introductory course on linear differential equations and some linear algebra. The essential material from such courses is outlined in Chapters 1–3, which can be glossed over by the mathematically sophisticated reader. Chapter 4 on higher order linear differential equations contains material on time lags and the Routh–Hurwitz criterion for oscillations that will merit review by most readers. Following Chapter 5 on numerical methods (mainly Runge–Kutta), problems in nonlinear neurodynamics are explored in an orderly sequence: multiple steady states and hysteresis, limit cycles, action potentials and bursting, chaos, neural synchronization, and Lyapunov functions. Finally, the cable or diffusion equation is presented as an avenue to dendritic potentials and compartmental neurons. Throughout, the mathematics is intermingled with the neuroscience of ionic currents, action potentials, motor control, and memory.

The book is accompanied by a disk containing many MatLab™ scripts that are intended to complement the text. These simulations will help to solidify and elucidate analytic results and concepts derived in the text, and they should also enhance the utility of the book for self-study. Beyond this, their inclusion brings the book closer to the realities of research in theoretical neuroscience, where a combination of analysis and simulation is to be found in much of the best work. MatLab was chosen because the scripts will run on the Macintosh, UNIX, or Windows platforms. Also, MatLab is the most common simulation language in my own area, vision research. Most scripts will run in the student version of MatLab, but a few towards the end of the book may require larger arrays than the student version permits. A brief Appendix introduces the MatLab scripts and their use with this book.

I have attempted to make the text as readable as possible by avoiding undue use of abstruse mathematical terminology. While this may annoy some who are mathematically sophisticated, it is my experience that most neuroscience students find this to be a more congenial approach to difficult mathematical concepts. In the same vein, I have frequently written in the first person plural, using 'we' and 'us'. I hope this will encourage the reader to view this book as a journey of exploration we are taking together.

A number of problems accompany each chapter. In addition, the text encourages the reader to explore different parameter ranges in many of the simulations, and this functions as a form of self-teaching problem. Given the MatLab simulations, it should be relatively easy for an instructor to create additional problems as desired.

Mathematics texts inevitably suffer from a number of typographical errors, particularly because equations are so difficult to proofread. I have checked the book several times but take full responsibility for any remaining errors. To aid the reader, I shall list known errors and corrections on the web site: http://spikes.bsd.uchicago.edu. Readers discovering errors are requested to e-mail me at hrw6@midway.uchicago.edu, and I will post them on the web.

In writing this book, I have done my symbolic derivations using Maple MathView™ software on the Macintosh. Use of this computer algebra and calculus software has been enormously valuable in minimizing errors in lengthy derivations. MathView™, Maple™, and Mathematica™ software are all of great value in conducting analytic manipulations, and readers familiar with any of these programs are encouraged to use them.

Many individuals have contributed directly or indirectly to *Spikes, decisions, and actions*. First were my parents, Hugh and Georgine Wilson, who always encouraged me to find learning exciting and to seek understanding. Two former professors at my alma mater, Wesleyan

University, deserve special mention. Tom Tashiro, a Shakespearian scholar, taught me about the relationship between data and theory through historical examples. Gilbert Burford, Professor of Physical Chemistry, showed me how rich and elegant the interplay between mathematics and science could be. At The University of Chicago, Stuart Rice, my thesis advisor, and Jack Cowan, my postdoctoral mentor, are the two individuals responsible for introducing me to mathematical approaches in biology and neuroscience. I am particularly grateful to Jack Cowan for introducing me to mathematical theories of neural functions, thus setting me on the career path I have followed since. I am indebted to Suzanne Weaver Smith from the University of Kentucky and to my lifelong friend Bob Morris from the University of Massachusetts for reading substantial portions of the book and providing invaluable feedback. Thanks are also due to Tony Marley of the McGill University Psychology Department for providing me with a very hospitable sabbatical environment in 1995–1996 during which this book was begun. The development of this book has benefitted from comments and questions by many former students involved with my nonlinear dynamics course over the years at The University of Chicago. Among these are Michael Hines, Jim Bergen, Bard Ermentrout, Charlie Smith, Bill Swanson, Sidney Lehky, Jeounghoon Kim, and Li-Ming Lin.

Finally, my deepest gratitude goes to my wife, Frances Wilkinson of McGill University, to whom this book is dedicated. I could never have succeeded in such a prolonged writing enterprise without her support, encouragement, and critical comments as work progressed.

Chicago   H.R.W.
1998

Contents

1   *Introduction*

1.1   Nonlinear dynamics in neuroscience

The nervous system of all higher animals is inherently both highly complex and highly nonlinear. The complexity is most obvious from consideration of the enormous number of neurons and synapses in the brain: the human brain is comprised of approximately 10^{12} neurons and over 10^{15} synapses! Brain nonlinearity becomes evident from equally striking observations. Nerve cells have a threshold for producing spikes or action potentials, the fundamental events of almost all neural signaling. Thus, weak stimulation has no effect yet several weak stimuli together produce a dramatic spike response. To cite another example, changing from a walk to a trot or gallop requires switching among dynamical modes (switching gears) in a nonlinear motor control network. More generally, all higher nervous systems make decisions between alternative courses of action, and decision making is an inherently nonlinear process. We do not act upon the average among alternatives; we act upon the winner among competing alternatives. *Spikes, decisions, and actions* is an exploration of the nonlinear mathematical principles by which brains generate spikes, make decisions, store memories, and control actions.

Despite the enormous complexity of the brain, a major goal of neuroscience (and indeed all science) is to understand and predict change: changes in neural firing rates caused by altered ionic concentrations; changes in behavior resulting from altered neural activity. Dynamics is the name given to the mathematics devoted to studying change, to predicting the future given knowledge of the present. The foundation of dynamics is the differential equation. Differential equations are based on a simple concept: knowledge of the present state of a system can be used to predict how it will evolve during the next instant of time. By repeating this predictive process from instant to instant, differential equations permit us to build up a picture of the future behavior of any system.

The motivation for introducing the neuroscientist to nonlinear dynamics should now be clear: it is the most powerful analytic tool available to us for understanding and predicting behaviors of complex systems. This became apparent as early as 1952, when Hodgkin and Huxley developed the highly nonlinear differential equations that predict the generation of action potentials in neurons (Hodgkin and Huxley, 1952). Since that time dynamical systems theory has been effectively employed in understanding and predicting the responses of vertebrate photoreceptors to light (Schnapf *et al.*, 1990), the behavior of neural networks that control swimming (Ekeberg, 1993), the neural dynamics underlying light adaptation in the retina (Wilson, 1997), and many other neural phenomena.

In addition to providing a powerful analytic tool, nonlinear dynamics will also introduce the neuroscientist to important new concepts and phenomena that simply cannot occur in linear systems. One of these is the limit cycle oscillation (Chapters 8 and 9), which may fairly be said to be the foundation of all biological and neural oscillatory behavior. Indeed, an understanding of limit cycles is necessary to both the scientist who would understand the ionic basis of action potentials and to the neuroscientist studying motor control at the network level. Systems with multiple equilibrium points (Chapter 6), which enable the system to categorize its inputs, make decisions, and perform short-term memory functions, are another concept unique to nonlinear dynamics. Finally, there is the notion of chaos, which is sufficiently fascinating to have captivated the popular press. Chaotic systems (Chapter 11) are ones which are perfectly predictable in principle but yet seem to behave in a highly unpredictable fashion, almost as though they possessed a modicum of free will! Chaos, too, is a dynamical phenomenon unique to nonlinear systems. Acquaintance with these and other nonlinear concepts will enrich the armamentarium of the contemporary neuroscientist whether or not her/his main focus is mathematical. In addition, nonlinear dynamical concepts are highly relevant to neuroscience-oriented discussions of philosophy of mind (e.g. Churchland, 1989), as will be emphasized by occasional asides in the text.

It has sometimes been claimed that nonlinear dynamical systems are just too complex to be adequately understood using available mathematical techniques. For example, Green (1976, p. 33) stated: '[Nonlinear] systems are sometimes almost inscrutable.' While this comment may have been justified earlier in the century, it is simply no longer true with respect to the types of nonlinearities inherent in the nervous system. A major goal of this book is to provide neuroscientists with the background necessary to appreciate the sophistication and processing power of neural nonlinearities. By restricting treatment to just those types of nonlinearities that are most relevant to neuroscience, it is possible to give a treatment of nonlinear dynamics that is succinct and yet sufficiently general to enable one to begin applying these powerful techniques in one's own research. Indeed, the title of this book, *Spikes, decisions, and actions*, reflects the importance of nonlinear dynamics at all levels from spike generation to the level of networks that make decisions and control actions.

Sometimes experimentalists who are resistant to theory claim that there is little to learn from models, because they believe we obtain no results other than those we consciously design into our models. While this may be more or less true for linear models due to their simplicity, it is demonstrably false in the case of nonlinear models. The Hodgkin–Huxley (1952) equations provide a classic example. More than 25 years elapsed before Rinzel (1978) and Best (1979) predicted from a mathematical analysis of these highly nonlinear equations that they should exhibit hysteresis. Even more counter-intuitive was the prediction that a brief *depolarizing* current pulse should permanently extinguish an ongoing spike train if delivered at the proper phase. Both predictions were experimentally verified by Guttman, Lewis, and Rinzel (1980) the next year. Furthermore, the Hodgkin–Huxley (1952) equations were published years before chaos was characterized by mathematicians. Yet in 1987 it was predicted and experimentally verified that the squid axon should exhibit chaos (Degn *et al.*, 1987). Thus, the Hodgkin–Huxley (1952) equations correctly predicted several novel phenomena that were clearly not conceived of by Hodgkin or Huxley in designing their equations.

One might argue that the ready availability of desktop microcomputers has made the study of nonlinear dynamics irrelevant: one need only approximate solutions of equations to study their behavior. This viewpoint rests on several major misconceptions, however. First, one is unlikely to have any idea what form the equations appropriate to a particular phenomenon might take without a grounding in nonlinear dynamics. Second, even when the relevant equations are already known, it is very difficult to determine how solutions depend on parameter values without knowledge of dynamical techniques. Finally, a knowledge of nonlinear dynamics is required if one is to be certain that one's computer approximations actually reflect the true dynamics of the system under study.

Having said this, it must certainly be acknowledged that computer simulations provide an extremely valuable and powerful tool for the neuroscientist. Accordingly, another goal of this book is to teach precisely those analytical techniques that will enhance the power and sophistication of neural simulations. To this end I have included numerous computer simulations of single neurons and small neural networks throughout the book. The goal of these is both to enable the reader to explore dynamical phenomena that are discussed in the book and to learn to apply analytical techniques to the understanding of network simulations. MatLab™ scripts for these simulations are contained on the accompanying disk. In addition, computers have enhanced the utility of some techniques and removed the drudgery from others. Accordingly, MatLab scripts are included to produce symbolic solutions to second order linear differential equations and to facilitate application of stability criteria to higher order systems. There is no point in doing by hand what a computer can do faster and more accurately.

Currently, 'connectionist' neural networks are very much in vogue, and a number of excellent texts already exist in this area (e.g. Hertz *et al.*, 1991). This book is not intended to compete with these but rather to complement them. Connectionism today is concerned with somewhat artificial neuron-like networks that can be trained (by back-propagation, for example) to associate and categorize stimuli. My emphasis will be on smaller, deterministic neural networks that are more biological in their behavior. Furthermore, the nonlinear analysis in this book provides much of the mathematical background upon which connectionist neural modeling has been built (see Chapter 14). It has been my experience that many students who are attracted to connectionist modeling have strong computer science backgrounds but rather little understanding of the nonlinear dynamics upon which their modeling is based. I hope that this book will help to rectify the situation.

1.2 Neuroscience and levels of abstraction

As in any branch of science, there are many different levels of abstraction at which a neuron or neural network may be described. At the most abstract level, a neuron may be described as a device that is either on or off (1 or 0). This was the description introduced by McCulloch and Pitts (1943) in their classic paper: 'A logical calculus of ideas immanent in nervous activity'. At a more detailed level, a neuron may be described by its spike rate, which varies continuously between zero (when the postsynaptic potential is below threshold) and some maximum value at which the spike rate saturates due to the absolute refractory period. A yet more detailed neural description is provided by the Hodgkin–Huxley (1952) equations, which describe the generation and shape of each individual action potential as a function of the underlying ionic currents. Finally, the most detailed descriptions incorporate the detailed geometry of the dendritic tree along with the spatial distribution of synapses and ion channels on the dendrites. This, of course, is not the end of the line: one could frame one's neural simulations at the level of quantum mechanical changes in ion channel gating molecules. However, contemporary neuroscience seldom goes so far: neural reductionism generally stops with the belief that quantum mechanical effects will either average out or else manifest themselves as statistical fluctuations.

Different levels of generality in the description of neurons are appropriate for different purposes, as is illustrated in Fig. 1.1. Generally speaking, the more detailed the description of individual neurons, the smaller the number of neurons that can be effectively modeled. As different types of neural experimentation produce data reflecting activity of widely varying neural populations, however, several different levels of

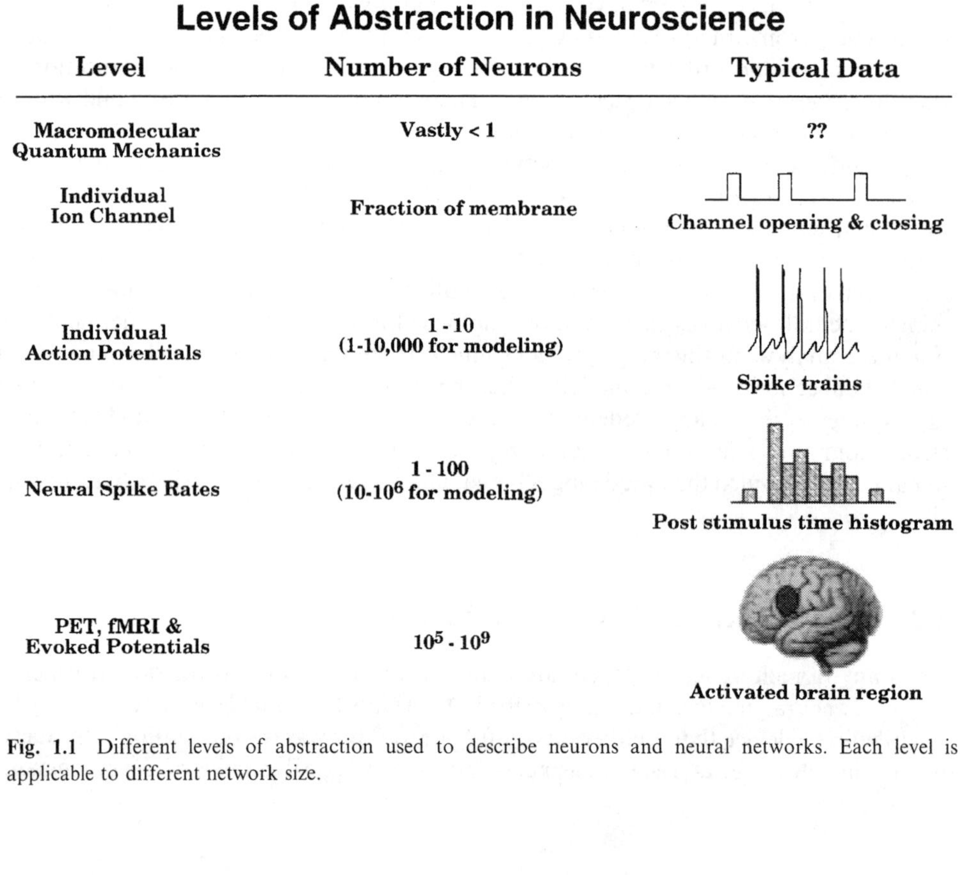

Levels of Abstraction in Neuroscience

Level	Number of Neurons	Typical Data
Macromolecular Quantum Mechanics	Vastly < 1	??
Individual Ion Channel	Fraction of membrane	Channel opening & closing
Individual Action Potentials	1 - 10 (1-10,000 for modeling)	Spike trains
Neural Spike Rates	1 - 100 (10-10^6 for modeling)	Post stimulus time histogram
PET, fMRI & Evoked Potentials	10^5 - 10^9	Activated brain region

Fig. 1.1 Different levels of abstraction used to describe neurons and neural networks. Each level is applicable to different network size.

abstraction are clearly relevant in theoretical neuroscience. For example, in my own work on visual system function, neural descriptions at the level of spike rates have proved to be extremely useful in simulating motion perception (Wilson and Kim, 1994), retinal light adaptation (Wilson, 1997), and aspects of cortical function (Wilson and Cowan, 1973). On the other hand, a scientist interested in intracellular recordings from individual neurons would surely require a much more detailed level of description. The material in this book encompasses three levels of neural description. The most general level is that of a neural network in which each neuron is represented by a continuous variable describing its varying spike rate. This level of description is adequate for simulating many aspects of categorization, decision making, short-term memory, and motor control. Description of a neuron by its spike rate is comparable to representation of responses by post-stimulus time histograms in the experimental literature (see Fig. 1.1). The next level of description focuses on the generation of spike trains produced by multiple ionic currents. This is the level of analysis of Hodgkin–Huxley and related equations. Finally, the most detailed level incorporates diffusion of ionic potentials along dendrites with complex geometry. Each of these levels of abstraction is appropriate for some theoretical purposes but inappropriate for others. An appreciation of the conditions under which each may be optimal will emerge from the chapters that follow.

1.3   Mathematical background

This book is intended to be useful to anyone who has completed one to two years of calculus and who is also familiar with some of the basics of vectors and matrices. A semester of linear differential equations would also be very helpful. The single most important concept from calculus is that of the exponential function and its derivative, which describes processes of growth and decay. To fix these ideas, suppose that a colony of microorganisms has population F at the beginning of an experiment. If the fraction of organisms that reproduce per unit time is k, then increases in the population will be described by the differential equation:

$$\frac{\mathrm{d}F}{\mathrm{d}t} = kF \tag{1.1}$$

This just states that the rate of change of the population, $\mathrm{d}F/\mathrm{d}t$, is proportional to the current population level F. As we shall see in the next chapter, eqn (1.1) is solved by the exponential function, which we shall now derive.

1.3.1   *The exponential function*

All scientists are familiar with the exponential function e^t, where $e = 2.71828\ldots$. However, it is frequently forgotten that e is actually determined by the solution to a differential equation. To see the central importance of e to differential equations, let us consider the differential equation satisfied by e^t:

$$\frac{de^t}{dt} = e^t \tag{1.2}$$

This equation states that e is the unique constant such that the function e^t is its own derivative. To see the consequences of this, first consider the definition of the derivative:

$$\frac{de^t}{dt} = \lim_{h \to 0} \frac{e^{t+h} - e^t}{h} \tag{1.3}$$

Using the observation that $a^{t+h} = a^t a^h$ for any positive $a, h,$ and t, we deduce:

$$\lim_{h \to 0} \frac{e^{t+h} - e^t}{h} = e^t \lim_{h \to 0} \frac{e^h - 1}{h} \tag{1.4}$$

Substitution of (1.4) and (1.3) into (1.2) yields the result:

$$\lim_{h \to 0} \frac{e^h - 1}{h} = 1 \tag{1.5}$$

Although the actual derivation is somewhat complex, it can now be shown that the value e satisfying (1.4) is given by:

$$e = \lim_{h \to 0} (1 + h)^{1/h} \tag{1.6}$$

The value of e may be approximated to any desired accuracy using (1.6). For example, $h = 1/10$ and $h = 1/1000$ yield:

$$e \approx \left(1 + \frac{1}{10}\right)^{10} = 2.5937$$

$$\tag{1.7}$$

$$e \approx \left(1 + \frac{1}{1000}\right)^{1000} = 2.7169$$

Thus, the value of e is determined by the differential equation (1.2). In the chapters that follow, the exponential function will be seen to provide the basis for analyzing virtually all differential equations of scientific interest.

1.3.2 *Taylor series*

A second very important concept is that of a Taylor series expansion. We shall frequently make use of the fact that near $t = a$, a function $f(t)$ may be approximated by:

$$f(t) \approx f(a) + h\frac{\mathrm{d}f(a)}{\mathrm{d}t} + \frac{h^2}{2}\frac{\mathrm{d}^2 f(a)}{\mathrm{d}t^2} + \text{(higher order terms)} \qquad (1.8)$$

In this expression, f and its derivatives are evaluated at $t = a$, and $h = t - a$ is the distance from a. The first two terms on the right of (1.8) describe the straight line tangent to f at $t = a$, while the next term represents the curvature of f. This formula will enable us to simplify nonlinear functions so as to gain insights into their properties, and the formula will also be instrumental in our development of computer approximations to the solutions of nonlinear differential equations.

To derive (1.8), suppose we wish to approximate a function $f(t)$ near $t = 0$. One of the simplest approximations is a polynomial in t. So, let us approximate $f(t)$ as:

$$f(t) \approx a + bt + ct^2 + \text{higher powers} \qquad (1.9)$$

To determine the coefficients a, b, c, etc., let us choose an approximation that has derivatives identical to those of f at point $t = 0$. First we evaluate both $f(t)$ and our polynomial at $t = 0$, with the result $f(0) = a$. This guarantees that $f(t)$ and our polynomial have the same value at $t = 0$. Next, we differentiate both $f(t)$ and the polynomial and then evaluate both at $t = 0$. This gives:

$$\frac{\mathrm{d}f}{\mathrm{d}t} = b + 2ct = b \quad \text{for } t = 0 \qquad (1.10)$$

Similarly, taking second derivatives of both sides of (1.9) and evaluating them at $t = 0$ gives:

$$\frac{\mathrm{d}^2 f}{\mathrm{d}t^2} = 2c \qquad (1.11)$$

If we now replace a, b, and c in (1.9) by the values obtained from (1.10) and (1.11), we obtain the polynomial approximation to $f(t)$:

$$f(t) \approx f(0) + \frac{\mathrm{d}f(0)}{\mathrm{d}t}t + \frac{1}{2}\frac{\mathrm{d}^2 f(0)}{\mathrm{d}t^2}t^2 + \text{higher order terms} \qquad (1.12)$$

This polynomial is called the Taylor series approximation to $f(t)$. It is easy to generalize this approach to show that the nth term must be:

$$n\text{th term} = \frac{1}{n!}\frac{\mathrm{d}^n f(0)}{\mathrm{d}t^n}t^n \qquad (1.13)$$

As one simple example, note from eqn (1.2) that the derivative of e^t is 1 at $t = 0$, and indeed all derivatives of e^t are also equal to 1 at $t = 0$. So the Taylor series for e^t to fourth order near $t = 0$ is just:

$$e^t \approx 1 + t + \frac{t^2}{2} + \frac{t^3}{6} + \frac{t^4}{24} \tag{1.14}$$

Although this gives the exact value of e^t only at $t = 0$, it is still reasonably accurate at $t = 1$, where it produces $e \approx 2.7083$, within 0.5% of the true value.

1.3.3 *Imaginary exponents*

A final aspect of exponential functions that will be of importance in our study of differential equations is expressed by **Euler's formula**:

$$e^{iat} = \cos(at) + i\sin(at) \tag{1.15}$$

where $i = \sqrt{-1}$. In other words, e with an imaginary exponent is equivalent to a combination of sine and cosine functions. Based on this equation, we shall see that sines and cosines provide a foundation for understanding neural oscillations.

A simple derivation of (1.15) can be sketched out based on Taylor series. Let us approximate $\exp(i\alpha t)$ near $t = 0$ by a fourth order Taylor's series:

$$\begin{aligned}
e^{i\alpha t} &\approx 1 + i\alpha t - \frac{(\alpha t)^2}{2} - i\frac{(\alpha t)^3}{6} + \frac{(\alpha t)^4}{24} \\
&= \left(1 - \frac{(\alpha t)^2}{2} + \frac{(\alpha t)^4}{24}\right) + i\left(\alpha t - \frac{(\alpha t)^3}{6}\right)
\end{aligned} \tag{1.16}$$

The second step is simply a grouping of the real and imaginary parts of the series separately. If you now compare the series for the real part of (1.15) with the Taylor series for $\cos(\alpha t)$, you will discover that they are identical. Similarly, the imaginary part of the Taylor series expansion in (1.15) is identical to the Taylor series for $\sin(\alpha t)$. This sketches a proof of (1.15), although more sophisticated proofs are available from complex variable theory.

1.4   Neurobiology background

In addition to the calculus concepts outlined above, this book assumes that the reader has some basic knowledge of neurobiology. This background is contained in several excellent introductory texts, such as *Neurobiology* by Shepherd (1994), *Neurons and Networks* by

Dowling (1992), or *Foundations of Neurobiology* by Delcomyn (1998). In this respect, *Spikes, decisions, and actions* is analogous to the many applied mathematics texts written for physics students. Those books choose their mathematical examples on the assumption that the reader has some understanding of springs, gravitation, etc. The examples I shall choose assume a basic knowledge of neuron morphology (axon, soma, synapse, etc.) and physiology (ion channels, resting potentials, action potentials) at the level offered by the books above.

For reference, a microscope drawing of a cortical neuron from Cajal (1911) is presented in Fig. 1.2. Briefly, other neurons make localized contacts with this particular neuron at chemical **synapses** located throughout the **dendrites** (or dendritic tree) and also on the cell body or **soma**. These synapses may be either excitatory (electrically depolarizing) or inhibitory (generally electrically hyperpolarizing). Electrical changes produced at all the synapses that are simultaneously active then propagate to the soma, thus producing a net **postsynaptic potential**. If the postsynaptic potential at the soma is sufficiently large to

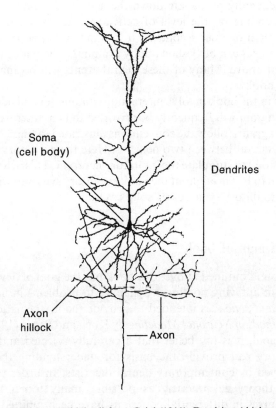

Fig. 1.2 Diagram of a cortical pyramidal cell from Cajal (1911). Dendrites (thick processes) emanate from the soma or cell body and are the site of synaptic contacts from other neurons. The thin axon is connected to the soma at the axon hillock, which is the site of action potential generation. Axons typically have many branches (several shown) and may cover very long distances before they contact other nerve cells at synapses.

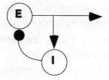

Fig. 1.3 Diagram of a two-neuron negative feedback network. Throughout the book excitatory synapses will be indicated by arrowheads, and inhibitory synapses will be indicated by solid circles.

exceed a threshold value, typically a depolarization of 10–15 millivolts (mV), the neuron generates a brief electrical pulse, known as a **spike** or **action potential** at its **axon hillock**. The axon hillock is the point of connection between the soma and the **axon**, which is a long thin process leading off from the soma and ultimately terminating in synapses on other neurons. Axons range in length from less than one millimeter to almost a meter, depending on the location and identity of the particular neuron. Spikes generated at the axon hillock subsequently propagate down the axon and produce chemical signals at the synapses. Depending on the level of activation, neurons can fire spikes at rates varying from less than one per second up to around 1000 per second (1000 Hz). This extremely brief synopsis can be fleshed out by examining the relevant chapters in any of the references cited above. Many of these neural events will be analyzed in dynamical terms later in this book.

Many examples in the book involve interactions among several neurons in a network. In these cases a diagram will frequently be provided to help interpret the mathematical equations. Such diagrams follow the conventions illustrated in Fig. 1.3. This figure shows a simple feedback circuit between two neurons. Here the excitatory neuron E sends an axon collateral (arrow) to stimulate the inhibitory neuron I. This neuron in turn provides feedback inhibition to E. Throughout the book excitatory synapses will be represented by arrowheads, and inhibitory synapses by solid circles.

1.5 Scope and plan of book

Given the background outlined above, this book has the goal of developing those skills that are essential in applying nonlinear dynamics to problems in neurobiology. Thus, *Spikes, decisions, and actions* is intended to provide the mathematical background for works such as *Methods in Neuronal Modeling* by Koch and Segev (1989). While this may be an ambitious goal, it is my belief that a carefully selected and coherent body of mathematical theory can provide the basis for understanding the vast majority of techniques employed by contemporary neural theorists. In order to provide this core of mathematical theory as concisely as possible, many topics typically taught in undergraduate courses in differential equations have been omitted. The instructor or advanced reader will note that there is no discussion of equations with time-varying coefficients or Laplace transforms. However, this does not reduce the utility of the book for neuroscientists. Laplace transforms have little relevance to nonlinear dynamical systems. Time-varying parameters, however, are really a version of nonlinear dynamics in disguise. The time variation may always be rewritten using multiplicative variables that satisfy their own differential equations. For example:

$$\frac{\mathrm{d}x}{\mathrm{d}t} = \frac{1}{1+t}x \tag{1.17}$$

is equivalent to the system of equations:

$$\frac{\mathrm{d}x}{\mathrm{d}t} = yx$$

$$\frac{\mathrm{d}y}{\mathrm{d}t} = -y^2 \tag{1.18}$$

with $y = 1$ at time $t = 0$. (The reader may verify this by substituting $y = 1/(1+t)$ into the second equation.) Equations in which the independent variable t does not appear explicitly in any of the parameters are said to be **autonomous systems**. (We will, of course, deal with the case where the stimulus to a neuron or network is time-varying.)

The plan of the book is straightforward. In order to understand nonlinear dynamics, one must first grasp certain fundamental aspects of linear dynamics. These are contained in Chapter 2, which discusses first order linear systems, and Chapters 3 and 4, which extend the analysis to second and higher order linear systems. Before delving into nonlinear differential equations, Chapter 5 develops techniques for simulating dynamical systems on the computer, the focus being on Runge–Kutta methods. Chapters 6 and 7 introduce the analysis of nonlinear neural systems with multiple equilibrium points, and Chapter 8 focuses on limit cycle oscillations in simple neural systems. With this background, Chapter 9 analyzes equations for action potential generation related to the Hodgkin–Huxley equations, and Chapter 10 extends this to bursting neurons controlled by limit cycles within limit cycles. Chapter 11 provides a brief introduction to neural chaos. Neural synchrony and motor control (with the lamprey swimming oscillator as a prime example) are covered in Chapters 12 and 13. Chapter 14 provides a brief introduction to Lyapunov function theory with applications to long-term memory. Chapter 15 provides an introduction to the one partial differential equation of greatest relevance to neuroscience: the diffusion or cable equation. This represents a straightforward extension of the foregoing material, as the method of separation of variables reduces the diffusion equation to ordinary differential equations. Applications to dendritic and action potential propagation, as well as compartmental neurons, are explored. Finally, Chapter 16 concludes with a resume of general principles underlying nonlinear neural dynamics.

MatLab scripts are included with this book to enable the reader to simulate all of the examples in the text without spending excessive time programming. In addition, many exercises at the end of chapters can be approached by making small modifications in the MatLab programs provided. MatLab was chosen because it is widely used by neuroscientists and because the same script will run on the Macintosh, Windows, and under UNIX. Although the MatLab simulations are not essential for readers with sophisticated mathematical backgrounds, they should be quite valuable to the neuroscience student exploring nonlinear dynamics for the first time.

I have always found nonlinear dynamics to be conceptually rich and fascinating, and I have truly enjoyed writing *Spikes, decisions, and actions*. I sincerely hope the reader will find the book exciting as well as informative.

2 *First order linear differential equations*

A differential equation describes the change in neural responses (or ionic concentrations, etc.) between the present time t and a time $(t + dt)$, which lies infinitesimally in the future, and it describes this change as some function of all the physiologically relevant variables at time t. The **order** of a differential equation is defined as the highest derivative present in the equation. As will be seen in the next chapter, this is equivalent to the number of coupled first order equations in a system of differential equations. In this chapter I introduce the simplest differential equation of major importance in science: the first order linear differential equation with constant coefficients. Solutions of this equation form the basis for understanding higher order equations, both linear and non-linear. Despite its linearity, however, the first order equation is still capable of describing the spike rate of a single neuron in response to stimulation, and we shall see that it can also describe interactions between postsynaptic potentials.

2.1 The fundamental first order equation

Let us begin our treatment of differential equations by considering the simplest, yet most fundamental of all differential equations in the sciences:

$$\frac{dx}{dt} = -\frac{1}{\tau}x \tag{2.1}$$

This equation states that the rate of change of the function $x(t)$ as a function of time t is equal to a constant times the function itself. The time constant, τ, can be shifted to the left side of the equation by simple multiplication, but the form (2.1) will be useful for our present purposes. The temporal units of τ are the same as t (i.e. milliseconds, ms, or fraction of a second). We can solve (2.1) by substituting an exponential function and then making use of its derivative from (1.2):

$$x(t) = A\,e^{at}, \quad \text{so} \quad \frac{dx}{dt} = Aa\,e^{at}$$

Substitution of this into (2.1) leads to the result:

$$Aa\,e^{at} = -\frac{A}{\tau}\,e^{at}$$

This equation is easily solved to give $a = -1/\tau$, so the solution to (2.1) is just $x = A\exp(-t/\tau)$. The minus sign on the right side of (2.1) results in a negative value for the exponent in the solution (τ is always positive by convention). As a negative exponent is generally the most appropriate physiological solution, it has been emphasized, although the same derivation obviously holds for positive values on the right-hand side. The constant A may be assigned any value we wish. To determine a unique value of A, we need to specify an **initial condition**, namely, a value for $x(t)$ at $t = 0$. Denoting this value by x_0 and substituting into the solution at $t = 0$, we see that the solution to (2.1) is:

$$x(t) = x_0\,e^{-t/\tau} \tag{2.2}$$

Let us now consider a somewhat more complex equation in which the right-hand side contains an arbitrary additive function $S(t)$. As we shall see shortly, this may be thought of as a time-varying stimulus applied to a neuron which responds with spike rate $x(t)$. The relevant equation is:

$$\frac{dx}{dt} = \frac{1}{\tau}(-x + S(t)) \tag{2.3}$$

Equation (2.1) is referred to as a **homogeneous** differential equation, because the right side only contains terms involving the unknown function x. Equation (2.3), on the other hand, is termed **inhomogeneous**, because the right-hand side contains an additional term that is independent of x. To solve (2.3), let us try to find a solution of the form:

$$x = A\,e^{-t/\tau} + H(t)\,e^{-t/\tau} \tag{2.4}$$

This is just the solution we obtained to (2.1) plus an additional function of time, $H(t)$, multiplied by the exponential that was obtained from solving (2.1).

Substitution of (2.4) into (2.3) yields:

$$e^{-t/\tau}\frac{dH}{dt} - \frac{1}{\tau}(A\,e^{-t/\tau} + H\,e^{-t/\tau}) = \frac{1}{\tau}(S - A\,e^{-t/\tau} - H\,e^{-t/\tau})$$

Cancellation of the second and third terms on both sides gives:

$$e^{-t/\tau}\frac{dH}{dt} = \frac{1}{\tau}S$$

This may be solved for $H(t)$ by integration with the result:

$$H(t) = \frac{1}{\tau}\int_0^t e^{t'/\tau}S(t')\,dt' \tag{2.5}$$

The last step results from isolating dH/dt on the left and then integrating both sides of the equation. This integration is carried out with respect to the 'dummy variable' t', which represents all past times between the start of stimulation at $t = 0$ and the present time t. Combining (2.2), (2.4), and (2.5), we have proved the theorem:

Theorem 1: The solution to the equation:

$$\frac{dx}{dt} = \frac{1}{\tau}(-x + S(t))$$

is

$$x(t) = A e^{-t/\tau} + \frac{1}{\tau} e^{-t/\tau} \int_0^t e^{t'/\tau} S(t') \, dt'$$

or

$$x(t) = A e^{-t/\tau} + \frac{1}{\tau} \int_0^t e^{-(t-t')/\tau} S(t') \, dt'$$

where A is chosen to satisfy the initial condition.

To obtain the second form for $x(t)$ in this theorem, the exponential has simply been moved inside the integral. Many readers will recognize the integral in the solution as a temporal **convolution integral**. This convolution adds up effects throughout the past history of stimulation, $S(t')$, weighting each past instant t' by an exponentially decaying function of the elapsed time between t' and the present, $\exp\{-(t - t')/\tau\}$. Thus, the influence of all previous stimulation is summed, but it dies out exponentially as we move further into the past.

To fix ideas, let us apply Theorem 1 to an example. For time in milliseconds, solve the equation:

$$\frac{dx}{dt} = \frac{1}{10}(-x + 50)$$

for an initial condition $x(0) = 0$. Using Theorem 1 we obtain:

$$x(t) = A e^{-t/10} + \frac{1}{10} \int_0^t e^{-(t-t')/10} 50 \, dt'$$

The integral is easily evaluated with the result:

$$\frac{1}{10} \int_0^t e^{-(t-t')/10} 50 \, dt' = 5 e^{-t/10} \int_0^t e^{t'/10} \, dt' = 50(1 - e^{-t/10})$$

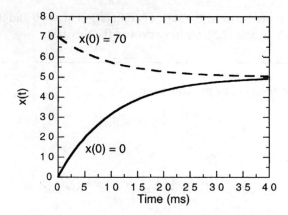

Fig. 2.1   Solutions of equation in text for two different initial conditions.

Therefore:

$$x(t) = A\,e^{-t/10} + 50(1 - e^{-t/10})$$

At $t = 0$, $x = A$, so the initial condition will be satisfied if $A = 0$. This produces the solution:

$$x(t) = 50(1 - e^{-t/10})$$

As a second case, let us solve the same equation again but with a different initial condition: $x(0) = 70$. Now $A = 70$ and

$$x(t) = 70\,e^{-t/10} + 50(1 - e^{-t/10}) = 50 + 20\,e^{-t/10}$$

Both of these solutions are plotted in Fig. 2.1, where it is easy to see that $x(t)$ approaches the value 50 with a decaying exponential time course. The time constant for this approach is 10 ms, and it is apparent that the solution has virtually reached its asymptotic value by $t = 40$ ms, i.e. within about 4 time constants. This is because $e^{-4} = 0.018$, within 2% of the asymptotic value.

To take another example, let us solve the following equation for $x(0) = 0$:

$$\frac{dx}{dt} = \frac{1}{20}(-x + 40\,e^{-t/20})$$

Using Theorem 1, we obtain:

$$x(t) = A\,e^{-t/20} + \frac{40}{20}\,e^{-t/20}\int_0^t e^{t'/20}\,e^{-t'/20}\,dt' = A\,e^{-t/20} + 2t\,e^{-t/20}$$

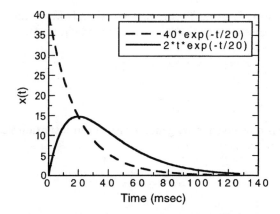

Fig. 2.2 $x(t)$ (solid line) in response to the time-varying stimulus plotted by the dashed line.

As $x(0) = 0$, the solution is:

$$x(t) = 2t\,e^{-t/20}$$

This function is graphed in Fig. 2.2. Note that the response $x(t)$ overshoots the stimulus $S(t)$ and then follows its exponential decay.

2.2 Cascades of first order equations

There are many physiological circumstances in which one first order differential equation provides the input to a second, the second to a third, and so forth. Frequently it is assumed on the basis of experimental evidence that each equation has the same time constant τ as the others. This is known as a **cascade** of equations, and it generally arises when there is a chain of chemical steps between an initial event and a final measured neural response. For example, the electrical response of photoreceptors known as rods in the primate retina are well described by a three-equation cascade. Let us derive the solution to such a cascade for the rod with stages x, y, and z obeying the equations:

$$\frac{dx}{dt} = -\frac{1}{\tau}x$$
$$\frac{dy}{dt} = \frac{1}{\tau}(-y + kx) \qquad\qquad (2.6)$$
$$\frac{dz}{dt} = \frac{1}{\tau}(-z + ky)$$

We will assume that several light quanta have just been captured so that $x(0) = 1$, but $y(0) = 0$, $z(0) = 0$. The constant k describes amplification by biochemical events in the rods. This is actually a third order differential equation (see Chapter 3), but it is so simple that we can solve it exactly using Theorem 1. We already know the solution to the first equation, which we can then substitute into the second equation:

$$x(t) = e^{-t/\tau} \quad \text{so}$$
$$\frac{dy}{dt} = \frac{1}{\tau}(-y + k e^{-t/\tau}) \tag{2.7}$$

The equation for y is now just a more general case of our last example, and we can again solve using Theorem 1 to get:

$$y(t) = k \frac{t}{\tau} e^{-t/\tau} \tag{2.8}$$

Substituting this expression for y in the final equation in (2.6) and again using Theorem 1 gives our final result:

$$z(t) = \frac{k^2}{2} \left(\frac{t}{\tau}\right)^2 e^{-t/\tau} \tag{2.9}$$

Thus our three-equation cascade in eqn (2.6) produces a response proportional to t^2 for small t. In comparing data with eqn (2.9) it is usually convenient to plot the data on double logarithmic coordinates, because these coordinates transform the rising phase of the response into a straight line with a slope of 2.0. Figure 2.3 illustrates such a plot of human rod electroretinogram (ERG) data (Hood and Birch, personal communication).

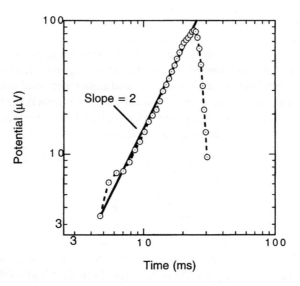

Fig. 2.3 Double logarithmic plot of human rod ERG data (from Hood and Birch, 1990). As shown by the solid line, the initial phase of the response has a slope of 2.0 in agreement with eqn (2.9).

As shown by the solid line, the rod ERG response has an initial slope of 2.0, so there are three stages in the rod biochemical response to light as described in (2.6) and solved in (2.9). This shows that a neuroscientist can sometimes infer aspects of the underlying biochemistry or circuitry from a theoretical analysis of the measured response.

We can generalize our treatment of cascades to N stages, each with time constant τ and amplification k to yield:

$$z_N(t) = \frac{k^{N-1}}{(N-1)!} \left(\frac{t}{\tau}\right)^{N-1} e^{-t/\tau} \tag{2.10}$$

where $z_N(t)$ is the response of the Nth stage of the reaction cascade. Note that we can allow different values of the amplification k so long as τ remains constant for all stages.

2.3 Responses of a simple model neuron

Let us now see how (2.3) can be modified to describe the response of a simple neuron to an external stimulus. This neuron will be represented by its spike rate as a function of time without describing the shape and timing of each individual spike. Before diving into the mathematics, however, a brief discussion of neural responses as a function of stimulus intensity is in order.

The particular physiological example I shall choose is from the visual system, but similar functional relationships with slightly different parameter values are common in the nervous system. Sclar, Maunsell, and Lennie (1990) measured the spike rate of visual neurons in response to stimuli of varying contrast or intensity. Recordings from several different levels of the visual system (lateral geniculate, striate cortex, middle temporal cortex) showed that all neurons could be described by a single equation in which only the parameters differed among visual areas. Albrecht and Hamilton (1982) have also found that this same equation provided a better fit to their data than several other candidate equations. The equation is known as the Naka–Rushton (1966) function in vision research and as the Michelis–Menton equation in chemical kinetics. This equation relates a stimulus intensity P, which may be thought of as the net postsynaptic potential reaching the site of spike generation, to response or spike rate $S(P)$ as follows:

$$S(P) = \begin{cases} \dfrac{MP^N}{\sigma^N + P^N} & \text{for } P \geq 0 \\ 0 & \text{for } P < 0 \end{cases} \tag{2.11}$$

In this equation M is the maximum spike rate for very intense stimuli, and σ determines the point at which $S(P)$ reaches half of its maximum. Hence, σ is termed the semi-saturation constant. Finally, N determines the maximum slope of the function, or how sharp the transition is between threshold and saturation. These points will become evident by inspection of Fig. 2.4, where $M = 100$, $\sigma = 50$, and N assumes several values within the range reported for visual neurons. In particular, Sclar *et al.* (1990) reported that lateral geniculate neurons were best fit by values of N averaging 1.4, visual cortical neurons had N values around 2.4, and middle temporal cortex neurons had N values around 3.0.

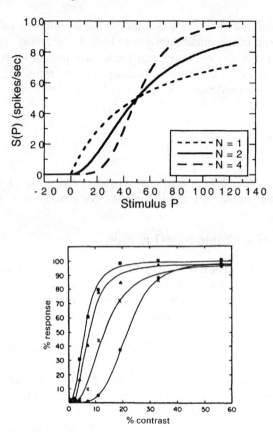

Fig. 2.4 Naka–Rushton function (2.11) plotted for three values of N with $\sigma = 50$ and $M = 100$ (top panel). The bottom panel shows spike rates of four different neurons along with fits of (2.11) to the response rate of each (reproduced with permission, Albrecht and Hamilton, 1982).

Similarly, Albrecht and Hamilton (1982) reported average values of $N = 3.4$ and $M = 120$. Representative data from the Albrecht and Hamilton (1982) study are plotted in Fig. 2.4 for comparison. In this book, we shall usually let $M = 100$ and $N = 2$ for mathematical convenience. This means that our Naka–Rushton function will have an accelerating nonlinearity near $x = 0$ and will have a maximum response rate of 100 spike/s. However, none of our conclusions depend on these particular choices. The semi-saturation constant σ will be varied to suit particular mathematical or physiological contexts. It is important to be aware that (2.11) represents the asymptotic or steady state firing rate of a neuron. As we shall see, neural responses will generally vary over time as they approach the rate determined by (2.11). Note that the general form of $S(P)$ involves a threshold for P near zero followed by a roughly linear region in which $S(P)$ increases proportionally to P. Finally, the spike rate saturates for large P. Many mathematically similar functions have been used in describing neurons, particularly the hyperbolic tangent, tanh. However, all of these functions have the same general sigmoidal, or S-like shape, which is exhibited by $S(P)$ in Fig. 2.4. As the Naka–Rushton function in (2.11) provides the best fit to physiological data (Albrecht and Hamilton, 1982), it will be used here.

We can now write down a differential equation describing the response of a single neuron to an arbitrary stimulus. If the Naka–Rushton function with $M = 100$, $N = 2$, and $\sigma = 40$ is designated by $S(P)$ and the neural response or spike rate is designated by R, then:

$$\frac{dR}{dt} = \frac{1}{\tau}(-R + S(P)) \tag{2.12}$$

Let us solve this equation as a function of t on the assumption that $R(0) = 0$. From Theorem 1, the solution may be seen to be:

$$R(t) = A\,e^{-t/\tau} + \frac{1}{\tau} \int_0^t e^{-(t-t')/\tau} S(P(t'))\,dt' \tag{2.13}$$

This equation describes the responses of cortical cells to a wide variety of time-varying stimuli. We can obtain exact results for the special case where P is a constant input. Then eqn (2.13) may be solved by simple integration yielding:

$$R(t) = (1 - e^{-t/\tau})S(P) \tag{2.14}$$

$R(t)$ is plotted for $P = 40$, 80, and 120 in Fig. 2.5, where $\tau = 15\,\text{ms}$. Note that $S(P)$ is similar to the solid curve in the top panel of Fig. 2.4. Thus, we have solved the equation for time evolution of the spike rate of a neuron that has the nonlinear stimulus–response relationship given by $S(P)$ but that nonetheless is governed by linear dynamics. The nonlinearity in this simple neuron is apparent in Fig. 2.5 from the fact that stimulus magnitudes differing by 40 produce a progressive compression in the asymptotic response levels. This example raises the important point that the solution of a differential equation can be a nonlinear function of its input, as $S(P)$ is indeed a nonlinear function of P, even

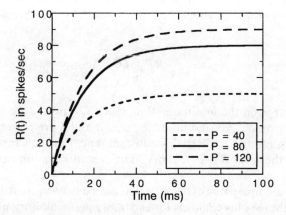

Fig. 2.5 Solution (2.14) to eqn (2.12) for stimulus levels $P = 40$, 80, and 120.

though the dynamics in (2.12) are linear in the response $R(t)$. When we speak of a differential equation as being linear, we really mean only that it is linear in the dynamical variables.

One more important point may be made concerning eqn (2.12). In the case where $S(P(t)) = S$, a constant independent of t, we can solve immediately for the **equilibrium state** of the neuron. At equilibrium, there is no variation with time, so $dR/dt = 0$. It follows from (2.12) that:

$$R = S \quad \text{when} \quad \frac{dR}{dt} = 0 \qquad (2.15)$$

The equilibrium state is also referred to as an **equilibrium point** or **steady state**, and these terms will be used interchangeably hereafter. The important concept to retain here is that an equilibrium point is defined mathematically as a state of the system where nothing changes with time. We shall subsequently see that these points are extremely important to our understanding of nonlinear neural systems. From the solution to (2.12) in (2.14), we can see that $R(t)$ approaches equilibrium exponentially as $t \rightarrow \infty$. Therefore, this equilibrium point is **asymptotically stable**. Equilibrium points and stability will be explored in detail in the next chapter.

2.4 Excitatory and inhibitory postsynaptic potentials

With the background gained thus far, we can now examine the role of ion channels in the generation of postsynaptic potentials. Ion channels are governed by Ohm's law, which states that $I = g(V - E)$, where I is the ionic current across the nerve membrane, g is the conductance (reciprocal of the resistance) usually in units of nano-Siemens (nS), and V is the voltage difference across the membrane in millivolts (mV). The effective voltage for any ionic current is the difference between the membrane potential V and the equilibrium potential E of the ionic species in question. For any ionic species, the **reversal potential** is identical to E, because that is the membrane potential at which the sign of the current I changes. The **equilibrium potential** E is determined by the **Nernst equation**, which states that:

$$E = \frac{RT}{zF} \ln \left(\frac{C_{\text{out}}}{C_{\text{in}}} \right) \qquad (2.16)$$

where z is the charge on the ion in question, and C_{out} and C_{in} are the respective concentrations of the ion outside and inside the cell. R and F are respectively the thermodynamic gas constant and the Faraday constant, and T is the temperature in degrees Kelvin. At 20°C the ratio $RT/F = 25\,\text{mV}$. The Nernst equation can be derived from thermodynamic principles.

Let us consider a simple patch of dendrite that has a passive current due to ionic leakage through the membrane plus channels for excitatory and inhibitory postsynaptic potentials (EPSPs and IPSPs). Due to the capacitive properties of the lipid bilayer of the

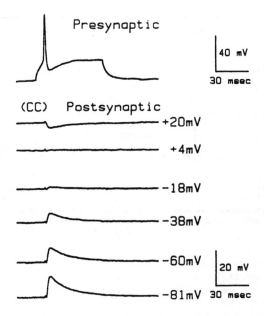

Fig. 2.6 Postsynaptic responses to a single presynaptic spike for various levels of the postsynaptic resting potential from +20 mV to −81 mV (reproduced with permission, Huettner and Baughman, 1988). The reversal or equilibrium potential is close to 0.0 mV.

membrane, the equation for the change of membrane potential V with time is:

$$\frac{dV}{dt} = -\frac{1}{\tau}\{g_l(V - E_l) + g_e(V - E_e) + g_i(V - E_i)\} \tag{2.17}$$

In this equation, g_l and E_l are the conductance and equilibrium potential for the leakage current; g_e and E_e refer to EPSP ion channels; and g_i and E_i are the conductance and equilibrium potential for the IPSP channels. The IPSPs simulated here are due to GABAa channels, which are present on all cortical neurons (Gutnick and Moody, 1995), including those of human neocortex (McCormick, 1989). The equilibrium potential for our excitatory synapse may be obtained from the data in Fig. 2.6, which shows EPSPs generated at a synapse for different values of the resting membrane potential V (Huettner and Baughman, 1988). It is apparent that the reversal potential occurs close to 0 mV, which is the value we shall adopt. Similarly, the data in Fig. 2.7 show that the early GABAa equilibrium potential (due to Cl⁻ ions) is about −75 mV in human neurons (McCormick, 1989). The average resting potential for human neurons is also about −75 mV (Avoli *et al.*, 1994). If we set $g_l = 1$ nS and $\tau = 12.5$ ms, (2.17) becomes:

$$\frac{dV}{dt} = -\frac{1}{12.5}\{(V + 75) + g_e V + g_i(V + 75)\} \tag{2.18}$$

When there is no transmitter release at either excitatory or inhibitory synapses, $g_e = 0$ and $g_i = 0$, and it is easy to see that $V = -75$ mV is the equilibrium value of (2.18). Suppose

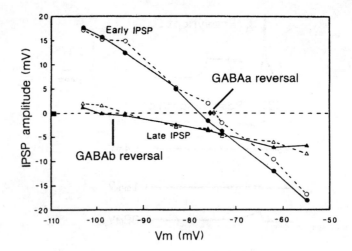

Fig. 2.7 IPSP amplitudes in a human cortical neuron plotted as a function of postsynaptic resting potential (reproduced with permission, McCormick, 1989). The GABAa reversal potential (early IPSP) occurs at about −75 mV. The late IPSP, due to a GABAb synapse, has a lower resting potential of about −95 mV.

now that an excitatory synapse becomes active. We represent this by letting $g_e = 2\,\text{nS}$ for a period of 1.0 ms. Equation (2.18) thus becomes:

$$\frac{\mathrm{d}V}{\mathrm{d}t} = -\frac{1}{12.5}\{(V + 75) + 2V\} = -\frac{1}{12.5}(3V + 75) \quad \text{for } 0 \le t \le 1$$
$$V(0) = -75$$

$$(2.19)$$

Thus,

$$\frac{\mathrm{d}V}{\mathrm{d}t} = \frac{3}{12.5}(-V - 25)$$

where it is assumed that V is at the resting potential before the EPSP. Using Theorem 1 the solution may be derived easily:

$$V(t) = A\,\mathrm{e}^{-0.24t} - \frac{3}{12.5}\,\mathrm{e}^{-0.24t}\int_0^t \mathrm{e}^{0.24t'}\,25\,\mathrm{d}t'$$

where $0.24 = 3/12.5$. Evaluation of the integral gives:

$$V(t) = -75\,\mathrm{e}^{-0.24t} - 25(1 - \mathrm{e}^{-0.24t}) = -50\,\mathrm{e}^{-0.24t} - 25 \qquad (2.20)$$

where A has been chosen so that $V(0)$ satisfies the initial condition. After 1.0 ms, $V(t) = -64.3\,\text{mV}$, so we see that this conductance change due to synaptic transmission has depolarized the neuron by 10.7 mV, which is just about the EPSP amplitude obtained for a membrane potential of −81 mV in Fig. 2.6. Equation (2.20) is plotted as a solid line

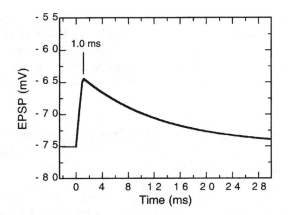

Fig. 2.8 Piecewise linear solution of (2.18) for an EPSP in which the conductance changes abruptly at 1.0 ms. The shape is very similar to the −81 mV record in Fig. 2.6.

for the first millisecond in Fig. 2.8. (The line appears almost straight only because 1.0 ms is such a short time relative to the time constant of 12.5 ms.)

After 1.0 ms, we assume that the effects of the neurotransmitter are terminated, so $g_e = 0$ again (synaptic conductance changes do not end quite so abruptly, but this is still a reasonable approximation). To solve for the decay in the postsynaptic potential, therefore, we must now solve the equation:

$$\frac{\mathrm{d}V}{\mathrm{d}t} = -\frac{1}{12.5}\{V + 75\} \quad \text{for } t > 1$$
$$V(1) = -64.3 \tag{2.21}$$

Note that our new initial condition is $V(1) = -64.3\,\text{mV}$, which is not the equilibrium value. Note also that this initial condition occurs at $t = 1$ ms, because that is the time at which the synaptic conductance reverts to its original value. To shift the initial condition from 0 to 1 ms, it is only necessary to replace t by $(t − 1)$ in the solution, so Theorem 1 produces the result:

$$V(t) = -64.3\,\mathrm{e}^{-0.08(t-1)} - 75(1 - \mathrm{e}^{-0.08(t-1)}) = 10.7\,\mathrm{e}^{-0.08(t-1)} - 75 \tag{2.22}$$

for $t \geq 1$ ms. This function is also plotted in Fig. 2.8 beginning at $t = 1$ ms. There is a striking similarity between our mathematical EPSP and the shape of the EPSP obtained in the −81 mV resting condition shown in Fig. 2.6. Note that the decay back to equilibrium in (2.22) is three times slower than the initial rise of the potential in (2.20). This is a mathematical consequence of membrane conductance changes during synaptic transmission, and it indicates that this problem is actually nonlinear! We have solved for the EPSP in this case using what is called a piecewise linear approximation, which is the simplest approximation to an inherently nonlinear dynamical problem.

Let us see just how nonlinear synaptic interactions typically are. Suppose that the dendrite is at the −75 mV resting potential, and synaptic activation causes a change from

$g_i = 0$ to $g_i = 12$ nS for 1.0 ms. Referring back to eqn (2.18), you will see that this will produce *no change in membrane potential*, because -75 mV is both the resting potential and the equilibrium potential for the GABAa synapse. Can an inhibitory synapse that has no effect by itself reduce the excitation produced by an EPSP? To answer this question, let us assume that an EPSP and an IPSP occur simultaneously in the dendrite. Now we must solve (2.18) with $g_e = 2$ nS and $g_i = 12$ nS for $0 \le t \le 1$:

$$\frac{dV}{dt} = -\frac{1}{12.5}\{(V+75) + 2V + 12(V+75)\} = -\frac{1}{12.5}\{15V + 975\} \qquad (2.23)$$

The solution is again found using Theorem 1:

$$V(t) = -10\,e^{-1.2t} - 65 \qquad (2.24)$$

Now the peak depolarization at $t = 1$ ms due to the EPSP has dropped to $V = -68$ mV. This is a drop from the peak of -64.3 mV when the EPSP occurred without a concurrent IPSP. Thus, an IPSP that has no effect on the membrane potential when it occurs alone has reduced the effect of a concurrent EPSP by about 35%. As a result of this effect, GABAa synapses are frequently termed **shunting synapses** because their effect is to shunt or short circuit the depolarizing current produced by EPSPs. This highly nonlinear interaction is essentially divisive rather than subtractive, although neural modelers sometimes assume that inhibition is inherently subtractive. Only when synapses are fairly far apart on the dendritic tree or on different dendritic branches do EPSPs and IPSPs interact in a manner approximating addition and subtraction.

2.5   Exercises

1. Solve the following equation for the initial condition $x(0) = 17$.

$$\frac{dx}{dt} = \frac{1}{13}(-x+5)$$

2. Solve the following equation for the response rate $R(t)$ of a neuron for each of the following values of the postsynaptic potential: $P = 10, 20$, and 30. Plot your results on a single graph for times up to 100 ms assuming that $R(0) = 5$ in each case.

$$\frac{dR}{dt} = \frac{1}{20}\left(-R + \frac{50P^4}{15^4 + P^4}\right)$$

3. Prove that eqn (2.9) follows from (2.6) and (2.8) by solving using Theorem 1. Now add a fourth stage, call it $w(t)$ to (2.7) governed by the equation:

$$\frac{dW}{dt} = \frac{1}{\tau}(-W + kz)$$

Derive the solution $w(t)$ for $w(0) = 0$. Does this agree with eqn (2.10)?

4. In this problem we will explore a piecewise linear approximation to action potential generation. There are two ionic channels, one for Na^+ ($E_{Na} = 50\,mV$) and one for K^+ ($E_K = -90\,mV$), resulting in the equation:

$$\frac{dV}{dt} = -4\{g_{Na}(V - 50) + g_K(V + 90)\}$$

where $g_K = 0.6$ is the K^+ resting conductance and $g_{Na} = 0.1$ is the Na^+ resting conductance. First solve for the equilibrium or resting potential and use this as the initial condition for the following. After threshold depolarization, the action potential may be approximated by three successive stages:

(a) A brief but large Na^+ conductance increase, so let $g_{Na} = 5.0$ and $g_K = 0.6$ for $0 \le t < 1$ ms.
(b) Na^+ conductance decreases to zero and K^+ conductance increases, so $g_{Na} = 0$ and $g_K = 2.0$ for $1 \le t < 4$ ms.
(c) Both conductances return to normal, so $g_{Na} = 0.1$ and $g_K = 0.6$ for $t \le 4$ ms. Calculate the piecewise linear solution to this problem and plot the result. This problem should give you an intuitive feel for the role of Na^+ and K^+ conductance changes in generating the action potential.

5. Generalize eqn (2.19) to the case where there are N simultaneous excitatory synaptic events. Assuming that there is no inhibition and that $g_e = 2\,nS$, the equation becomes:

$$\frac{dV}{dt} = -\frac{1}{12.5}\{(V + 75) + 2NV\} \quad \text{for } 0 \le t \le 1$$
$$V(0) = -75$$

Obtain the analytical solution for $0 \le t \le 1$ ms (you need not solve for the decay phase). Plot the peak EPSP value $V(1)$ for $1 \le N \le 12$. Do EPSPs add linearly, or does the biophysics produce saturation effects?

6. Solve eqn (2.18) for the case where there is one excitatory synaptic event but N concurrent inhibitory events. Assuming that $g_e = 2\,nS$, and $g_i = 12\,nS$ the equation becomes:

$$\frac{dV}{dt} = -\frac{1}{12.5}\{(V + 75) + 2V + 12N(V + 75)\}$$
$$V(0) = -75$$

Obtain the analytical solution for $0 \le t \le 1$ ms (you need not solve for the decay phase). Plot the peak EPSP value $V(1)$ for $0 \le N \le 8$. Discuss the results of this shunting inhibition in terms of the linearity or nonlinearity of synaptic effects.

3   *Two-dimensional systems and state space*

The first order differential equations discussed in the previous chapter are limited in their applicability, as a single equation can only describe responses of a single neuron or a single postsynaptic potential, etc. We are mainly interested in studying systems in which many neurons interact to produce complex behavior. This leads us naturally to consider what happens when several first order differential equations are coupled together to produce a neural network. As all of the dynamical features of multi-component linear systems can occur when only two component processes interact, this chapter will focus on an exhaustive classification of the behavior of such systems. (As will be seen later, however, certain *nonlinear* phenomena, such as chaos, can only occur in systems with three or more components.) As two-component systems can describe responses of two interacting model neurons, their responses can be regarded as defining two axes in a plane. This insight leads to the concept of the **state space** of a dynamical system, which will be developed here and employed throughout the book.

3.1   Second order equations and normal form

By definition, a second order differential equation is one which contains only first and second derivatives of the dependent variable with respect to time t, but no higher derivatives. Without doubt, the most famous second order differential equation ever written is Newton's (1687) second law of motion, $f = ma$, where f is force, m is mass, and a is acceleration. In a typical textbook example of a weight at the end of a spring, the restorative force $f = -kx$, where x is the position relative to the resting length of the spring, and the air resistance contributes a second force term that is proportional to velocity. As acceleration is the rate of change of velocity, $a = \mathrm{d}v/\mathrm{d}t$, and velocity is the rate of change of position, $v = \mathrm{d}x/\mathrm{d}t$, Newton's second law for a spring can be written as:

$$m\frac{\mathrm{d}^2 x}{\mathrm{d}t^2} = -kx - r\frac{\mathrm{d}x}{\mathrm{d}t} \quad \text{or}$$

$$m\frac{\mathrm{d}^2 x}{\mathrm{d}t^2} + r\frac{\mathrm{d}x}{\mathrm{d}t} + kx = 0 \tag{3.1}$$

Given our focus on neuroscience, it is interesting to note that this same equation provides a simplified description of the contraction of muscle fibers, as will be discussed below.

Let us now proceed to solve (3.1). Assuming k and r are constants, we may proceed in the same manner as (2.1) in the last chapter by substituting an exponential function $x = a \exp(\lambda t)$:

$$m\frac{d^2 x}{dt^2} + r\frac{dx}{dt} + kx = m\lambda^2 a\,e^{\lambda t} + r\lambda a\,e^{\lambda t} + ka\,e^{\lambda t} = 0 \tag{3.2}$$

Factoring (3.2), we obtain the following **characteristic equation** for λ:

$$m\lambda^2 + r\lambda + k = 0 \tag{3.3}$$

This quadratic equation is readily solved to yield:

$$\lambda = \frac{-r \pm \sqrt{r^2 - 4mk}}{2m} \tag{3.4}$$

Note that (3.4) yields two values, λ_1 and λ_2, that solve (3.1). This means that the general form of $x(t)$ will be a linear combination of two exponentials:

$$x(t) = a\,e^{\lambda_1 t} + b\,e^{\lambda_2 t} \tag{3.5}$$

where a and b are constants. The particular values of a and b must be determined from the **initial conditions**. As there are two constants, there must be two initial conditions for this second order problem, and these are generally taken to be the values of both x and the velocity (dx/dt) at $t = 0$.

Rather than pursue this example further, however, let us now see how (3.1) can be recast as a system with two interacting components. All we need do is define a new variable, y, which we equate to the velocity, dx/dt, and then substitute this into the first equation in (3.1). The resulting two-equation system is:

$$\frac{dx}{dt} = y$$
$$\frac{dy}{dt} = \frac{1}{m}(-kx - ry) \tag{3.6}$$

Thus, we see that Newton's second law can be re-described by a pair of equations that incorporate interactions between position and velocity. When a higher order differential equation is written as a system of coupled first order differential equations, as in (3.6), the equation will be said to be in **normal form**. This is a more general definition of normal form than is employed in some other texts, but it will suffice here. Almost all higher order differential equations of scientific interest can be written in normal form using the same stratagem of defining new variables to represent the various derivatives of the function. To take one third order example, the equation:

$$\frac{d^3 x}{dt^3} + a\frac{d^2 x}{dt^2} + b\frac{dx}{dt} + cx = g(t) \tag{3.7}$$

can be written in normal form as the equivalent system of first order equations:

$$\frac{dx}{dt} = y$$

$$\frac{dy}{dt} = z \qquad (3.8)$$

$$\frac{dz}{dt} = -az - by - cx + g(t)$$

Nonlinear differential equations can always be written in normal form so long as they are linear in the highest order derivative. Differential equations are termed **quasi-linear** if the term containing the highest order derivative is linear. Virtually all equations in both theoretical physics and neuroscience fall into this category. Let us consider just one example, the famous van der Pol (1926) equation, which was developed as the first mathematical model of the heartbeat:

$$\frac{d^2x}{dt^2} - a(1 - x^2)\frac{dx}{dt} + bx = 0$$

This is converted to normal form in the same manner as our linear examples:

$$\frac{dx}{dt} = y$$

$$\frac{dy}{dt} = a(1 - x^2)y - bx \qquad (3.9)$$

A slightly modified version of this equation (the FitzHugh–Nagumo equation) has been proposed as a simple approximation to the Hodgkin–Huxley equations and will be discussed in Chapter 8. All quasi-linear equations can be cast into normal form, and it is thus the most general form for representing differential equations. It is important to note, however, that nonlinear differential equations in normal form cannot always be converted back into a single higher order equation.

Normal form equations will be very important to our study of neurodynamics for two reasons. First, the most powerful methods for solving nonlinear differential equations numerically by computer require that the equations be cast in normal form (see Chapter 5). More importantly, it is natural to think of neural problems in terms of their components, typically either neurons, ionic species, or ion channels. In these cases, one writes one first order equation for each component and studies the resultant normal form system. As a simple example, suppose that we have two neurons, one excitatory (E), and one inhibitory (I), that are connected to form an inhibitory feedback loop, as illustrated in Fig. 1.3. If each neuron is described by its spike rate as in (2.12), this two neuron network would be described by:

$$\frac{dE}{dt} = \frac{1}{\tau_E}[-E + S(p - bI)]$$

$$\frac{dI}{dt} = \frac{1}{\tau_I}[-I + S(aE)] \qquad (3.10)$$

where S is the Naka–Rushton function defined in (2.11). The constants a and b here represent weights of the synapses from $E \rightarrow I$ and from $I \rightarrow E$ respectively, while p is the excitatory input to the E neuron. Nonlinear neural equations like these have been analyzed in detail (Wilson and Cowan, 1972), and we shall study them in Chapters 6–8 after we have developed some further background.

3.2 Solution of second order systems

In the last section we saw that all quasi-linear second order systems can be written in normal form. Restricting ourselves to linear systems at present, a general second order system may be written in vector notation as:

$$\frac{d}{dt}\begin{pmatrix} x \\ y \end{pmatrix} = \begin{pmatrix} a_1 & a_2 \\ a_3 & a_4 \end{pmatrix}\begin{pmatrix} x \\ y \end{pmatrix} + \begin{pmatrix} b_1 \\ b_2 \end{pmatrix}$$

or

$$\frac{d\vec{X}}{dt} = \overset{\leftrightarrow}{A}\vec{X} + \vec{B} \quad \text{where} \quad \vec{X} = \begin{pmatrix} x \\ y \end{pmatrix} \tag{3.11}$$

(The double headed arrow over the A will always designate a square matrix.) All of the as and bs in (3.11) are assumed to be constants. In analyzing any system of differential equations such as this, the first step is to ascertain the location of the **equilibrium point** or **steady state**. This occurs where there is no change in the system with time, so the temporal derivatives must all vanish simultaneously at the steady state. In vector form, the solution is:

$$\frac{d\vec{X}_{eq}}{dt} = \begin{pmatrix} 0 \\ 0 \end{pmatrix}$$

so

$$\vec{X}_{eq} = -\overset{\leftrightarrow}{A}^{-1}\vec{B} \tag{3.12}$$

where A^{-1} is the inverse of matrix A. (Rather than calculating A^{-1} outright, it is more efficient and accurate computationally to solve for X_{eq} using a technique known as Gaussian elimination with back substitution, see Press *et al.*, 1986.) Solving (3.12) is equivalent to solving two simultaneous linear equations. To simplify this, the disk accompanying this book contains the MatLab™ script **Equilibrium.m** for solving (3.12) using a Gaussian elimination algorithm. The user simply types in the values of the four as and the two bs, and MatLab solves (3.12) for the coordinates of the steady state. As one example using the program:

$$\overset{\leftrightarrow}{A} = \begin{pmatrix} -9 & -5 \\ 1 & -3 \end{pmatrix} \quad \vec{B} = \begin{pmatrix} 7 \\ 1 \end{pmatrix}$$

so

$$\vec{X}_{eq} = \begin{pmatrix} \frac{1}{2} \\ \frac{1}{2} \end{pmatrix} \tag{3.13}$$

In this example, as in any linear system, there will be a unique steady state, as sets of simultaneous linear equations have (except in pathological cases) a unique solution. Because of this fact, (3.11) can be rewritten by translating the steady state to the origin. This is simply done by subtracting the coordinates of the steady state solution from x and y, with the result:

$$\vec{X}' \equiv \vec{X} - \vec{X}_{eq} \quad \text{so} \quad \frac{d\vec{X}'}{dt} = \overleftrightarrow{A}\vec{X}' \tag{3.14}$$

This transformation eliminates the vector B from the equation without otherwise affecting the dynamics. Once (3.14) has been solved for X', X_{eq} is simply added to produce the final solution.

We can now derive the general solution to (3.14). As in the simple case of (3.1) above, where we obtained the characteristic equation (3.3) by substituting an exponential function for x, we can solve (3.14) by substituting a vector of exponentials for X with arbitrary coefficients a and b:

$$\vec{X}' \equiv \begin{pmatrix} a\,e^{\lambda t} \\ b\,e^{\lambda t} \end{pmatrix}, \quad \text{so} \quad \frac{d\vec{X}'}{dt} = \lambda\vec{X}' = \overleftrightarrow{A}\vec{X}' \tag{3.15}$$

The right-hand equality here may be rearranged to give:

$$\{\overleftrightarrow{A} - \lambda\overleftrightarrow{I}\}\vec{X}' = \begin{pmatrix} 0 \\ 0 \end{pmatrix}, \quad \text{where} \quad \overleftrightarrow{I} = \begin{pmatrix} 1 & 0 \\ 0 & 1 \end{pmatrix} \tag{3.16}$$

I is known as the **identity matrix**. One trivial solution of this equation is $X' = 0$, which is clearly of no scientific interest. This, however, is the unique solution to the equation if the matrix $\{A - \lambda I\}$ has an inverse (the reader should convince herself of this by applying the inverse of this matrix to both sides of the equation). Therefore, (3.16) can have non-trivial solutions only if $\{A - \lambda I\}$ does not have an inverse. As is derived in linear algebra texts, this can occur only if the determinant of this matrix vanishes:

$$|\overleftrightarrow{A} - \lambda\overleftrightarrow{I}| = 0 \tag{3.17}$$

The determinant in (3.17) is just a quadratic polynomial, so (3.17) generates the **characteristic equation** for the system (3.14). To continue our earlier example, if we take A from (3.13), eqn (3.17) yields:

$$|\overleftrightarrow{A} - \lambda\overleftrightarrow{I}| = \left| \begin{pmatrix} -9 - \lambda & -5 \\ 1 & -3 - \lambda \end{pmatrix} \right| = (-9 - \lambda)(-3 - \lambda) + 5 = 0 \tag{3.18}$$

The solutions to this quadratic equation are $\lambda = -4, -8$. These solutions of the characteristic equation are called **eigenvalues**. We have thus derived a theorem applicable to all cases where the two eigenvalues are not identical (that unusual case is treated later):

Theorem 2: The solution to the second order differential equation:

$$\frac{\mathrm{d}\vec{X}}{\mathrm{d}t} = \overleftrightarrow{A}\vec{X} + \vec{B}$$

is obtained by first solving for the equilibrium state:

$$\vec{X}_{\mathrm{eq}} = -\overleftrightarrow{A}^{-1}\vec{B}$$

Next, one determines the two **eigenvalues**, λ_1 and λ_2, of the characteristic equation:

$$|\overleftrightarrow{A} - \lambda\overleftrightarrow{I}| = 0$$

where I is the identity matrix. Assuming $\lambda_1 \neq \lambda_2$, the solution is:

$$\vec{X} = \begin{pmatrix} a_1\,\mathrm{e}^{\lambda_1 t} \\ b_1\,\mathrm{e}^{\lambda_1 t} \end{pmatrix} + \begin{pmatrix} a_2\,\mathrm{e}^{\lambda_2 t} \\ b_2\,\mathrm{e}^{\lambda_2 t} \end{pmatrix} + \vec{X}_{\mathrm{eq}}$$

where a_1, a_2, b_1, and b_2 are constants determined by substitution back into the original equation and by the initial conditions.

The task of solving for the coefficients a_1, a_2, b_1, and b_2 is really quite mechanical and we will therefore let MatLab^{TM} do the work for us. However, it is important to have an understanding of the operations involved. We will need four equations to solve for the four unknown parameters. To obtain the first two equations, we substitute X back into the original differential equation, which produces two algebraic equations in a_1, a_2, b_1, and b_2. This procedure is known as obtaining the **eigenvectors** of the A matrix. Two further equations are required to solve for the remaining two coefficients, and these equations are supplied by the initial conditions.

Let us see how this works out in practice by considering an example. We have seen from (3.15) that substitution of a vector of exponential functions into a second order differential equation in normal form reduces it to a matrix equation of the form:

$$\lambda\vec{X} = \overleftrightarrow{A}\vec{X}$$

In this equation λ is an **eigenvalue** of the matrix A. From the example in (3.18) we have found that $\lambda = -8$ is one of the eigenvalues of the matrix:

$$\overleftrightarrow{A} = \begin{pmatrix} -9 & -5 \\ 1 & -3 \end{pmatrix} \tag{3.19}$$

Therefore, the general solution X for $\lambda = -8$ must satisfy:

$$-8 \begin{pmatrix} a_1\,e^{-8t} \\ b_1\,e^{-8t} \end{pmatrix} = \begin{pmatrix} -9 & -5 \\ 1 & -3 \end{pmatrix} \begin{pmatrix} a_1\,e^{-8t} \\ b_1\,e^{-8t} \end{pmatrix} = \begin{pmatrix} -9a_1\,e^{-8t} & -5b_1\,e^{-8t} \\ a_1\,e^{-8t} & -3b_1\,e^{-8t} \end{pmatrix}$$

Equating the top entries in both vectors gives the result:

$$-8a_1\,e^{-8t} = -9a_1\,e^{-8t} - 5b_1\,e^{-8t}$$

Therefore, $a_1 = -5b_1$. The reader should verify that equating the bottom entries in the vectors gives exactly the same result. This means that either a_1 or b_1 can assume an arbitrary value, and we must therefore use one initial condition to specify that value. Thus, the X vector satisfying our equation is:

$$\vec{X} = b_1 \begin{pmatrix} -5\,e^{-8t} \\ e^{-8t} \end{pmatrix}$$

where b_1 is determined from the initial conditions. This is called an **eigenvector** of the matrix A. If we repeat this procedure with the other eigenvalue, $\lambda = -4$, of A, we will obtain a relationship between a_2 and b_2 defining the second eigenvector of A, which completes the solution.

The procedure for solving linear second order differential equations using Theorem 2 is very important, and fortunately it is sufficiently mechanical so that it can be done entirely by computer. The MatLab program **LinearOrder2.m** produces a complete solution to any two-component, linear system for any initial condition using Theorem 2. For simplicity, it is assumed that the equilibrium state has been translated to the origin. The program then plots the solutions (the state space plot will be discussed later) and prints out a symbolic solution in terms of exponentials that is accurate to about 10^{-9}. All the user need do is type in the A matrix and an initial condition vector B. As an example, we can solve the following differential equation defined by the matrix A in (3.19) for the initial condition X_0 using **LinearOrder2.m**:

$$\frac{d\vec{X}}{dt} = \begin{pmatrix} -9 & -5 \\ 1 & -3 \end{pmatrix} \vec{X} + \begin{pmatrix} 7 \\ 1 \end{pmatrix} \quad \text{with} \quad \vec{X}_0 = \begin{pmatrix} 0 \\ 0 \end{pmatrix}$$

First we solve for the equilibrium point using **Equilibrium.m** to obtain the result:

$$\vec{X}_{eq} = \begin{pmatrix} \frac{1}{2} \\ \frac{1}{2} \end{pmatrix}$$

Next we subtract this from the initial condition vector X_0, to obtain a new initial condition $(X_0 - X_{eq})$. This translates the equilibrium point to the origin. Now we can use the program **LinearOrder2.m** by entering the A matrix and this new initial condition.

The result is:

$$\vec{X} = \begin{pmatrix} -1.25\,e^{-8t} + 0.75\,e^{-4t} \\ 0.25\,e^{-8t} - 0.75\,e^{-4t} \end{pmatrix}$$

Finally, we add X_{eq} to this to obtain the final result:

$$\vec{X} = \begin{pmatrix} -1.25\,e^{-8t} + 0.75\,e^{-4t} \\ 0.25\,e^{-8t} - 0.75\,e^{-4t} \end{pmatrix} + \begin{pmatrix} \frac{1}{2} \\ \frac{1}{2} \end{pmatrix}$$

You can verify that $X(0) = 0$, and $X(\infty) = X_{eq}$. As you can see, MatLab^{TM} removes the drudgery from this calculation!

3.3 Negative feedback in the retina

Now let us apply our knowledge of linear second order differential equations to a real problem in neuroscience: negative feedback on cone photoreceptors in the primate retina. There is extensive evidence that the cones stimulate horizontal cells, while horizontal cells provide inhibitory (or negative) feedback onto the cones (Burkhardt, 1993, 1995). Schnapf *et al.* (1990) recorded from single primate cones and constructed a linear feedback circuit to explain their results, which are plotted in Fig. 3.1. The cone current in response to a step change in luminance shows a transient overshoot followed by a slight undershoot before reaching its equilibrium value. In slightly modified and simplified form, the Schnapf *et al.* (1990) circuit can be adapted to describe cone (C) and horizontal (H) cell interactions by the following differential equations:

$$\frac{dC}{dt} = \frac{1}{\tau_C}(-C - kH + L)$$
$$\frac{dH}{dt} = \frac{1}{\tau_H}(-H + C)$$

(3.20)

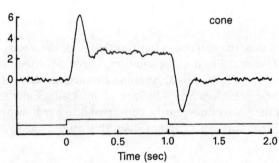

Fig. 3.1 Electrical response (current) of a primate cone photoreceptor to a 1.0 s light stimulus turned on at $t = 0$ (reproduced with permission, Schnapf *et al.*, 1990). The response at stimulus offset is the mirror image of that at stimulus onset.

L is the light-induced stimulus due to phototransduction, and k is a constant representing the strength or gain of the inhibitory feedback from the horizontal cell. (Schnapf *et al.* (1990) related their circuit to feedback within the cone itself rather than feedback from the horizontal cells, but the same equations describe either case.) Appropriate values for the time constants and the gain are $\tau_C = 0.025$ s, $\tau_H = 0.08$ s, and $k = 4$. Also, we shall assume $C = 0$ and $H = 0$ are the initial conditions. Thus, (3.20) becomes:

$$\frac{dC}{dt} = 40(-C - 4H + L)$$
$$\frac{dH}{dt} = 12.5(-H + C)$$

(3.21)

In analyzing (3.21) let us first determine the equilibrium state on the assumption that the light level L is constant. Equating the time derivatives to zero, the equilibrium is determined by:

$$-C - 4H + L = 0$$
$$-H + C = 0$$

The solutions to this are easily seen to be:

$$C = \frac{L}{5} \quad \text{and} \quad H = C$$

(3.22)

Thus, the subtractive feedback in this neural circuit reduces L by the factor $1/5$ in the steady state. If $L = 10$, the equilibrium state is $C = 2$, $H = 2$. Next, we shift the steady state to the origin by subtracting the equilibrium values from the initial conditions. The **homogeneous** equation thus obtained from (3.21) is:

$$\frac{d}{dt}\begin{pmatrix} C' \\ H' \end{pmatrix} = \begin{pmatrix} -40 & -160 \\ 12.5 & -12.5 \end{pmatrix} \begin{pmatrix} C' \\ H' \end{pmatrix}$$

(3.23)

where C' and H' denote the variables after translation of the steady state to the origin. We may now use Theorem 2 to find the solution. The initial conditions minus the equilibrium values are $C' = -2$, $H' = -2$. Running **LinearOrder2.m** with these initial conditions produces the eigenvalues $\lambda = -26.25 \pm 42.56i$. Using Euler's formula (1.15), the imaginary parts of the eigenvalues may be converted into a real combination of sine and cosine functions. The MatLab program does this automatically, and produces the solution:

$$C(t) = -2\,e^{-26.25t}\cos(42.56t) + 8.17\,e^{-26.25t}\sin(42.56t) + 2$$
$$H(t) = -2\,e^{-26.25t}\cos(42.56t) - 1.23\,e^{-26.25t}\sin(42.56t) + 2$$

(3.24)

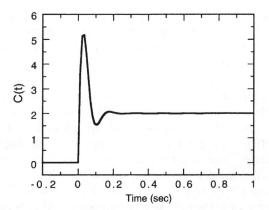

Fig. 3.2 Solution $C(t)$ to (3.21) as represented by (3.24). The response at stimulus onset ($t = 0$) is very similar to that of the primate cone in Fig. 3.1.

The equilibrium values (2, 2) have been added in accord with Theorem 2. $C(t)$ is plotted as a function of time in Fig. 3.2. The similarity between the solution (3.24) and the physiological data in Fig. 3.1 is apparent. The response at the termination of light stimulation has not been plotted in Fig. 3.2, but it is mathematically the mirror image of the response to stimulus onset, and this is evident from the experimental data.

3.4 Stability and state space

Based on Theorem 2, all possible solutions of second order differential equations with constant coefficients may now be grouped into a small number of categories. This categorization is based on the fact that the characteristic equation is quadratic and therefore must have exactly two roots. We have already seen that the unique equilibrium point for a linear differential equation can always be translated to the origin. Thus, it might be expected that solutions can be characterized by their behavior near the origin. An important concept here is that of a trajectory. A **trajectory** is the entire time course of the solution of a differential equation from $t = 0$ to $t = \infty$, given a particular initial condition. Thus, any differential equation defines an infinite number of trajectories, each corresponding to a different initial condition. Additionally, note that an equilibrium point is itself a trajectory, since a trajectory starting at equilibrium must by definition remain there for all eternity!

We may now define the concepts of stability, asymptotic stability, and instability of an equilibrium point. An equilibrium point of a system of differential equations is **asymptotically stable** if all trajectories starting within a region containing the equilibrium point decay to that point exponentially as $t \to \infty$. Conversely, the equilibrium is **unstable** if at least one trajectory beginning in a region containing the point leaves that region permanently. Finally, an equilibrium is **stable** or **neutrally stable** if nearby trajectories remain nearby as $t \to \infty$ but do not approach asymptotically.

Trajectories for second order differential equations may be conveniently plotted in a two-dimensional space in which the variables of the equations define the axes. Such a space is known as a **state space** or a **phase space**, terms that will be used interchangeably. Analytical solution of the system equations yields explicit functions of t, such as $C(t)$ and $H(t)$ in (3.24). If time is regarded as a parameter, then it is possible to plot the entire course of each system trajectory as a curve in the phase space. Phase space plots reveal the relationship between the variables at all points on the trajectory. Both the time dependence of solutions and the phase space are plotted by **LinearOrder2.m**, and experimentation with that program will provide a great deal of insight into the nature of phase space trajectories. Note that the phase space plot contains arrows indicating the local directions in which trajectories flow at various points in the space.

All possible solutions to the second order differential equation described in Theorem 2 may now be categorized and their phase space trajectories illustrated. From the fact that the characteristic equation is quadratic with real coefficients, the two eigenvalues must either both be real, both be pure imaginary, or else be a complex conjugate pair. The possibilities are enumerated below, and typical equations are given along with solutions for initial conditions $x(0) = 1$, $y(0) = 1$.

A **spiral point** results when the eigenvalues are a complex conjugate pair. Thus, the solutions are in the form of exponentials multiplied by a sine and a cosine. The spiral point is asymptotically stable if the real part of the eigenvalues is negative. A typical trajectory of an asymptotically stable spiral point is plotted in the upper left of Fig. 3.3, which makes

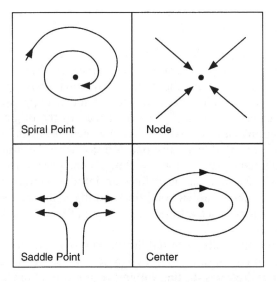

Fig. 3.3 Typical phase plane trajectories for the four characteristic equilibrium points of linear dynamical systems: spiral point, node, saddle point, and center. The horizontal and vertical axes represent the two variables describing each system. Equilibrium points are depicted by a dot in the center of each plot. Trajectories illustrated for the spiral point and node are asymptotically stable. Saddle points are always unstable, while centers are stable but not asymptotically stable.

clear that the trajectory is in fact a spiral. The following equation provides an example of a system with an asymptotically stable spiral point:

$$\frac{d}{dt}\begin{pmatrix} x \\ y \end{pmatrix} = \begin{pmatrix} -2 & -16 \\ 4 & -2 \end{pmatrix}\begin{pmatrix} x \\ y \end{pmatrix}$$

solution : (3.25)

$$x(t) = e^{-2t}\cos(8t) - 2e^{-2t}\sin(8t)$$

$$y(t) = e^{-2t}\cos(8t) + 0.5e^{-2t}\sin(8t)$$

The equilibrium point is a **node** if the eigenvalues are both real and have the same sign. The node is asymptotically stable if the eigenvalues are negative, and it is unstable if they are positive. Typical trajectories at a node are plotted in Fig. 3.3, and an illustrative example is:

$$\frac{d}{dt}\begin{pmatrix} x \\ y \end{pmatrix} = \begin{pmatrix} -2 & 4 \\ 0 & -3 \end{pmatrix}\begin{pmatrix} x \\ y \end{pmatrix}$$

solution : (3.26)

$$x(t) = 5e^{-2t} - 4e^{-3t}$$

$$y(t) = e^{-3t}$$

A **saddle point** occurs when both eigenvalues are real but have opposite signs. Because one eigenvalue is positive, all saddle points are unstable. As illustrated in Fig. 3.3, trajectories approach a saddle point along one axis (y in this example) but diverge from it along a different axis (x in this case). Typical equations generating a saddle point are:

$$\frac{d}{dt}\begin{pmatrix} x \\ y \end{pmatrix} = \begin{pmatrix} 2 & -1 \\ 0 & -3 \end{pmatrix}\begin{pmatrix} x \\ y \end{pmatrix}$$

solution : (3.27)

$$x(t) = 0.8e^{2t} + 0.2e^{-3t}$$

$$y(t) = e^{-3t}$$

The final possibility is that the pair of eigenvalues are pure imaginary, and this condition defines a **center**. Euler's formula (1.15) in this case dictates that all trajectories must be a sum of a sine and a cosine of the same frequency. Because of this, all trajectories around a center will be strictly periodic oscillations. Because any periodic function repeats itself, phase space trajectories around a center will always be closed circular or ellipsoidal shapes like those plotted in the lower right of Fig. 3.3. A typical example of a system with a center is:

$$\frac{d}{dt}\begin{pmatrix} x \\ y \end{pmatrix} = \begin{pmatrix} 1 & -2 \\ 5 & -1 \end{pmatrix}\begin{pmatrix} x \\ y \end{pmatrix}$$

solution :   (3.28)

$$x(t) = \cos(3t) - 0.33\sin(3t)$$
$$y(t) = \cos(3t) + 1.33\sin(3t)$$

As the nature of the solutions and their stability are determined by the eigenvalues of the characteristic equation, all trajectories for all possible initial conditions will have the same qualitative behavior in any given linear system. However, this holds true only for linear systems, and it must be qualified when nonlinear systems are discussed. Indeed, the power and elegance of phase space representations can only be fully appreciated in nonlinear dynamics.

3.5   Critical damping and muscle contraction

Theorem 2 covers all possible second order linear differential equations except for one: the case where both roots of the characteristic equation are identical. The case of two identical roots is generally called **critical damping** for historical reasons deriving from physics. Critical damping is an exceptional case, as the probability that the coefficients of the characteristic equation, if chosen at random, would generate identical eigenvalues is zero. Nevertheless, it is easy to construct mathematical examples in which the two eigenvalues are identical, and these will be solved now. Physiologically, critical damping is significant because it represents the simplest approximation to the dynamics of muscle contraction. In addition, the cascades of equations treated in the previous chapter are another case of critical damping (because the time constants of all stages are identical).

Let us motivate critical damping by considering a simple model of muscle contraction. Figure 3.4 plots data on the force generated by a cat soleus muscle as a function of length at two different levels of motorneuron activation (Rack and Westbury, 1969). Although the overall curves are nonlinear, above an equilibrium length, x_0, the force generated is nearly linear over the considerable range indicated by the solid lines. No force is generated for $x < x_0$, a state where the muscle is relaxed. The equilibrium length shifts to smaller values as the level of motorneuron activity increases. Neural specification of x_0 is believed to be the way in which the central nervous system determines the desired length and force exerted by each muscle in the body, and this is termed the 'equilibrium point hypothesis'. Over the linear range, the force of contraction is of the form $\alpha^2(x - x_0)$, where the length of the muscle x is always assumed to be greater than or equal to x_0. The force of contraction is generated as actin–myosin bonds are formed, after which a configurational change in the myosin head causes the muscle fiber to shorten (see Rothwell, 1994). There is also some frictional resistance to contraction within the muscle due to the

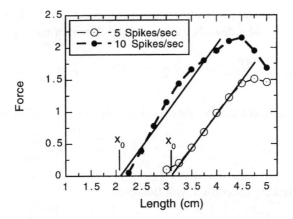

Fig. 3.4 Force generated by cat soleus muscle as a function of muscle length in centimeters (reproduced with permission, Rack and Westbury, 1969). Changing the driving spike frequency from 5 to 10 Hz shifts the linear portion (solid lines) of the force curve to shorter lengths. The stimulating spike rate determines x_0, the equilibrium length at which the force drops to zero.

breaking of actin–myosin bonds following contraction, and this friction is proportional to the velocity of contraction. Using Newton's second law (3.1) to describe muscle contraction leads to the equation:

$$\frac{d^2x}{dt^2} + 2\alpha \frac{dx}{dt} + \alpha^2(x - x_0) = 0 \tag{3.29}$$

It is easy to see that the equilibrium state occurs when $x = x_0$, so we can eliminate x_0 by defining the new variable $x' = x - x_0$. The coefficients in (3.29) have been carefully chosen so that the resulting characteristic equation, obtained by substituting $x' = e^{\lambda t}$, will have identical eigenvalues:

$$\lambda^2 + 2\alpha\lambda + \alpha^2 = 0 \tag{3.30}$$

so $\lambda = -\alpha$. This is the most general formulation of the one case not covered by Theorem 2. One solution will certainly be:

$$x'(t) = a e^{-\alpha t} \tag{3.31}$$

where a is determined by the initial condition. However, there must also be a second solution, just as there are always two exponentials when the eigenvalues differ.

Whatever form the second solution takes, it must certainly be some other function of t. Let us therefore guess that it involves the same exponential as (3.31) and try to find a second solution of the form:

$$x'(t) = h(t) e^{-\alpha t} \tag{3.32}$$

where $h(t)$ must be determined. Substituting (3.32) into (3.29) (with the x_0 term removed by the change of variables described above) and differentiating gives:

$$e^{-\alpha t} \frac{d^2 h}{dt^2} - 2\alpha e^{-\alpha t} \frac{dh}{dt} + \alpha^2 e^{-\alpha t} h + 2\alpha \left(e^{-\alpha t} \frac{dh}{dt} - \alpha e^{-\alpha t} h \right) + \alpha^2 e^{-\alpha t} h = 0 \quad (3.33)$$

You can see that almost all the terms cancel conveniently, so (3.33) reduces to:

$$e^{-\alpha t} \frac{d^2 h}{dt^2} = 0 \quad (3.34)$$

This is easily solved by integration with the result:

$$h = bt \quad (3.35)$$

where b is an arbitrary constant of integration. Substitution of this result into (3.32) shows that the general solution to (3.29) is:

$$x(t) = a e^{-\alpha t} + bt e^{-\alpha t} + x_0 \quad (3.36)$$

where a and b are determined by the initial conditions. This special case, omitted from Theorem 2, is called **critical damping**, because the smallest independent variation in the coefficient of λ in (3.30) produces either two different negative real eigenvalues or else a complex conjugate pair of eigenvalues with negative real parts. Suppose, that is, that the characteristic equation for a dynamical system was:

$$\lambda^2 + (2\alpha + \delta)\lambda + \alpha^2 = 0 \quad (3.37)$$

where δ represents a small parameter variation. The eigenvalues will produce a spiral point for small $\delta < 0$; a node for $\delta > 0$, and critical damping only when $\delta = 0$.

Thus, critical damping forms the boundary between an asymptotically stable node and an asymptotically stable spiral point.

Because of variability in nature, real physiological systems will almost never exhibit critical damping (for aficionados, critical damping forms a set of measure zero in dynamics). However, values of coefficients that are close to those producing critical damping in (3.29) will produce responses very similar to the critical damping condition. Furthermore, critical damping produces the fastest approach to equilibrium for a linear differential equation because the two identical eigenvalues are smallest in this case, thus making the exponential decay fastest. Regarding muscles as an instance of critical damping, this suggests that evolution has optimized motor control for speed of response.

Returning now to our example of muscle contraction, let us solve (3.29) for one of the simplest motor control systems in mammals: the control of rapid or saccadic eye movements. The open and solid symbols in Fig. 3.5 respectively plot angular positions of the eye during 15° and 30° saccades (Clark and Stark, 1974). These eye movements are obviously quite rapid, being completed within about 50 ms of their initiation.

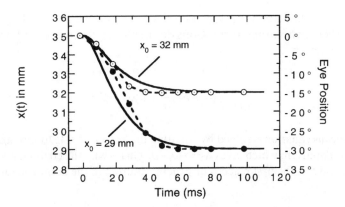

Fig. 3.5 Data on saccadic eye rotations in degrees as a function of time in milliseconds (Clark and Stark, 1974). Saccades of 15° and 30° amplitude are plotted by solid and open circles respectively. The two solid curves plot solutions to (3.29) given by (3.36) for two different values of the equilibrium length x_0.

Eye rotations in the horizontal plane are controlled by a pair of muscles, one of which rotates the eye to the right and the other to the left. Each muscle is approximately 35 mm long when the eye is looking straight ahead, and shortening of the appropriate muscle by either 3 or 6 mm will produce a rotation of approximately 15° or 30° to the appropriate side. Applying (3.29) first to a 15° rotation, therefore, the muscle will initially be at length $x(0) = 35$ mm, and the equilibrium length will be 3 mm shorter, so $x_0 = 32$ mm. Because the eye is initially at rest, the remaining initial condition is that the velocity be zero: $dx/dt = 0$. Based on the data in Fig. 3.5, $\alpha = 0.09$, which is 1/11 ms. Given these values, the solution (3.36) to (3.29) is:

$$x(t) = 3\,e^{-0.09t} + 0.27t\,e^{-0.09t} + 32 \qquad (3.38)$$

The reader should verify that this satisfies the initial conditions. This result is plotted in Fig. 3.5 for $x_0 = 32$ mm. The agreement of the critically damped solution with the saccade data is reasonably good given that the biomechanics of eye movements have certainly been oversimplified here. The figure also shows the solution for $x_0 = 29$ mm, which corresponds reasonably well to the 30° saccade data. The solution in this case has been relegated to the Exercises.

3.6 Responses to time-varying inputs

There is only one more topic that we must consider to complete our analysis of second order linear differential equations. This is the case where the system is driven by a time-varying input or **forcing function**. We saw in Theorem 1 of the previous chapter that the solution to a first order linear equation was the sum of the solution without forcing plus a **convolution integral** that produced a weighted sum of the history of stimulation. As we shall see, this result generalizes to the second order case as well.

To simplify the derivation, the second order form of the differential equation will be employed rather than normal form. Consider the fairly general second order equation with forcing function $S(t)$:

$$\frac{d^2x}{dt^2} + (a+b)\frac{dx}{dt} + abx = S(t) \tag{3.39}$$

The **homogeneous equation** is solved by equating the left-hand side to zero and solving the characteristic equation, which gives $\lambda = -a, -b$. This is why the coefficients were written in terms of a and b in the first place. The general solution of the homogeneous equation is therefore:

$$x(t) = A e^{-at} + B e^{-bt} \tag{3.40}$$

To find a specific solution to (3.39) that involves $S(t)$, we can adopt the same strategy that was used above to solve the critical damping problem. This is termed the **variation of parameters approach**. Thus, we attempt to find a solution where constants A and B in (3.40) are replaced by two functions of t, $g(t)$ and $h(t)$:

$$x = g e^{-at} + h e^{-bt} \tag{3.41}$$

As this is only one equation in the two unknown functions g and h, a second equation must be chosen to fully define the solutions. To obtain a second equation in g and h, differentiate (3.41):

$$\frac{dx}{dt} = -ag e^{-at} - bh e^{-bt} + e^{-at}\frac{dg}{dt} + e^{-bt}\frac{dh}{dt} \tag{3.42}$$

To obtain a second equation determining the solution, let us require that:

$$e^{-at}\frac{dg}{dt} + e^{-bt}\frac{dh}{dt} = 0$$

so: (3.43)

$$\frac{dx}{dt} = -ag e^{-at} - bh e^{-bt}$$

We can now substitute dx/dt from (3.43) and x from (3.41) into (3.39) to give:

$$-a e^{-at}\frac{dg}{dt} - b e^{-bt}\frac{dh}{dt} = S \tag{3.44}$$

In obtaining this result, dx/dt in (3.43) has been differentiated a second time. If the first relationship in (3.43) is used to eliminate dg/dt from (3.44), the result is:

$$\frac{dh}{dt} = \frac{1}{a-b} e^{bt} S(t) \tag{3.45}$$

This is readily integrated to give the solution for $h(t)$ so long as $a \neq b$:

$$h(t) = \frac{1}{a-b} \int_0^t e^{bt'} S(t') \, dt' \tag{3.46}$$

Solving the first equation in (3.43) for dh/dt and substituting into (3.44) yields a similar solution for $g(t)$:

$$g(t) = \frac{1}{b-a} \int_0^t e^{at'} S(t') \, dt' \tag{3.47}$$

Substitution (3.46) and (3.47) into (3.41) produces the solution to (3.39) for any forcing function $S(t)$. This result may be stated as a theorem:

Theorem 3: The solution to the equation

$$\frac{d^2 x}{dt^2} + (a+b) \frac{dx}{dt} + abx = S(t)$$

when $a \neq b$ is given by:

$$x(t) = A e^{-at} + B e^{-bt} + \frac{1}{b-a} \int_0^t e^{-a(t-t')} S(t') \, dt' + \frac{1}{a-b} \int_0^t e^{-b(t-t')} S(t') \, dt'$$

where A and B are arbitrary constants chosen to satisfy the initial conditions.

Thus, we see that Theorem 3 is a straightforward generalization of Theorem 1 to the case of a second order system. In both cases the forcing function appears in convolution integrals involving the exponential decay eigenfunctions we found in solving the homogeneous equation. In the case of critical damping, $a = b$, so Theorem 3 will not apply. However, the same approach using variation of parameters can be shown to solve the problem in that case as well.

You may find the choice of (3.43) as a second equation relating $g(t)$ and $h(t)$ to be a somewhat capricious if clever trick. In a sense this is true. However, a second equation relating g and h is essential if the problem is to be solved. Furthermore, the equation in Theorem 3 does have a unique solution, as is proved in more extended treatments of linear differential equations. Therefore, we can be certain that our choice of (3.43) has allowed us to derive the unique solution to (3.39).

Congratulations! You have now covered those aspects of the theory of first and second order linear differential equations that are most relevant to contemporary neuroscience. Although mathematics courses in linear differential equations certainly include more material, the additional topics are generally of less scientific interest. Furthermore, we will soon cover some of these topics, such as differential equations with time-varying coefficients in the much more general context of nonlinear dynamics, which is where they truly belong.

3.7 Exercises

1. Put each of the following three equations in normal form:

$$\frac{d^2x}{dt^2} - 7\frac{dx}{dt} + 11x = 5$$

$$\frac{d^2x}{dt^2} - 2\frac{x^2}{1+x^2}\left(\frac{dx}{dt}\right)^3 + 11x^2 = 8$$

$$\frac{d^4x}{dt^4} - 3x^2\frac{d^3x}{dt^3} + 7\frac{dx}{dt}\frac{d^2x}{dt^2} + x - 2x^3 = 0$$

2. For the following second order differential equation find integer values (if possible) of a and b that produce each of the following types of equilibrium points: (a) spiral point, (b) node, (c) saddle point, (d) center, (e) critically damped asymptotically stable point. For (a), (b), and (e) find values for both an asymptotically stable equilibrium and for an unstable equilibrium. In each case give the exact solution for the initial condition $x(0) = 2$; $y(0) = -1$. (Note: there are many solutions in each case, and you need to find only one. The goal is to help you understand state space and the various types of equilibria.)

$$\frac{d\vec{X}}{dt} = \begin{pmatrix} -2 & a \\ 5 & b \end{pmatrix}\vec{X}$$

3. At low light levels the parameter values of the retinal cones change and the feedback from the horizontal cells becomes weaker. A modified version of (3.20) that is appropriate for low light conditions is:

$$\frac{dC}{dt} = \frac{1}{\tau_C}(-C - 0.5H + L)$$

$$\frac{dH}{dt} = \frac{1}{\tau_H}(-H + C)$$

with $\tau_C = 100$ ms and $\tau_H = 500$ ms. Solve for the steady state as a function of L. Now obtain an exact solution for $L = 3$ and the initial conditions $C(0) = 0$, $H(0) = 0$.

4. Solve (3.29) for the motor control of a saccadic eye movement through an angle of 30°. In particular, let $x_0 = 29$ mm, $x(0) = 35$ mm, and $\alpha = 0.1$. Write down the explicit solution and plot the result for $0 \le t \le 150$ ms.

5. The equation below describes a critically damped spring model of muscle contraction, with x_0 as the equilibrium length specified by the central nervous system:

$$\frac{dx}{dt} = v$$

$$\frac{dv}{dt} = \frac{1}{\tau}\left(-\frac{(x - x_0)}{\tau} - 2v\right)$$

Obtain the exact solution of these equations for $\tau = 0.08$ s and $x_0 = 4$ mm given the initial conditions $x(0) = 7$ mm, $v(0) = 0$. (Hint: first convert the equation back to a single second order differential equation.)

6. Solve the following equation with forcing function $3t^2$ for $x(t)$:

$$\frac{d^2x}{dt^2} + 5\frac{dx}{dt} + 6x = 3t^2$$

The initial conditions are $x(0) = 1$ and $dx(0)/dt = 0$. Plot the solution for $0 \le t \le 2$.

4   *Higher dimensional linear systems*

Our goal in theoretical neuroscience is to be able to describe and analyze systems with many components, be they neurons, dendrites, or ionic currents. The material in Chapters 2 and 3 has enabled us to examine systems with one or two components, such as the feedback circuit related to retinal cones. Fortunately, we now have all the background needed to generalize to linear systems with an arbitrarily large number of components. The reason for this is that all possible types of solutions found in high order linear systems with constant coefficients are simple generalizations of the solutions of second order systems.

In dealing with higher order systems, it is generally not possible to find all of the eigenvalues of the system analytically, because polynomial equations beyond fourth order do not have general analytical solutions. Fortunately, MatLab provides fast and accurate numerical approximations to the eigenvalues of quite large systems. Furthermore, we shall sometimes be interested only in whether or not all eigenvalues have negative real parts, as negative real parts guarantee that the steady state is asymptotically stable. There is a very powerful criterion due to Routh and Hurwitz that tests for negative real parts.

The Routh–Hurwitz criterion can also be used to determine the strength of synaptic connections to guarantee that a neural network will generate oscillations. This can be extremely valuable when we know from physiology experiments that a network does oscillate, yet we lack detailed quantitative information on the synaptic weights. The same approach also elucidates the effect of delays in neural feedback loops, which can sometimes produce undesirable oscillations possibly related to some forms of motor control disorders.

4.1   Solutions of *N*th order linear systems

Let us consider a system of N interacting neurons or N ionic species. If the interactions are linear, the system can always be described by a differential equation in normal form:

$$\frac{\mathrm{d}\vec{X}}{\mathrm{d}t} = \overleftrightarrow{A}\vec{X} \tag{4.1}$$

This can be solved in the same manner as was used in deriving Theorem 2, namely, by substituting a vector of exponentials and then solving the resulting characteristic equation. This approach leads to the following theorem:

Theorem 4: The solution to the Nth order differential equation (4.1) is obtained by finding the N roots, λ_1 to λ_N, of the characteristic equation:

$$|\overleftrightarrow{A} - \lambda \overleftrightarrow{I}| = 0$$

where I is the identity matrix. The components of the solution vector are then of the form:

$$\sum_{n=1}^{N} A_n e^{\lambda_n t}$$

where values of the constants A_n are determined by solving for the eigenvectors and by the initial conditions. If a single eigenvalue, say $\lambda = \omega$, occurs k times, then the solutions associated with this eigenvalue will be of the form:

$$A_1 e^{\omega t} + A_2 t\, e^{\omega t} + A_3 t^2\, e^{\omega t} + \cdots + A_k t^{k-1}\, e^{\omega t}$$

As in the case of second order systems, roots that occur in complex conjugate or pure imaginary pairs will introduce sine and cosine terms, and multiple roots generate polynomials multiplied by exponentials as in critical damping.

As a first example, consider the three-neuron system illustrated in Fig. 4.1. Each neuron excites one neuron (arrows) and is in turn inhibited by the other (line with black circle). If we take the excitatory synaptic strength to be 7, the inhibitory synaptic strength to be –10, and the self decay rate to be –5, the three coupled linear equations describing this system would be:

$$\frac{d}{dt}\begin{pmatrix} E_1 \\ E_2 \\ E_3 \end{pmatrix} = \begin{pmatrix} -5 & -10 & 7 \\ 7 & -5 & -10 \\ -10 & 7 & -5 \end{pmatrix} \begin{pmatrix} E_1 \\ E_2 \\ E_3 \end{pmatrix} \tag{4.2}$$

The eigenvalues of this equation are obtained using the MatLab^{TM} function **eig()** by typing the following in the command window: eig([–5, –10, 7; 7, –5, –10; –10, 7, –5]). This produces the eigenvalues $\lambda = -8, -3.5 \pm 14.7i$, so the neural responses E_i will all have the form:

$$E_i = A\,e^{-8t} + B\,e^{-3.5t}\sin(14.7t) + C\,e^{-3.5t}\cos(14.7t) \tag{4.3}$$

A, B, and C can be determined by substitution back into (4.2) to solve for the eigenvectors along with the initial conditions. Even without bothering to determine these constants, however, we have already discovered that the equilibrium point of this three-neuron system is an asymptotically stable spiral point. In fact, it will not be necessary to solve for all the constants in solutions of higher order equations such as (4.3), because we will almost always be dealing with nonlinear problems where the important information is contained in the eigenvalues.

4.2 Routh–Hurwitz criterion and oscillations

For large linear dynamical systems, two questions frequently arise: (1) is the equilibrium point asymptotically stable? (2) Under what conditions will the system produce an oscillation? For example, if we knew that the neural network in Fig. 4.1 produced oscillations under physiological conditions, we might wish to discover what values of the excitatory and inhibitory synaptic strengths were required to sustain such behavior. Both of these questions are answered by the Routh–Hurwitz criterion and related theorems. In approaching this problem, let us first examine a necessary but not sufficient condition for asymptotic stability of the steady state at the origin. (Recall that any steady state can be translated to the origin by subtraction.)

Theorem 5: The stability, asymptotic stability, or instability of the equilibrium point of (4.1) is determined by the roots of the characteristic equation:

$$|\overset{\leftrightarrow}{A} - \lambda \overset{\leftrightarrow}{I}| = 0$$

The equilibrium will be asymptotically stable if all the roots of the characteristic equation have negative real parts. Writing the characteristic equation as:

$$\lambda^N + a_1 \lambda^{N-1} + a_2 \lambda^{N-2} + \cdots + a_{N-1}\lambda + a_N = 0$$

A necessary (but not sufficient) condition for all roots to have negative real parts is:

$$a_k > 0 \quad \text{for } 1 \leq k \leq N$$

This theorem provides a quick check for the possibility of asymptotic stability. Note that all coefficients must be positive; zero values are excluded. Theorem 5 will not be proved, because the proof focuses on the mathematical nature of polynomials and has no further connections with dynamical systems theory.

If the coefficients of the characteristic equation satisfy Theorem 5, the next step is to apply the much more complex but definitive Routh–Hurwitz test, which I shall state as a theorem. In order to simplify the notation, let us state the theorem for a fifth order system. It readily generalizes to any higher order.

Theorem 6 (Routh–Hurwitz theorem): Given the coefficients a_k of the characteristic equation in Theorem 5, compute the following series of determinants for order $N = 5$:

$$\Delta_1 = a_1, \qquad \Delta_2 = \begin{vmatrix} a_1 & 1 \\ a_3 & a_2 \end{vmatrix}$$

$$\Delta_3 = \begin{vmatrix} a_1 & 1 & 0 \\ a_3 & a_2 & a_1 \\ a_5 & a_4 & a_3 \end{vmatrix}, \qquad \Delta_4 = \begin{vmatrix} a_1 & 1 & 0 & 0 \\ a_3 & a_2 & a_1 & 1 \\ a_5 & a_4 & a_3 & a_2 \\ 0 & 0 & a_5 & a_4 \end{vmatrix}$$

$$\Delta_5 = a_5 \Delta_4$$

The system is asymptotically stable if and only if:

$$\Delta_k > 0 \quad \text{for } 1 \le k \le N$$

Furthermore, if $\Delta_N = 0$, then $\lambda = 0$ is one eigenvalue.

The proof of this theorem is an exercise in the theory of complex variables as applied to the roots of polynomials and may be found in Willems (1970).

For reference, the most general form of the Routh–Hurwitz determinants for a system of order N is:

$$\Delta_k = \begin{vmatrix} a_1 & 1 & 0 & 0 & 0 & \cdots \\ a_3 & a_2 & a_1 & 1 & 0 & \cdots \\ a_5 & a_4 & a_3 & a_2 & \cdots & \cdots \\ a_7 & a_6 & \cdots & \cdots & \cdots & \cdots \\ \vdots & \vdots & \vdots & \vdots & \vdots & \vdots \\ a_{2k-1} & a_{2k-2} & \cdots & \cdots & \cdots & \cdots \end{vmatrix} \tag{4.4}$$

Note that all of the lower order determinants referred to in Theorem 5 are obtained by starting in the upper left and considering the successively larger square matrices that are $2 \times 2, 3 \times 3, 4 \times 4$, and so on. The convention in (4.4) is that $a_j = 0$ if $j > N$ or if $j < 0$. The theorem has been formulated here so that $a_0 = 1$, where a_0 is the coefficient of λ^N. Many texts make a_0 explicit, although it is trivial to divide through by a_0 to obtain the characteristic equation used here.

To take an example, Δ_5 in a seventh order system would be:

$$\begin{vmatrix} a_1 & 1 & 0 & 0 & 0 \\ a_3 & a_2 & a_1 & 1 & 0 \\ a_5 & a_4 & a_3 & a_2 & a_1 \\ a_7 & a_6 & a_5 & a_4 & a_3 \\ 0 & 0 & a_7 & a_6 & a_5 \end{vmatrix} \tag{4.5}$$

Theorem 6 was extremely valuable in determining the stability of higher order systems before the advent of programs such as MatLab that will compute the eigenvalues directly. Why, then, would one employ this theorem today? One answer is that the Routh–Hurwitz criterion is easy to program and may well yield a more accurate result than the numerical methods employed to find eigenvalues of higher order systems. This is because the derminants required by Theorem 6 involve simple multiplication, addition, and subtraction; and the final stability test is just an inequality. More importantly, there is one application of the Routh–Hurwitz Theorem that can be extremely useful in studying neural systems. Many neural control systems have evolved to generate oscillations. Two major examples of such systems are neuronal control of locomotion and respiration. Accordingly, it will be very useful to determine conditions under which dynamical equations will generate periodic solutions. For linear systems a criterion for the existence of oscillations is produced as a by-product of Theorem 6. This is the Routh–Hurwitz criterion for oscillations:

Theorem 7 (Routh–Hurwitz criterion for oscillations): For the Nth order system in (4.1) compute the Routh–Hurwitz determinants in Theorem 6. One pair of eigenvalues will be purely imaginary and the system will therefore produce a sinusoidal oscillation if and only if:

$$\Delta_k > 0 \text{ and } \Delta_{N-1} = 0 \text{ for } 1 \le k \le N - 2$$

Note from Theorem 6 that the final Routh–Hurwitz determinant $\Delta_N = 0$ as well if the conditions of Theorem 7 are met. This theorem can also be usefully applied to nonlinear dynamical systems, as we shall see in Chapter 8.

In addition to determining whether a system will produce an oscillation, Theorem 7 can be used to determine parameter values where neural networks will oscillate. To see this, let us go back to the network in Fig. 4.1 and use Theorem 7 to determine whether there is some strength of the inhibitory synaptic connections in the network that will produce oscillations. Recall that with an inhibitory strength of -10 in (4.2) the origin was found to be an asymptotically stable spiral point, and therefore no oscillations could occur. Let us now treat the inhibitory synaptic strength as an unknown parameter designated by $-g$ (for gain). Now (4.2) may be rewritten as:

$$\frac{\mathrm{d}}{\mathrm{d}t} \begin{pmatrix} E_1 \\ E_2 \\ E_3 \end{pmatrix} = \begin{pmatrix} -5 & -g & 7 \\ 7 & -5 & -g \\ -g & 7 & -5 \end{pmatrix} \begin{pmatrix} E_1 \\ E_2 \\ E_3 \end{pmatrix} \tag{4.6}$$

How can we solve for g so that an oscillation results? Referring to Theorem 7, we see that this will occur if the second Routh–Hurwitz determinant $\Delta_2 = 0$. The lower order determinants must, of course, be greater than zero, or the equilibrium point will be unstable. To carry out the computation, we would first determine the coefficients of the characteristic equation for (4.6) and then form the Routh–Hurwitz determinants. Finally,

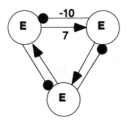

Fig. 4.1 A cyclical three-neuron network in which each neuron excites the next and is inhibited by the previous neuron.

we would have to solve the algebraic equation for $\Delta_2 = 0$ to find the value of g that causes oscillations.

Fortunately, all of this tedious calculation can be done using MatLab script **Routh_ Hurwitz.m**. Before using **Routh_Hurwitz.m**, it is first necessary to type the coefficients of the A matrix into the function **Hopf.m** (representing the unknown as G), and then save this function. Now run **Routh_Hurwitz.m**, and you will be prompted for a 'Guess' or first approximation at the solution, which should be a positive real number in this case. The program first finds a value of G that causes the Routh–Hurwitz determinant $\Delta_{N-1} = 0$ to satisfy Theorem 7. The value is $G = 17$ in this case. The **Routh_Hurwitz.m** script then uses this G value in the A matrix to calculate all of the Routh–Hurwitz determinants and find the eigenvalues. Finally, it indicates whether the equilibrium point is asymptotically stable, unstable, or a center. In this example, the eigenvalues are -15 and $\pm 20.78i$, so the neural responses all have the general form:

$$E_k = A\,e^{-15t} + B\cos(20.78t) + C\sin(20.78t) \qquad (4.7)$$

where $k = 1, 2, 3$. So the solution to (4.6), which is described in a three-dimensional state space, decays onto a two-dimensional surface within that space where the solution oscillates sinusoidally. Note that the frequency of this oscillation is reported in radians; conversion to Hz requires division by 2π. Thus, Theorem 7 has been used to determine the inhibitory synaptic strength that will cause oscillations in the network depicted in Fig. 4.1.

4.3 Oscillatory control of respiration

Oscillations abound in the neural control of movements. Walking, running, and swimming are all governed by networks of motorneurons that generate periodic responses causing the various muscles involved to contract in an appropriate and repetitive sequence. Here we shall examine a neural oscillation that most of us take for granted: the control of respiration. Anatomical and physiological data (Cohen, 1968) suggest that respiration might be controlled by a mutually inhibitory network similar to that illustrated in Fig. 4.2. Neuron 1 in this network inhibits neurons 2 and 3 but has no effect on

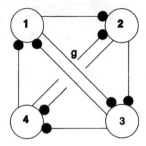

Fig. 4.2 Four-neuron network for respiratory control involving sequential neural disinhibition.

neuron 4, from which it receives inhibition. These interactions are repeated for the other neurons. This is a sequential network that involves inhibition but also disinhibition. To appreciate this, note that inhibition of neurons 2 and 3 by neuron 1 shuts off all inhibition to neuron 4.

To analyze this network and determine the inhibitory strengths that will produce a respiratory oscillation, let us assume that the inhibitory synaptic weights between adjacent neurons are -5 and that each neural response will decay to zero at a rate of -3 in the absence of stimulation. What strength $-g$ must the diagonal inhibition in the network have to generate periodic behavior? The system of four coupled linear neural equations is:

$$\frac{d}{dt}\begin{pmatrix} E_1 \\ E_2 \\ E_3 \\ E_4 \end{pmatrix} = \begin{pmatrix} -3 & 0 & -g & -5 \\ -5 & -3 & 0 & -g \\ -g & -5 & -3 & 0 \\ 0 & -g & -5 & -3 \end{pmatrix}\begin{pmatrix} E_1 \\ E_2 \\ E_3 \\ E_4 \end{pmatrix} \tag{4.8}$$

Type the A matrix into **Hopf.m** (again using G for the unknown), save this function, and run **Routh_Hurwitz.m**. The program reveals that $G=3$ will generate an oscillation of the general form:

$$E_k = A\,\mathrm{e}^{-t} + B\,\mathrm{e}^{-11t} + C\cos(5t) + D\sin(5t) \tag{4.9}$$

where $k = 1, 2, 3, 4$. Assuming that t is in seconds the oscillation frequency is $5/(2\pi) = 0.8$ Hz, about right for respiration. Here again the Routh–Hurwitz criterion in Theorem 7 has found a parameter value that produces a neural oscillation. As the first two terms in (4.9) die out with increasing t, the oscillation occurs on a two-dimensional surface in the four-dimensional state space of the system. These examples will generalize to nonlinear neural networks, because Theorem 7 can be employed in conjunction with the Hopf Bifurcation Theorem (see Chapter 8).

4.4 Feedback with delays

A final, very important example of oscillations in neural systems is related to delays in feedback loops. The negative feedback loop between horizontal cells and cones

(Chapter 3) generates an asymptotically stable spiral point. Can this network also produce an oscillation? Consider the very general linear feedback network between an excitatory neuron E and an inhibitory neuron I. The equations are:

$$\frac{dE}{dt} = \frac{1}{\tau_E}(-E - aI)$$

$$\frac{dI}{dt} = \frac{1}{\tau_I}(-I + bE)$$

(4.10)

For $a > 0, b > 0$, eqn (4.10) describes a negative feedback loop as illustrated in Fig. 1.3. The A matrix is easily be seen to be:

$$\overleftrightarrow{A} = \begin{pmatrix} -1/\tau_E & -a/\tau_E \\ b/\tau_I & -1/\tau_I \end{pmatrix}$$

(4.11)

so the eigenvalues are:

$$\lambda = -\frac{1}{2}\left(\frac{1}{\tau_E} + \frac{1}{\tau_I}\right) \pm \frac{\sqrt{(\tau_E - \tau_I)^2 - 4ab\tau_E\tau_I}}{2\tau_E\tau_I}$$

(4.12)

As real(λ) < 0, all solutions to eqn (4.10) must be decaying exponential functions of time, and so oscillations are impossible in this two component feedback loop. The reason for this is that both E and I decay exponentially with their respective time constants, as required for physiological plausibility. (It is, of course, possible to generate oscillations in an idealized second order system such as a spring without any frictional resistance.)

Does this analysis indicate that linear feedback systems can never oscillate? To answer this, suppose that physiological conditions caused a delay (for example, an axonal conduction time delay) in the feedback loop. To represent such a delay exactly in eqn (4.10), however, becomes extremely complex. In fact, differential equations with delays are infinite-dimensional dynamical systems! You can convince yourself of this fact from the following argument. An N-dimensional system requires N initial conditions, one for each variable at time $t = 0$. If there is a delay in the system of say 5 ms, a continuum of values must be specified between $t = -5$ ms and $t = 0$ to specify the initial state of the system. Hence, dynamical systems with true delays become extraordinarily complex, and the interested reader is referred to discussions by Glass and Mackey (1988), MacDonald (1989), and Milton (1996).

As a simplified approach to the problem, suppose we take our cue from the fact that delays increase the dimensionality of a dynamical system. This increase is certainly infinite, but let us be modest and introduce just one additional differential equation to approximate the delay by defining the variable Δ for delay. Before seeing the effect this has on eqn (4.10), let us see how an additional equation alters the response of a single first order equation. Consider therefore:

$$\frac{dx}{dt} = \frac{1}{\tau}(-x + 1)$$
$$\frac{d\Delta}{dt} = \frac{1}{\delta}(-\Delta + x)$$

(4.13)

where δ will approximate the delay time lag in milliseconds, and the initial conditions $x(0) = 0$ and $\Delta(0) = 0$. Note that the equation for Δ has been constructed so that $\Delta = x$ in the steady state, a requirement that any delay must meet. The first equation was solved in Chapter 2, and the second can also be solved using Theorem 1 with the results:

$$x(t) = 1 - e^{-t/\tau}$$
$$\Delta(t) = 1 - \frac{\tau e^{-t/\tau}}{\tau - \delta} + \frac{\delta e^{-t/\delta}}{\tau - \delta}$$

(4.14)

assuming $\tau \neq \delta$ (the solution involves critical damping otherwise). Now let $\tau = 10$ ms and $\delta = 5$ ms and examine the solutions plotted on the left of Fig. 4.3. It is clear that the response Δ lags behind $x(t)$, which is required of a delay. The figure also plots $x(t)$ with a true 5 ms delay to show that $\Delta(t)$ with $\delta = 5$ ms provides a modest approximation to the delay. If a more accurate approximation is desired, one can always include a chain of additional delay stages in eqn (4.13). For example, with four delay stages one would set $\delta = 1.25$ ms in this case, and the approximation to a true delay is greatly improved as shown on the right side of Fig. 4.3. In the limit of an infinite number of stages our approximation would be exact (see below). Remember, however, that in all computer simulations neural time delays are *de facto* represented by a finite number of stages simply because computers can only calculate a result at a finite number of time points. Thus, computer simulations reduce to embellishments of the delay approximation in eqn (4.13).

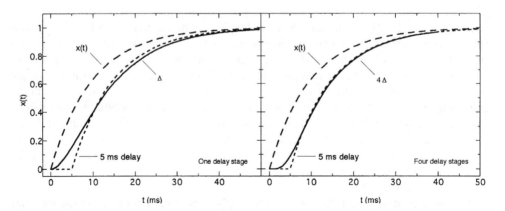

Fig. 4.3 Solid lines plot approximations to a 5 ms time delay by the introduction of one (left) or four (right) additional differential stages Δ as in (4.13). For comparison, an actual 5 ms delay of $x(t)$ (long dashes) is also shown (short dashes). Additional stages increase the accuracy of the approximation.

For most purposes the addition of one or two delay stages in the manner just described will permit a satisfactory approximation of delay effects on dynamics. (As emphasized by MacDonald (1989), however, a very large number of stages is sometimes necessary to explain all aspects of a true time delay.) Let us return to the feedback system in (4.10) and introduce a delay before the inhibition I begins to exert its effect. If we let $a = 2$, $b = 8$, and the time constants be 10 ms and 50 ms for E and I respectively, the equations describing the system become:

$$\frac{dE}{dt} = \frac{1}{10}(-E - 2\Delta)$$
$$\frac{dI}{dt} = \frac{1}{50}(-I + 8E) \qquad (4.15)$$
$$\frac{d\Delta}{dt} = \frac{1}{\delta}(-\Delta + I)$$

The presence of the delay stage Δ can be represented in a simple neural diagram like Fig. 4.4. Is there any value of the delay time δ that will produce an oscillation? The matrix for (4.15) is:

$$\overset{\leftrightarrow}{A} = \begin{pmatrix} -1/10 & 0 & -1/5 \\ 8/50 & -1/50 & 0 \\ 0 & 1/\delta & -1/\delta \end{pmatrix} \qquad (4.16)$$

The Routh–Hurwitz criterion in Theorem 7 can be used to solve for δ by entering matrix A into the MatLab function **Hopf.m** (always with G as the unknown), saving this function, and then running **Routh_Hurwitz.m**. You will find that $\delta = 7.61$ ms will cause solutions to (4.15) to be periodic with a frequency of 0.133, which is $0.133/2\pi$ cycles/ms or 21.2 Hz. Thus even a short feedback lag in eqn (4.15) can lead to rapid oscillations. Oscillations caused by delays in neural transmission may well be one cause of the tremors exhibited by patients with multiple sclerosis, a disease in which axonal transmission is known to be slowed down (Beuter *et al.*, 1993). Indeed, Mackey and Milton (1987) coined the term 'dynamical diseases' to refer to physiological systems that become dysfunctional due to alterations such as increased time delays. One cautionary note: not all feedback loops are guaranteed to oscillate for some value of δ in eqn (4.15). Whether an oscillation can occur or not is dependent on the feedback gain as well, an issue

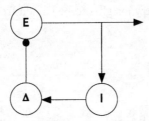

Fig. 4.4 Negative feedback loop with a delay stage Δ introduced between the I and E neurons. The network is described by (4.15).

explored in the problems. If a given feedback system cannot oscillate, **Routh_Hurwitz.m** cannot find a solution.

For readers with a more advanced background, here is a sketch of the proof that introduction of an increasing number of first delay stages N as in (4.13) becomes an exact mathematical description of a delay in the limit $N \to \infty$. It was shown in Chapter 2 that a cascade of $(N+1)$ first order stages with identical time constants τ produces the response $R(t)$:

$$R(t) = \frac{1}{\tau N!} \left(\frac{t}{\tau}\right)^N e^{-t/\tau} \tag{4.17}$$

When integrated this function has unit area for all $N \geq 1$, and the peak always occurs at $\tau_{delay} = N\tau$. For any desired delay, therefore, let us require that $\tau = \tau_{delay}/N$. This is known as the gamma function approximation to delays (MacDonald, 1989). Now, in the limit as $N \to \infty$, (4.17) is a function with unit area, infinitesimal width, and infinite height centered at τ_{delay}. This unusual function is the Dirac δ function (Dirac, 1958) and is a form of 'generalized function' (Lighthill, 1958). The convolution of $\delta(t - \tau_{delay})$ with any function $f(t)$ produces exactly $f(t - \tau_{delay})$, which is what we wanted to prove. Limitations in using the gamma function with small N to approximate delays are discussed by MacDonald (1989).

4.5 Synopsis

This chapter has been short because almost all attributes of linear second order systems generalize to higher order systems. The major differences between second order and higher order linear systems are practical: eigenvalues and eigenvectors are no longer as simple to compute, although MatLab^{TM} is an enormous aid here. Some scientists apparently believe that ready access to computers has obviated the need to understand and utilize analytical approaches to dynamical systems. This perception is misguided for two reasons. First, a scientist cannot hope to create and analyze a neural model derived from her/his research without a sophisticated appreciation of the underlying mathematics. Second, without analytical techniques, attempts to find parameter values that generate particular types of solutions, such as the oscillations in our simulation of respiration, can degenerate into laborious trial-and-error computer simultions. Beginning in the next chapter, we shall integrate computer simulations with nonlinear analysis, emphasizing the complementary insights to be gained from each.

4.6 Exercises

1. Consider the following equation:

$$\frac{d\vec{X}}{dt} = \begin{pmatrix} -3 & 2 & b \\ b & -2 & 7 \\ -4 & b & -4 \end{pmatrix} \vec{X}$$

(a) Determine the value of b that will cause the system to oscillate. (b) Write down the general form of the analytical solution in terms of exponentials, etc. (Just write down the combination of functions with their eigenvalues that will be involved; you need not solve for the eigenvectors.)

2. Extend the example of delays in the feedback system in eqn (4.15) by introducing a second delay between the excitatory neuron and the inhibitory neuron in addition to the delay already present. This will produce a fourth order system. Assume that the constant δ is the same for both delay stages, and solve for δ to produce an oscillation. How does the total feedback delay, 2δ, compare with the delay calculated for just one stage in the chapter?

3. Consider the following example of a feedback system in which there is a delay in neural transmission from the excitatory neuron E to the inhibitory neuron I. The strength of the feedback inhibition depends on a parameter g. Determine how the delay depends on g by solving for δ to produce oscillations for the following values: $g = 10, 15, 25$. Also determine the frequency of the oscillation in Hz (not radians) in each case assuming the time constants are in ms. What trends do you observe as the inhibitory feedback becomes stronger?

$$\frac{dE}{dt} = \frac{1}{15}(-E - gI)$$

$$\frac{d\Delta}{dt} = \frac{1}{\delta}(-\Delta + E)$$

$$\frac{dI}{dt} = \frac{1}{40}(-I + \Delta)$$

5   *Approximation and simulation*

Systems of linear differential equations with constant coefficients can be solved exactly in terms of sine, cosine, and exponential functions, with the occasional polynomial thrown in when several eigenvalues are identical. Why, then, is it necessary to approximate the solutions of differential equations on a computer? There are several answers. First, if the input or stimulus to the system is a sufficiently complex function of time, it may be impossible to evaluate integrals such as those in Theorems 1 and 3 exactly. Indeed, this is true when a sinusoidal stimulus provides input to the Naka–Rushton function in eqn (2.11). So approximation is necessary if we wish to determine the response of this system. A second case is that of linear systems that represent interactions among many components. Although we can in principle write down the solution in terms of sines, cosines, and exponentials as indicated by Theorem 4, in practice it may be more efficacious to simulate the response for the particular initial conditions of interest.

The final, and most important, reason for employing approximation methods is to obtain solutions to systems of nonlinear differential equations. As will be seen in subsequent chapters, virtually all of the truly interesting neural problems are inherently nonlinear. Nonlinear differential equations do not generally have solutions that can be written down in terms of known functions like exponentials. So the only way to make detailed predictions about the temporal evolution of nonlinear neural systems is by resorting to simulation. Therefore, let us examine some accurate approximation methods of general utility.

5.1   Euler's method

Let us focus our attention on the problem of solving a first order differential equation of the form:

$$\frac{\mathrm{d}x}{\mathrm{d}t} = F(x, t) \tag{5.1}$$

where F can be a linear or nonlinear function of x and t. At $t=0$ the initial condition is $x(0) = x_0$. In most approximation methods the time variable is divided into a series of very small steps spaced duration h apart. For any time t_{N+1} and t_N this produces the relationship:

$$t_{N+1} - t_N = h \quad \text{so} \quad t_N = Nh \tag{5.2}$$

A finite number of time steps must be used to reduce the problem to a finite number of calculations suitable for a computer.

The problem of approximating x may now be formulated thus: given the value of $x(t_N)$, how do we approximate the next value, $x(t_{N+1}) = x(t_N + h)$? Euler's insight was to use the Taylor series approximation to $x(t)$ but limit the expansion to the first term in the polynomial. Thus:

$$x(t_N + h) \approx x(t_N) + h\frac{dx}{dt} \qquad (5.3)$$

where dx/dt is evaluated at t_N. As the value of the derivative in this expression is given explicitly by (5.1):

$$x(t_N + h) \approx x(t_N) + hF(x(t_N), t_N) \qquad (5.4)$$

Equation (5.4) is Euler's approximation to the exact solution of (5.1). Using (5.2) and (5.4), we can start at $t = 0$ and use the value $x(0) = x_0$ to calculate the value of $x(h)$. The process is then iterated to estimate successive values of x.

As an example, let us take an equation that was solved exactly in Chapter 2:

$$\frac{dx}{dt} = \frac{1}{20}(-x + 40\,e^{-t/20}) \quad \text{with} \quad x(0) = 0 \qquad (5.5)$$

Euler's approximation from (5.4) is:

$$x(t + h) \approx x(t) + \frac{h}{20}(-x(t) + 40\,e^{-t/20})$$

As h is divided by the time constant 20 ms above, h should be chosen to be some reasonably small fraction of this time constant. As a first choice, let $h = 4$ ms or one-fifth of the time constant. The result of simulating $x(t)$ for 40 ms (i.e. 10 time steps) is plotted in Fig. 5.1, where only part of the ordinate is shown to emphasize differences. Euler's

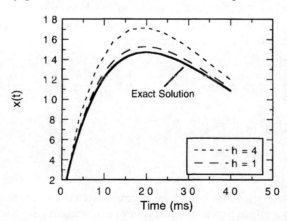

Fig. 5.1 Illustration of Euler's method with two different time steps h compared with the exact solution to (5.5).

method progressively overestimates the exact solution in this case and produces a value of 11.94 after 40 ms. Compared to the exact value of 10.83, this represents an error of 10%, not too impressive. The situation is improved by reducing the step size to $h = 1$ ms, which produces an estimate of $x = 11.10$ at 40 ms, an error reduction to 2.5%. Thus, the approximation has improved by a factor of 4.0 by reducing the time step to a quarter of the original value, but this improvement is at the cost of having to do calculations for 40 time iterations instead of 10. This is always the issue in approximating the solution to differential equations: the simulation will become more and more accurate as h is made smaller, but this multiplies the number of computations and hence the amount of computer time for the simulation. Life is too short to use excruciatingly small time steps!

5.2   Runge–Kutta methods

Although Euler's method can be used for simple problems if the step size h is sufficiently small, it is rarely used in practice, because more accurate results may be obtained with the same amount of computation using other methods. Furthermore, around a steady state that is a center, Euler's method is actually guaranteed to produce an unstable spiral solution regardless of step size h, a phenomenon investigated in the problems. The reason for the inaccuracy of Euler's method is that (5.4) only uses a straight line approximation to the solution at each time to obtain the approximation at the next time step, a consequence of truncating Taylor's series in (5.3) after the linear term in h. One might guess that greater accuracy could be obtained by retaining additional terms in the Taylor series approximation. This guess is correct, and it forms the basis for Runge–Kutta approximation methods. There are several different orders of Runge–Kutta methods, the order being determined by the highest power retained in the Taylor series approximation. The fourth order Runge–Kutta method is the most common, and we shall use it for neural simulations later in this book. To simplify the algebra, however, let us examine the second order case. In addition, let $F(x)$ in (5.1) be independent of t, which will simplify notation and highlight the conceptual aspects of the derivation.

To derive the second order Runge–Kutta approximation, let us first expand the Taylor series in (5.3) to include the second order term:

$$x(t_N + h) \approx x(t_N) + h\,\frac{dx}{dt} + \frac{h^2}{2}\frac{d^2x}{dt^2} \qquad (5.6)$$

Substituting for dx/dt using (5.1) gives:

$$x(t_N + h) \approx x(t_N) + hF + \frac{h^2}{2}\frac{dF}{dt} \qquad (5.7)$$

This strategy seems promising, but it now becomes necessary to evaluate dF/dt. If F is a well-defined mathematical function, the differentiation might be carried out analytically. (Note that F depends on t implicitly through its dependence on $x(t)$ even if t does not appear explicitly in F.) However, this can lead to very messy expressions, especially when higher order Runge–Kutta methods are considered. For this reason the Runge–Kutta method employs a very clever trick to avoid any explicit calculation of dF/dt. Assuming for simplicity that F depends only on x and not t, the chain rule for differentiation yields:

$$\frac{dF}{dt} = \frac{dF}{dx}\frac{dx}{dt} = F\frac{dF}{dx}$$

so

$$x(t_N + h) \approx x(t_N) + hF + \frac{h^2}{2}F\frac{dF}{dx} \tag{5.8}$$

To avoid having to compute dF/dx, let us see if we can find an approximation of the form:

$$x(t_N + h) \approx x + ahF + bhF(x + chF) \tag{5.9}$$

What we have done here is to replace the h^2 term in (5.8) by a function of a function, $F(x + chF)$. Three constants, a, b, and c, have been introduced into (5.9), and these must now be chosen to make (5.9) identical to (5.8). As h is very small, we may now use Taylor's series to expand the last term in (5.9), retaining only the first order term:

$$F(x + chF) \approx F(x) + h\frac{dF}{dx}\frac{dx}{dh} = F(x) + h\frac{dF}{dx}cF \tag{5.10}$$

Substituting this into (5.9) gives:

$$x(t_N + h) \approx x + (a + b)hF + bch^2F\frac{dF}{dx} \tag{5.11}$$

Looking at (5.8) and (5.11), one can see that the two will be identical if:

$$a + b = 1 \quad \text{and} \quad bc = \tfrac{1}{2} \tag{5.12}$$

Note that these conditions on a, b, and c do not uniquely specify the solution, so one more equation is needed. As there is no mathematical reason for preferring one choice over any other, aesthetic considerations are usually indulged by setting $a = b$. Therefore, (5.12) yields: $a = 1/2$; $b = 1/2$; and $c = 1$. Using these values in (5.9) gives:

$$x(t_N + h) \approx x + \frac{h}{2}F + \frac{h}{2}F(x + hF) \tag{5.13}$$

We have now shown that (5.13) is identical to the second order Taylor expansion in (5.6) and (5.7), but (5.13) only involves the function F and none of its derivatives.

Equation (5.13) is the second order Runge–Kutta approximation. Although (5.13) may seem strange, it actually has a very simple and intuitive interpretation. Recall from (5.1) that $F = dx/dt$. Substitution of this relationship back in the two occurrences of

F in (5.13) yields:

$$x(t_N + h) \approx x + \frac{h}{2}\frac{\mathrm{d}x}{\mathrm{d}t} + \frac{h}{2}F\left(x + h\frac{\mathrm{d}x}{\mathrm{d}t}\right)$$

so

$$x(t_N + h) \approx x + \frac{h}{2}\left\{\frac{\mathrm{d}x_N}{\mathrm{d}t} + \frac{\mathrm{d}x_{N+1}}{\mathrm{d}t}\right\} \qquad (5.14)$$

This indicates that the second order Runge–Kutta approximation to $x(t + h)$ is similar to the Euler method in (5.4), except that two estimates of the slope are obtained and averaged. The first estimate is obtained at t_N, and the second one is obtained based on Euler's approximation of $x(t_N + h)$.

Let us look at this graphically using a very simple equation:

$$\frac{\mathrm{d}x}{\mathrm{d}t} = -\frac{x}{10} \quad \text{with} \quad x(0) = 1 \qquad (5.15)$$

We shall take $h = 4$ ms, which is large but will emphasize the principle of the Runge–Kutta method. This proceeds through four steps (a–d) to approximate $x(h) = x(4)$ from $x(0)$.
(a) Calculate the slope $F(x(0))$ (depicted at (1) in Fig. 5.2 as a heavy line):

$$F(x(0)) = \frac{\mathrm{d}x(0)}{\mathrm{d}t} = -\frac{1}{10} \qquad (5.16)$$

(b) Use this estimate to compute a first approximation of $x_1(4)$ using the equivalent of Euler's formula (5.4) (open point at (2) in Fig. 5.2):

$$x_1(4) \approx x(0) + hF(x(0)) = 1 + 4\left(-\frac{1}{10}\right) = 0.6 \qquad (5.17)$$

(c) Now use $x_1(4)$ to compute a second estimate of the slope $F(x(4))$ (heavy line at (3) in Fig. 5.2):

$$F(x_1(4)) = \frac{\mathrm{d}x_1(4)}{\mathrm{d}t} = -\frac{0.6}{10} = -0.06 \qquad (5.18)$$

(d) As indicated in (5.14) $F(x(0))$ and $F(x_1(4))$ from (5.16) and (5.18) are now averaged to obtain the second order Runge–Kutta estimate of $x(4)$ (solid point at (4) in Fig. 5.2):

$$x(4) \approx x(0) + \frac{h}{2}\{F(x(0)) + F(x_1(4))\} = 1 + \frac{4}{2}(-0.10 - 0.06) = 0.68 \qquad (5.19)$$

These four steps are shown in Fig. 5.2 along with both the Euler approximation (dotted line) and the second order Runge–Kutta solution (dashed line). These approximations are compared with the exact solution (solid curve), which is $x(t) = \exp(-t/10)$. At the end of one 4 ms time step the true result is 0.67, and the Runge–Kutta approximation is 0.68,

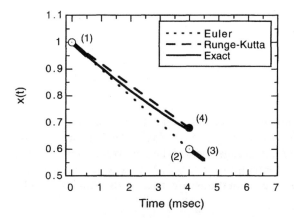

Fig. 5.2 Graphical representation of the second order Runge–Kutta method. Numbers 1–4 refer to the four steps embodied in the approximation (5.13) or (5.14).

which is an error of 1.5%. In contrast, Euler's method, which was used in (5.17 gives an estimate of 0.6, which is in error by 10.5%. To be fair, the Runge–Kutta method required twice as many calculations as Euler's method, so the Euler approximation should be computed using two steps of 2 ms each for comparison. The result is easily calculated to be 0.64, which is an improvement over 0.60, but still is in error by 4.7%. Thus, even for the same number of calculations, the second order Runge–Kutta result with a 4 ms step size is 3.0 times more accurate than Euler's method with a 2 ms step size.

We can also apply the second order Runge–Kutta routine to (5.5) and compare it with Euler's method. With a step size $h = 2$ ms, the error after 40 ms is 0.1%, which is much better than the 2.5% error of Euler's method using a step of $h = 1$ ms. Here again, Runge–Kutta is superior to Euler's method for the same number of calculations. This is the reason Euler's method is never used in serious work: it is a waste of effort!

5.3 Errors in approximate solutions

The previous examples demonstrate that the second order Runge–Kutta approximation is more accurate than Euler's method even when Euler's method uses half the step size so that the same number of computations are required. This raises the question: just how accurate is the Runge–Kutta approximation? This might seem to be impossible to answer, as the exact solution will generally be unknown, so there is only the approximation itself to rely on. However, we can still obtain a useful estimate of the error. To do so, recall that the second order Runge–Kutta method in (5.6) is based on truncation of the Taylor series after two terms. This means that the error in the computation will mainly depend on the next term, which is proportional to h^3. To obtain an estimate of the error, suppose we approximate a solution twice: once with step size h and once with half the step size, $h/2$. There will be twice as many steps in the latter case, but the simulation will certainly be more accurate. If $x(t + h)$ is the true solution after a step h, and $x_1(t + h)$ and $x_2(t + h)$ represent the two approximations using either one step h or two steps each of size $h/2$, we have approximately:

$$x(t+h) \approx x_1(t+h) + h^3\Omega$$
$$x(t+h) \approx x_2(t+h) + 2\left(\frac{h}{2}\right)^3 \Omega \tag{5.20}$$

The constant Ω that appears here is related to the maximum value of the third derivative of $x(t)$ in the interval according to the Taylor series formula. Although Ω is unknown, it will be the same number regardless of the step size chosen. Note that $(h/2)^3$ is multiplied by 2 in the second equation, as two $h/2$ steps were required to cover the interval h. If both of these approximations are treated as equalities and $\Delta = x_2 - x_1$, simple algebra gives:

$$\Delta \equiv x_2(t+h) - x_1(t+h) = h^3\Omega - 2\left(\frac{h}{2}\right)^3 \Omega \tag{5.21}$$

so

$$\Delta = \tfrac{3}{4}h^3\Omega$$

As Δ is measured by simulating the solution twice, once with step h and once with step $h/2$, (5.21) can be substituted for the error term containing Ω in (5.20):

$$2\left(\frac{h}{2}\right)^3 \Omega = \frac{\Delta}{3} \quad \text{so} \quad x(t+h) \approx x_2(t+h) + \frac{\Delta}{3}$$

This means that the estimated error is given by:

$$\text{Error} \approx \frac{\Delta}{3} \tag{5.22}$$

Thus, the error of the second order Runge–Kutta method using the smaller step size $h/2$ is approximately $\Delta/3$. It is important to note, however, that this is the error for a single step. Error will, of course, be cumulative in any simulation where multiple steps are involved, so it is prudent in neural simulations to take Δ itself as an estimate of the overall error involved in the computation.

As an example, suppose we use second order Runge–Kutta to simulate (5.5) for step sizes of both $h = 2$ ms and $h = 4$ ms. After 40 ms (20 steps for x_2 and 10 steps for x_1), $x_1(40) = 10.778$ and $x_2(40) = 10.816$, so the estimated error is:

$$\text{Error} \approx \frac{\Delta}{3} = 0.0127 \tag{5.23}$$

In this case the exact solution is 10.827, which is indeed within the estimated range.

It is also instructive to apply this same error estimation procedure to Euler's method. Recalling that Euler's method in (5.3) truncates the Taylor series after the first term:

$$x(t + h) \approx x_1(t + h) + h^2\Omega$$

$$x(t + h) \approx x_2(t + h) + 2\left(\frac{h}{2}\right)^2\Omega \tag{5.24}$$

Repeating the procedure used to obtain (5.22) reveals that the error in Euler's method is approximately:

$$\text{Error} \approx \Delta \tag{5.25}$$

When applied to (5.15) and Fig. 5.2, Euler's method with a 4 ms step produced the estimate $x_1(4) = 0.60$, while Euler's method with two 2 ms steps yielded $x_2(4) = 0.64$. Thus, $\Delta = 0.04$, the error estimate by (5.25). As the exact value is $x(4) = 0.67$, the computed error estimate is again appropriate. From a comparison of (5.22) with (5.25), one might expect the error from Euler's method to be about three times larger than that from second order Runge–Kutta for the same number of calculations, which is correct.

5.4 Fourth order Runge–Kutta

The improvements in accuracy obtained with second order Runge–Kutta approximations as compared to Euler's method suggest that the former technique might be profitably extended to higher orders. This has indeed been done, and third, fourth, and fifth order Runge–Kutta formulas are tabulated in various books. Each of these is derived in a manner analogous to the derivation of (5.13) from (5.6), except that the initial Taylor series expansion is extended to third, fourth, or fifth order terms. The contemporary standard in research is generally the fourth order Runge–Kutta scheme, and accordingly we shall standardize on it in the rest of this book. The fourth order Runge–Kutta method provides a significant improvement in accuracy over either second or third order methods for the same amount of computation, and it requires only four computations at each step. In contrast, fifth order Runge–Kutta requires six computations per step and thus is less efficient for a minimal gain in accuracy.

The formulas for the fourth order Runge–Kutta method are summarized below. Given $x(t)$ defined by eqn (5.1), the approximation to $x(t + h)$ based on terms in the Taylor series polynomial up to h^4 is:

$$x(t + h) \approx x(t) + \frac{h}{6}(K_1 + 2K_2 + 2K_3 + K_4)$$

where

$$K_1 = F(x(t), t)$$

$$K_2 = F\left(x(t) + \frac{h}{2}K_1, t + \frac{h}{2}\right)$$

$$K_3 = F\left(x(t) + \frac{h}{2}K_2, t + \frac{h}{2}\right) \qquad (5.26)$$

$$K_4 = F(x(t) + hK_3, t + h)$$

From the previous discussion, one can see that this formula again has a simple graphical interpretation. First we calculate dx/dt at time t, which is K_1. Next we use K_1 in Euler's formula (5.4) to move half way through the time interval to $t + h/2$, where we again calculate dx/dt, obtaining K_2. Repeating this procedure again gives us K_3, and K_3 is then used with Euler's formula to obtain an estimate of the slope K_4 at the end of the interval. Finally, a weighted average of these four estimates of dx/dt is used to compute $x(t + h)$.

As an example, let us again approximate (5.15) with $h = 4$ ms. Simple calculation gives:

$$\frac{dx}{dt} = -\frac{x}{10} \quad \text{with} \quad x(0) = 1$$

so

$$K_1 = F(1) = -0.10$$
$$K_2 = F(0.8) = -0.08$$
$$K_3 = F(0.84) = -0.084 \qquad (5.27)$$
$$K_4 = F(0.664) = -0.0664$$

so

$$x(4) \approx x(0) + \frac{4}{6}(-0.4944) = 0.6704$$

The slopes $K_1 - K_4$ are plotted at the points where they are calculated in Fig. 5.3 along with a comparison to the exact result. As the exact solution is $x = \exp(-t/10)$, $x(4) = 0.67032$. Thus, the fourth order Runge–Kutta estimate is accurate to 0.01% here.

As the fourth order Runge–Kutta routine is based on the Taylor series expansion up to fourth order, an estimate of the error is obtained from a comparison of simulations with step sizes of h and $h/2$, called $x_1(t)$ and $x_2(t)$ respectively. Using the approach developed earlier, the fourth order error estimate is:

$$\text{Error} \approx \frac{\Delta}{15} \equiv \frac{x_2(t) - x_1(t)}{15} \qquad (5.28)$$

Both Euler's method and the various Runge–Kutta routines can be generalized to a differential equation of any order that is cast into normal form. Taking the case of two coupled differential equations as an example, the fourth order Runge–Kutta approximation requires application of (5.26) to both equations simultaneously. The relevant

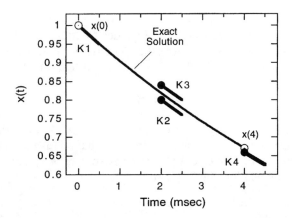

Fig. 5.3 Graphical representation of the four values of dx/dt, K1–K4, calculated in the fourth order Runge–Kutta approximation (5.28) to eqn (5.15). The exact solution and the estimate of x(4) are also shown.

formulae thus become:

$$\frac{dx}{dt} = F(x, y, t) \quad \text{and} \quad \frac{dy}{dt} = G(x, y, t)$$

so

$$x(t + h) \approx x(t) + \frac{h}{6}(K_1 + 2K_2 + 2K_3 + K_4)$$

$$y(t + h) \approx y(t) + \frac{h}{6}(L_1 + 2L_2 + 2L_3 + L_4)$$

where

$$
\begin{aligned}
K_1 &= F(x, y, t); & L_1 &= G(x, y, t) \\
K_2 &= F\left(x + \frac{h}{2}K_1, y + \frac{h}{2}L_1, t + \frac{h}{2}\right); & L_2 &= G\left(x + \frac{h}{2}K_1, y + \frac{h}{2}L_1, t + \frac{h}{2}\right) \\
K_3 &= F\left(x + \frac{h}{2}K_2, y + \frac{h}{2}L_2, t + \frac{h}{2}\right); & L_3 &= G\left(x + \frac{h}{2}K_2, y + \frac{h}{2}L_2, t + \frac{h}{2}\right) \\
K_4 &= F(x + hK_3, y + hL_3, t + h); & L_4 &= G(x + hK_3, y + hL_3, t + h)
\end{aligned}
$$

(5.29)

The accompanying MatLab script, **RungeKutta4.m**, implements (5.29) for any number of equations (see Appendix for details). In calculating the error in (5.29) it is apparent that there will be one value of Δ for x and another for y. It is prudent to use the larger of these as an estimate of the overall error of the simulation.

5.5 Variable step size routines

There is one further embellishment to the fourth order Runge–Kutta routine that bears mentioning. As the error in (5.28) is computed from the Taylor polynomial term in h^5, which is the first term ignored in fourth order Runge–Kutta simulations, it is possible to use Δ to estimate a new value of h, h_{new}, in order to attain a desired improvement in accuracy. It must be recognized, however, that Δ represents the error accumulated over N steps of the simulation. If the calculation is carried out to a final time $T = Nh$, then:

$$\frac{\Delta}{15} = Nh^5 M = h^4 TM \tag{5.30}$$

where M is related to the maximum value of d^5x/dt^5. Thus, although the error per step in fourth order Runge–Kutta is proportional to h^5, the cumulative error is proportional to h^4. If we denote the new desired error by ϵ, then the new value h_{new} that should give this size error obeys the same relationship as (5.30):

$$\frac{\epsilon}{15} = h_{new}^4 TM \tag{5.31}$$

where M is approximately the same as in (5.30). By now taking the ratio of (5.31) to (5.30) and solving for h_{new}, we obtain the formula:

$$h_{new} \leq h \left| \frac{\epsilon}{\Delta} \right|^{1/4} \tag{5.32}$$

The $\leq$ sign has been introduced to emphasize both that this is an approximate result and that ϵ is presumably the maximum desirable error. When one employs (5.32) to reduce error, h_{new} should be reduced to at least $h/2$ in order to have a significant effect.

In the neural simulations following in later chapters, it is generally adequate to compute a solution with both h and $2h$ as step sizes so that (5.28) may be used to estimate the error in the h step size simulation. If this is too large, then (5.32) may be used to estimate h_{new} with which to repeat the simulation. A final simulation with $2h_{new}$ as the step size may be used to check that the new error calculated from (5.28) is indeed within the desired tolerance. If necessary, the procedure can be repeated. A more sophisticated approach is to use (5.32) at each step in the Runge–Kutta simulation to adjust the step size h to maintain a desired degree of accuracy throughout the computation (an adaptive step size routine). This is generally more efficient than using the smallest h necessary for a desired degree of accuracy, but it is no more accurate. Fixed step size methods will therefore be adopted here, and the interested reader is referred to Press *et al.* (1986) for further consideration of adaptive step size methods. It is also worth mentioning that MatLab implements an adaptive step size routine in the **ode45()** procedure, which the interested reader is invited to explore.

5.6 Exercises

1. In Chapter 2 we developed an equation for a simple neuron using the Naka–Rushton function $S(P)$ defined in (2.11). Simulate the solution of the following differential equation using **RungeKutta4.m**:

$$\frac{dR}{dt} = \frac{1}{0.02}(-R + S(P))$$

$$S(P) = \begin{cases} \dfrac{100P^2}{25 + P^2} & \text{for } P \geq 0 \\ 0 & \text{for } P < 0 \end{cases}$$

$$P(t) = 20 \sin(2\pi 10 t)$$

where both the time constant and t are in seconds. Let $R(0) = 0$ for your initial condition. Run your simulation for 1.0 s and plot the results. Use a time step h of 0.004 s (i.e. 4 ms). Obtain an error estimate of your result by repeating the simulation with $h = 2$ ms. You have just modeled the response of a simple cell in monkey visual cortex (Movshon *et al.*, 1978)!

2. The program **Euler.m** implements an Euler approximation to the simple equations:

$$\frac{dx}{dt} = 2y$$

$$\frac{dy}{dt} = -2x$$

Assume the initial condition is $x(0) = 2$, $y(0) = 0$, and write down the exact solution. Next use **Euler.m** to simulate these equations for time steps $h = 0.1$, 0.01, and 0.001. Does the Euler approximation produce a trajectory that is stable, asymptotically stable, or unstable? Does the stability of the Euler approximation agree with that of the exact solution? Why or why not? (Hint: sketch a phase plane trajectory for the exact solution and discuss how the Euler approximation deviates from this.) This problem will also show you how to plot phase plane solutions for any two variables in a simulation.

3. Derive the error estimate (5.28) for the fourth order Runge–Kutta procedure.

4. The following equation is known as the van der Pol equation:

$$\frac{d^2x}{dt^2} + (x^2 - 5)\frac{dx}{dt} + 9x = 0$$

This was the first mathematical model proposed for cardiac rhythms, and it has also been used to simulate brain waves. Solve this equation using **RungeKutta4.m** (convert to normal form first) with the initial conditions $x(0) = 1$ and $dx/dt = 0$ at $t = 0$. Carry out the simulation up to time $t = 10$. First choose your time step $h = 0.05$. Next, make an estimate of the error of your approximation. Using the formulas in the chapter, reduce h until your error estimate is less than 0.001. Be sure to indicate the value of h used, how you arrived at it, and your final error estimate. Plot both $x(t)$ and the phase plane $(x, dx/dt)$ for your final simulation.

6   *Nonlinear neurodynamics and bifurcations*

We have completed our survey of linear dynamical systems and have developed simulation methods suitable for application to either linear or nonlinear systems. This chapter begins our exploration of nonlinear dynamics in neural systems. The chapter will develop the basic approach, which relates the local properties of nonlinear systems to those of associated linear systems. Three different ways in which two neurons can interact will be discussed next: (1) negative feedback in a divisive gain control, (2) mutual excitation in short-term memory, and (3) mutual inhibition in neural decision making. It will also be shown that neural adaptation can lead to memory loss, a topic that will introduce bifurcation theory.

Neurons will be described by the temporal variation of their spike rates in this chapter rather than at the level of individual spikes. There are several reasons for this. First, many principles of neurodynamics can be effectively studied at the spike rate level of description, and the mathematics is considerably simplified. Second, spike trains are examples of nonlinear oscillations mathematically, and so this level of description must await development of that topic in Chapter 8. Finally, experimental data reported as histograms and many theoretical problems in neuroscience are in fact described as spike rates.

6.1   Steady states and isoclines

Let us begin with a general mathematical description of a two interacting neurons:

$$\frac{dx_1}{dt} = F(x_1, x_2)$$
$$\frac{dx_2}{dt} = G(x_1, x_2)$$

(6.1)

F and G can be any of a wide range of nonlinear functions of x_1 and x_2, but they must have certain properties to insure that (6.1) will have unique solutions. It will be sufficient to require that in the region of physiological interest F and G are both finite and continuous. We can even tolerate a finite number of finite discontinuities in F or G or their derivatives (as was the case in our piecewise linear approximation to an EPSP in Chapter 2). These requirements do not limit us to any significant extent, as all physiologically plausible systems satisfy these constraints.

The dynamical behavior of (6.1) will generally be quite complex and will require Runge–Kutta simulation in order to follow the temporal evolution in detail. However, we can learn an enormous amount about the solutions by studying the nature of the **equilibrium points** or **steady states** of the system. As $dx_1/dt = 0$ and $dx_2/dt = 0$ at a steady state, the steady states of (6.1) must satisfy the equations:

$$F(x_1, x_2) = 0$$
$$G(x_1, x_2) = 0 \qquad (6.2)$$

If this were a linear system, these would be simultaneous linear equations. For a nonlinear system, however, each of the equations in (6.2) will describe some curve in the (x_1, x_2) state space of the system, and the intersections of these curves will determine the equilibrium points. Equations (6.2) describe the **isoclines** of the system, a term which literally means 'equal slope'. Thus, $F = 0$ describes the locus of points in state space where $dx_1/dt = 0$, so all trajectories must cross this isocline parallel to the x_2 axis. Similarly, on the isocline where $G = 0$, $dx_2/dt = 0$, and trajectories will parallel the x_1 axis. F and G are simultaneously zero where the isoclines intersect, and this defines an equilibrium point. It should be noted here that some authors use the term **nullcline** instead of isocline to emphasize that the slope is null or zero along these particular isoclines. However, isoclines are almost never discussed today except for the zero slope case, so the term isocline will only be used here in the restrictive sense defined by (6.2).

6.2   A divisive gain control

To fix these ideas, let us consider a nonlinear gain control network that employs feedback inhibition. This network was originally developed as a cortical divisive gain control to explain certain aspects of psychophysics related to orientation selective cells in the visual cortex (Wilson and Humanski, 1993). More recently, a similar feedback circuit has been used to describe amacrine cell feedback onto bipolar cells in primate and human retinas (Wilson, 1997; see Chapter 7). The simple negative feedback network to be described here is like the cone, horizontal cell feedback network discussed in Chapter 3, except that the inhibition divides the stimulus to the first neuron. As pointed out in Chapter 2, shunting inhibitory synapses are well approximated by division.

Letting B represent the bipolar cell response to light level L and A the amacrine cell response, the equations are:

$$\frac{dB}{dt} = \frac{1}{\tau_B}\left(-B + \frac{L}{1+A}\right)$$
$$\frac{dA}{dt} = \frac{1}{\tau_A}(-A + 2B) \qquad (6.3)$$

This equation has well-behaved solutions in all regions that exclude $A = -1$, where the right-hand side of the first equation becomes infinite and the solution is not defined. As the light inputs L can only be positive or zero, however, it is easy to show that if the initial conditions lie in the first quadrant or at the origin, then the system must stay in the first quadrant for all future times. That is, if $A(0) \geq 0$, $B(0) \geq 0$, and $L \geq 0$ solutions can never leave the first quadrant. To prove this, note that whenever $B = 0$, $dB/dt \geq 0$ because A,

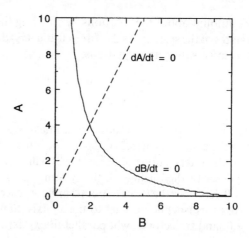

Fig. 6.1 Phase plane and isoclines (6.4) for eqn (6.3) plotted for $L = 10$. The unique steady state in the first quadrant is located where the isoclines intersect.

$L \geq 0$. Therefore B must become positive (or remain zero), but it cannot become negative. Similarly, $dA/dt \geq 0$ whenever $A = 0$, so A can never become negative. Thus, all trajectories starting in the first quadrant must remain there.

The isoclines are now obtained by setting the derivatives in (6.3) equal to zero, giving:

$$B = \frac{L}{1 + A} \quad \text{and} \quad A = 2B \tag{6.4}$$

These are plotted in the state space of this system in Fig. 6.1. The one intersection of the isoclines in the first quadrant defines the steady state, which can be obtained analytically in this case by solving (6.4):

$$A = 2B; \quad B = \frac{-1 + \sqrt{1 + 8L}}{4} \tag{6.5}$$

Note that the second root or equilibrium point does not lie in the first quadrant, so it is irrelevant to the neurobiology of the problem. Neural systems with multiple steady states in the first quadrant will be considered shortly. In the example plotted in Fig. 6.1, $L = 10$, so the steady state is at: $B = 2$; $A = 4$.

6.3   Stability of steady states

Having found the equilibrium point of (6.3), let us now determine its nature (node, spiral point, etc.) and stability characteristics. For a linear system this simply requires obtaining the eigenvalues from the characteristic equation, but how should we proceed with a nonlinear system? The answer turns out to be simple: expand the nonlinear functions in (6.1) or (6.3) in a Taylor series evaluated at the equilibrium point and retain only the linear terms. This produces an associated linear equation in the vicinity of the equilibrium state:

$$
\frac{d}{dt}\begin{pmatrix} x_1 \\ x_2 \end{pmatrix} = \begin{pmatrix} \left.\frac{\partial F}{\partial x_1}\right|_{Eq} & \left.\frac{\partial F}{\partial x_2}\right|_{Eq} \\ \left.\frac{\partial G}{\partial x_1}\right|_{Eq} & \left.\frac{\partial G}{\partial x_2}\right|_{Eq} \end{pmatrix}\begin{pmatrix} x_1 \\ x_2 \end{pmatrix} \tag{6.6}
$$

where the partial derivatives in the matrix are evaluated at the equilibrium point and x_1 and x_2 are the values of the variables relative to the equilibrium values. This matrix of first partial derivatives is called the **Jacobian** or **Jacobian matrix**. The Jacobian for any non-linear system of N equations will always be $N \times N$ in size.

The higher order Taylor series terms in the expansion of F and G will be at least quadratic in x_1 and x_2, so the linear terms in the Jacobian will dominate when x_1 and x_2 are sufficiently close to equilibrium. Therefore, you might expect that the linear stability analysis of (6.6) would also apply to (6.1) near equilibrium. This is basically correct and is captured in the following theorem, which applies to nonlinear systems of any order:

Theorem 8: Given the nonlinear system described by the equation:

$$
\frac{d\vec{X}}{dt} = \vec{F}(\vec{X})
$$

and an equilibrium point at X_{Eq}, which is a solution to:

$$
\vec{F}(\vec{X}_{Eq}) = \vec{0}
$$

calculate the Jacobian to produce an associated linear equation:

$$
\frac{d\vec{x}}{dt} = \overleftrightarrow{A}\vec{x}
$$

where

$$
\overleftrightarrow{A} = \begin{pmatrix} \frac{\partial F_1}{\partial x_1} & \frac{\partial F_1}{\partial x_2} & \cdots \\ \frac{\partial F_2}{\partial x_1} & \frac{\partial F_2}{\partial x_2} & \cdots \\ \vdots & \vdots & \frac{\partial F_N}{\partial x_N} \end{pmatrix}
$$

where all partial derivatives are evaluated at X_{Eq}. Then sufficiently near X_{Eq}: (a) if all eigenvalues of the linear system have negative real parts, the nonlinear system is asymptotically stable; and (b) if the linear system has at least one eigenvalue with a positive real part, the nonlinear system is unstable. In addition, the type of equilibrium point for the nonlinear system, i.e. spiral point, node, or saddle point, will be the same as that for the associated linear equation.

Two special cases are explicitly excluded by Theorem 8. First, if A has a pair of pure imaginary roots and therefore is a stable center (but not asymptotically stable), the theorem does not apply. Second, if any root of A is zero, the theorem again fails to apply. The higher order terms in the Taylor expansion of F become critical when the associated linear system has any roots with zero real part. It is also important to note that Theorem 8 does not specify just how close to equilibrium one must be for the theorem to apply. Determining this requires the use of Lyapunov functions, which will be introduced in Chapter 14.

Let us apply Theorem 8 to the divisive gain control in (6.3). For $L = 10$, the steady state occurs at $B = 2$; $A = 4$. Let time constants be 10 ms. Calculation of the Jacobian now gives:

$$\overset{\leftrightarrow}{A} = \begin{pmatrix} -\dfrac{1}{10} & -\dfrac{1}{(1+A)^2} \\ \dfrac{1}{5} & -\dfrac{1}{10} \end{pmatrix} = \begin{pmatrix} -\dfrac{1}{10} & -\dfrac{1}{25} \\ \dfrac{1}{5} & -\dfrac{1}{10} \end{pmatrix} \tag{6.7}$$

where the right-hand equality results from evaluation of the Jacobian at the steady state. Using **LinearOrder2.m**, the eigenvalues are $\lambda = -0.1 \pm 0.089\mathrm{i}$. Thus, the equilibrium is an asymptotically stable spiral point for both the associated linear system and for the nonlinear system in (6.3). Given this analysis of (6.3), it is now appropriate to simulate the solution using Runge–Kutta methods. Results of simulations using the MatLab script **DivFB.m** are plotted in Fig. 6.2. As can be seen, the solution with initial conditions $A = 0$, $B = 0$ asymptotically approaches the unique steady state in the first quadrant with a damped oscillation, as was predicted by analysis of (6.7). The simulation suggests that trajectories starting far from the steady state will approach it asymptotically, although this cannot be determined using Theorem 8. Figure 6.2 also shows that the transient overshoot of the response is relatively larger when the light intensity $L = 100$ than it is when $L = 10$. This is a manifestation of the nonlinear dynamics.

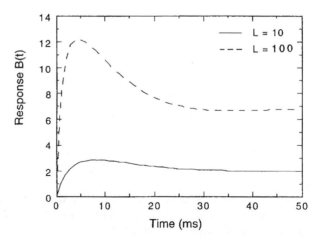

Fig. 6.2 Responses $B(t)$ in (6.3) to two stimulus levels, $L = 10$ and $L = 100$.

6.4 A short-term memory circuit

In the previous example, we analyzed a nonlinear, two-neuron network that had a single equilibrium point in the first quadrant. However, many nonlinear systems are interesting precisely because they have multiple equilibria, so let us examine such a system now. To motivate the mathematical discussion, we shall examine a physiological example involving short-term memory. Consider an experiment in which a monkey is first presented with a briefly flashed stimulus, a red or green light in this case. The stimulus then disappears, but the monkey is required to wait during a delay until it receives a second signal indicating that it is time to respond. The monkey must then make one of two responses indicating whether it remembers a red or green stimulus, and it is rewarded if the response is correct. This is known as a delayed matching task, because the monkey must remember the stimulus during the delay in order to be rewarded. Fuster (1995) has shown that single neurons in the temporal and prefrontal cortex of monkeys can be switched on by the brief stimulus presentation, and these neurons will then continue to fire at a higher rate for 20 s or more after the stimulus disappears. For example, the neuron shown in Fig. 6.3 increased its firing rate following presentation of a red stimulus and maintained its increased firing rate for 16 s after the stimulus vanished, when the monkey made a correct choice and was rewarded. This particular neuron was selective for the red stimulus, as it never responded to the green one.

Let us examine a very simple neural network with responses like those of prefrontal neurons during this delayed response task. The system consists of just two neurons which are mutually excitatory and whose spike rates are described by the Naka–Rushton function in (2.11) and Fig. 2.3 with a maximum spike rate of 100/s and $N = 2$. Assuming that the neurons have identical properties and connection strengths for simplicity, they will be described by the equations:

$$\frac{dE_1}{dt} = \frac{1}{\tau}\left(-E_1 + \frac{100(3E_2)^2}{120^2 + (3E_2)^2}\right)$$
$$\frac{dE_2}{dt} = \frac{1}{\tau}\left(-E_2 + \frac{100(3E_1)^2}{120^2 + (3E_1)^2}\right)$$

(6.8)

This problem can be tackled in the same fashion as the previous one: first plot the isoclines and find the equilibrium points, then use Theorem 8 to determine the stability characteristics of the equilibria.

The state space and isoclines of (6.8) are plotted in Fig. 6.4. The isoclines intersect at three points rather than just one, so this system must have three steady states. Solving the isocline equations for steady states will generally require numerical approximations using MatLab. However, in this case the symmetry of the problem simplifies things considerably. From the symmetry of (6.8), it can be inferred that $E_1 = E_2$ at equilibrium, so the steady states will obey the simplified equation:

$$E_1 = \frac{100(3E_1)^2}{120^2 + (3E_1)^2}$$

Spikes, decisions, and actions

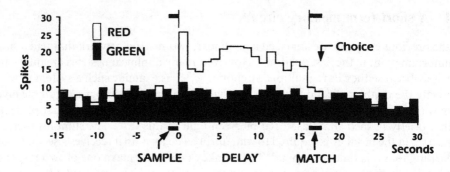

Fig. 6.3 Responses of a neuron in monkey inferiortemporal cortex during a short-term memory task (reproduced with permission, Fuster, 1995). Following a 1.0 s presentation of a red sample, this neuron fires at more than twice its resting level for 16 s until the signal to make a match appears and the monkey makes a choice to receive a reward. The same neuron did not increase its response when the sample was green.

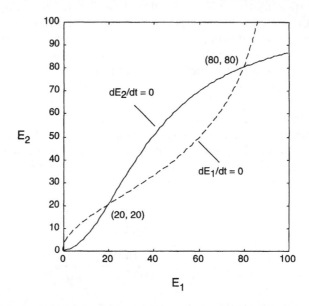

Fig. 6.4 Isoclines of (6.8) intersect at three points, thus producing steady states at $(0, 0)$, $(20, 20)$ and $(80, 80)$.

so

$$9E_1^3 - 900E_1^2 + 120^2E_1 = 0 \tag{6.9}$$

Although this is a cubic equation, one can see that $E_1 = 0$ is one solution, so the origin must be one equilibrium point. The other two are found by solving the quadratic equation that results after factoring E_1 out of (6.9), which produces the result:

$$E_1 = \frac{900 \pm \sqrt{(900)^2 - 36(120)^2}}{18} = 50 \pm 30$$

So the remaining two equilibria are at $(20, 20)$ and $(80, 80)$ as shown in Fig. 6.4.

The next task is to apply Theorem 8 to each of the three steady states of (6.8) in turn. To do this we shall need to know the derivative of the Naka–Rushton function $S(x)$ from (2.11), where $N = 2$. From basic calculus:

$$S(x) = \frac{M(ax)^2}{\sigma^2 + (ax)^2}$$

so (6.10)

$$\frac{dS}{dx} = \frac{2M\sigma^2 a^2 x}{\left(\sigma^2 + (ax)^2\right)^2}$$

Letting $\tau = 20$ ms, the Jacobian matrix for (6.8) will therefore have the following form:

$$\overleftrightarrow{A} = \begin{pmatrix} -\dfrac{1}{20} & \dfrac{1}{20}\left[\dfrac{200(120)^2 9E_2}{\left((120)^2+(3E_2)^2\right)^2}\right] \\ \dfrac{1}{20}\left[\dfrac{200(120)^2 9E_1}{\left((120)^2+(3E_1)^2\right)^2}\right] & -\dfrac{1}{20} \end{pmatrix}$$ (6.11)

The Jacobian must be evaluated at each equilibrium point, after which **Linear Order2.m** can be used to obtain the eigenvalues. This produces the following matrices and stability characteristics at the three singular points:

$$\begin{pmatrix} 0 \\ 0 \end{pmatrix}: \overleftrightarrow{A} = \begin{pmatrix} -0.05 & 0 \\ 0 & -0.05 \end{pmatrix}, \lambda = -0.05, \; -0.05 \quad \text{Asymptotically stable node.}$$

$$\begin{pmatrix} 20 \\ 20 \end{pmatrix}: \overleftrightarrow{A} = \begin{pmatrix} -0.05 & 0.08 \\ 0.08 & -0.05 \end{pmatrix}, \lambda = +0.03, \; -0.13 \quad \text{Unstable saddle point.} \quad (6.12)$$

$$\begin{pmatrix} 80 \\ 80 \end{pmatrix}: \overleftrightarrow{A} = \begin{pmatrix} -0.05 & 0.02 \\ 0.02 & -0.05 \end{pmatrix}, \lambda = -0.07, \; -0.03 \quad \text{Asymptotically stable node.}$$

Therefore, all trajectories must diverge from $(20, 20)$ and trajectories near either $(0, 0)$ or $(80, 80)$ will approach these steady states asymptotically. Try running MatLab script **STMemory.m** with a range of different initial conditions to convince yourself that trajectories will indeed converge to either $(0, 0)$ or $(80, 80)$.

6.5 Hysteresis, bifurcation, and memory

This short-term memory network exhibits an important nonlinear phenomenon known as **hysteresis**. The term hysteresis is derived from a Greek term meaning 'to lag behind'. In the present context, this means that the present state of our neural network is determined not just by the present state of stimulation but also by the history of stimulation. Hysteresis is most easily exhibited if (6.8) is modified to include an external stimulus K, assumed to be the same for each of the two neurons. Now (6.9), which defines the steady states, becomes:

$$E_1 = \frac{100(3E_1 + K)_+^2}{120^2 + (3E_1 + K)_+^2} \qquad (6.13)$$

where the subscripted plus sign indicates that the expressions in parentheses evaluate to zero for negative arguments. The steady state values of E_1 now depend on K, and MatLab has been used to solve (6.13) with the results plotted in Fig. 6.5. There is a range of K values between A and B for which three steady states exist. Two are asymptotically stable (solid lines), and they are separated by an unstable saddle point (dashed line), as was found in (6.12) and Fig. 6.4. If we begin stimulating the network with $K < A$ and increase K, the network will remain in the lower or resting steady state throughout the AB range. If, however, we begin stimulation with $K > B$ and decrease K, the network will stay in the upper asymptotically stable state as K traverses the region AB. Thus, over the stimulus range $A < K < B$, the equilibrium state that the system is in depends on the previous history of stimulation. Furthermore, if K is varied slowly back and forth across the range shown in the diagram, the neural response will trace out the loop shown by the arrows in the figure. This is known as a **hysteresis loop**.

Let us consider in more detail the reasons that this memory network exhibits hysteresis. When $K = A$, the unstable saddle point coalesces with the upper asymptotically stable equilibrium, and the two vanish when $K < A$. Similarly, when $K = B$, the lower asymptotically stable state and the saddle point state coalesce, and both vanish when $K > B$. This appearance or disappearance of a pair of equilibria is known as a **bifurcation**, which literally means a splitting in two. At a **bifurcation point** two equilibrium points (or one equilibrium and one nonlinear oscillation, see Chapter 8) are either created together, or else they merge and vanish together. The mathematical reason that a bifurcation always involves the creation or disappearance of a pair of steady states is that two roots of

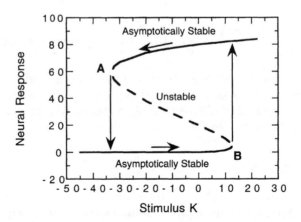

Fig. 6.5 Hysteresis loop and bifurcations generated by (6.8) in the presence of stimulus K in (6.13). Between A and B two steady states are asymptotically stable nodes, while the intervening one is an unstable saddle point. If K is swept back and forth across range AB, the system will trace out the hysteresis loop shown by the arrows.

(6.13) become complex valued at the bifurcation point and therefore no longer exist on the real plane. Although bifurcations involving two equilibrium points always occur in an asymptotically stable–unstable pair, we shall see in Chapter 8 that there are bifurcations of oscillatory solutions from steady states as well.

Hysteresis is the mathematical basis of the short-term memory capacity of this two-neuron network. The system will remain in the lower asymptotically stable equilibrium where neither neuron is firing until there is a stimulus $K > B$, after which the neural response will rapidly jump to the upper asymptotically stable state as depicted by the vertical arrow at B in Fig. 6.5. This constitutes triggering of the network's short-term memory. If the stimulus is now turned off, so $K = 0$, the neurons will remain in the upper asymptotically stable state and thus continue to fire, because the upper state exists and is asymptotically stable for $K = 0$. Thus, the network 'remembers' via hysteresis that a relevant stimulus has occurred. To shut the network off, there must be a negative or inhibitory stimulus ($K < A < 0$) to erase the short-term memory activity once triggered. This is hysteresis in short-term memory.

6.6 Adaptation, forgetting, and catastrophe theory

The short-term memory network in (6.8) has a physiological shortcoming: after the network has been triggered to its active state, both E neurons will continue to fire at a rate of 80 spikes/s until the response is actively inhibited. In the absence of such inhibition, activity will continue forever. So, the network does not incorporate the physiological fact that neurons (like muscles) slowly adapt or fatigue when they continue to fire at high rates for long periods of time. As will be seen in a moment, the consequence for the animal is forgetting. Let us extend our analysis of short-term memory by incorporating neural adaptation into the network.

The ionic mechanisms underlying neural adaptation will be explored in Chapter 10, but the reduction of spike rates caused by adaptation can be easily incorporated here. Studies of both single neurons in visual cortex (Bonds, 1991) and the perceptual consequences of visual pattern adaptation (Wilson and Humanski, 1993) indicate that adaptation causes a slow increase in the constant σ of the Naka–Rushton function (2.11). This in turn reduces the firing rate of the neurons. So, in addition to the two neural activity equations in (6.8), adaptation requires the introduction of two variables, A_1 and A_2. The resulting system of four equations is:

$$\frac{dE_1}{dt} = \frac{1}{\tau}\left(-E_1 + \frac{100(3E_2)_+^2}{(120 + A_1)^2 + (3E_2)_+^2}\right)$$

$$\frac{dE_2}{dt} = \frac{1}{\tau}\left(-E_2 + \frac{100(3E_1)_+^2}{(120 + A_2)^2 + (3E_1)_+^2}\right) \qquad (6.14)$$

$$\frac{dA_1}{dt} = \frac{1}{\tau_a}(-A_1 + 0.7E_1)$$

$$\frac{dA_2}{dt} = \frac{1}{\tau_a}(-A_2 + 0.7E_2)$$

Assume again that $\tau = 20\,\text{ms}$, but because neural adaptation is a very slow process, a reasonable value for the adaptation time constant is $\tau_a = 4000\,\text{ms}$, i.e. 4.0 s! The firing rate in (6.14) will continuously decrease as A_1 and A_2 increase. A_1 and A_2, in turn, will increase when $E_1 > 0$ or $E_2 > 0$ respectively, reaching the equilibrium values:

$$A_1 = 0.7E_1 \quad \text{and} \quad A_2 = 0.7E_2 \tag{6.15}$$

Let us set $E_1 = E_2$ due to the symmetry of the network and solve for the equilibrium states of (6.14), which obey the equation:

$$E_1\left((120 + 0.7E_1)^2 + 9E_1^2\right) = 900E_1 \tag{6.16}$$

$E_1 = 0$ is clearly one solution of (6.16), so the resting state $(0, 0, 0, 0)$ is one equilibrium point of (6.14). After factoring out E_1 in (6.16) and solving the resulting quadratic equation, both roots turn out to be complex. This means that there is only one equilibrium point for (6.14), and this is at the origin. Does this mean that the system no longer has interesting behavior and cannot function as a short-term memory network? The answer is no, and the key to understanding this lies in the fact that the adaptation variables have a time constant that is 200 times larger than the neural time constants, so A_1 and A_2 will increase very slowly. They change so slowly, in fact, that we can analyze the system using the approximation that E_1 and E_2 come to equilibrium very rapidly relative to any change in A_1 and A_2 so that E_1 and E_2 are essentially always at the equilibrium determined by the current values of A_1 and A_2. Thus, we can understand the behavior of this system by analyzing the isoclines and equilibrium points of just the first two equations in (6.14) while treating A_1 and A_2 as parameters. This general approach, in which system behavior on a fast time-scale is treated as being in equilibrium relative to behavior on a very slow time-scale, is a widespread and very important method in nonlinear analysis.

Let us set $A_1 = A_2$ due to symmetry and consider how the isoclines in Fig. 6.4, which describe system (6.14) when $A_1 = 0$, vary as A_1 increases. The isocline equations are now:

$$
\begin{aligned}
E_1 &= \frac{100(3E_2)^2}{(120 + A_1)^2 + (3E_2)^2} \\[2mm]
E_2 &= \frac{100(3E_1)^2}{(120 + A_1)^2 + (3E_1)^2}
\end{aligned}
\tag{6.17}
$$

Figure 6.6 shows the isoclines and equilibrium points for $A_1 = 24$ and $A_1 = 36$. For $A_1 = 24$, the system still has three steady states, and stability analysis shows that both the origin and $(64, 64)$ are asymptotically stable nodes, while $(36, 36)$ is an unstable saddle point. Compared to Fig. 6.4, however, the firing rate in the asymptotically stable state has decreased, and the saddle point has moved closer. When A_1 increases to 36, one node and the saddle point have coalesced and vanished leaving only the origin as an asymptotically stable steady state. The node and saddle point coalesce when $A_1 = 30$. (The isocline equations still have three solutions, but two have become complex conjugates and thus have moved off the phase plane.) To get a feel for this change, run the MatLab script

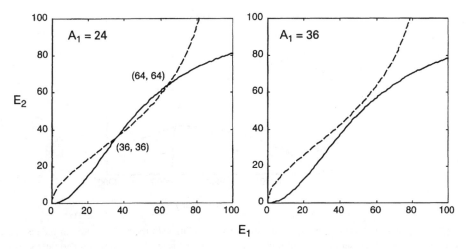

Fig. 6.6 Isoclines (6.17) for system (6.14) plotted for two values of A_1. On the left there are three steady states, but only the steady state at the origin remains on the right.

IsoclineMovie.m, which produces an animation showing the gradual change of the isoclines in (6.17) as A_1 increases from 0 to 40.

Thus, the adaptation variables A_1 and A_2 slowly move the equilibrium point corresponding to short-term memory activity toward the saddle point until they meet and vanish from the phase plane. To corroborate this analysis, let us simulate (6.14) using the script **STMadapt.m**. In this simulation, a stimulus $K = 50$ is presented for 200 ms and is then turned off. As illustrated in Fig. 6.7, however, the ensuing neural activity outlasts the stimulus for about 5000 ms until adaptation results in loss of stimulus memory, at which point neural activity rapidly ceases. A larger adaptation time constant τ_a in (6.14) would increase the duration of short-term memory but would also increase the simulation time. Thus, neural adaptation can lead to forgetting in short-term memory tasks, and indeed monkeys do forget and make mistakes in delayed response tasks when the delay lasts too many seconds. Correlated with this behavioral change, firing rates of prefrontal neurons drop back to baseline levels (Fuster, 1995).

This short-term memory network with adaptation is a very important example of the insights that can be gained by using arguments from symmetry and taking advantage of widely different time constants. Doubtless, no two neurons ever have exactly the same characteristics, but just as obviously, there is only enough genetic information to specify classes of neurons, so it is reasonable to assume that different members of the same class will be very similar. By letting $E_1 = E_2$ and $A_1 = A_2$, analysis of (6.14) was reduced from four to two dimensions.

In (6.14) there is an extreme difference in time scales, because fluctuations in neural spike rates are much more rapid than neural adaptation. When there is an extreme difference in time scales, the rapidly changing variables are almost always very near equilibrium, but the positions of the equilibria change very slowly. In this example, the spike rates E_1 and E_2 vary rapidly and hence arrive at an equilibrium point determined by the

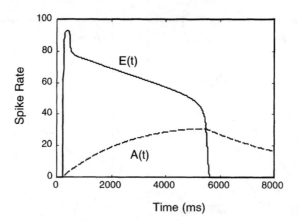

Fig. 6.7 Response of (6.14) to a brief, 200 ms stimulus coinciding with the narrow peak on the upper left. Recurrent excitation maintains activity of both $E(t)$ cells at a high level, but activity slowly decays as neural adaptation $A(t)$ builds up. After 5000 ms a sudden loss of neural activity occurs at a bifurcation.

current value of the adaptation variables A_1 and A_2. These adaptation variables change 200 times more slowly than the spike rates, so they can be viewed as parameters that slowly change the structure of the equilibrium points of the system. As illustrated in Fig. 6.7, the response of the system follows the slowly changing equilibrium points for more than 5000ms. Then a **catastrophe** occurs: one asymptotically stable equilibrium joins with the unstable saddle point and vanishes, so the neural response rapidly drops to zero. Reference to Fig. 6.5 shows that the adaptation variable functions like a slowly varying input driving the system, once excited, back through the bifurcation point at A.

The mathematical notion of a catastrophe or bifurcation also underlies the geology of plate tectonics and earthquakes. As pressure builds up on tectonic plates, they compress only slightly for a long time, so the distance between points on opposite sides of a fault line changes little. At some point, however, the pressure becomes great enough to overcome frictional forces, and the plates rapidly slip to a new equilibrium position, thus producing an earthquake, which can be a true catastrophe in the vernacular sense! The mathematical concepts analogous to those in this neural short-term memory example underlie geophysical catastrophes as well.

6.7 Competition and neural decisions

So far we have analyzed two nonlinear neural networks: one for divisive gain control and one for short-term memory. The former involved a negative feedback loop, while the latter incorporated mutual excitation. A further possible interaction between two neurons is mutual inhibition, which will be examined here. As we shall see, the state space of two mutually inhibitory neurons is similar to that of the memory network in having two asymptotically stable steady states separated by an unstable saddle point. However, each steady state in this case is defined by activity in one neuron and complete inhibition of the other, so this network makes one of two mutually exclusive decisions based on the relative strengths of inputs to the two neurons.

Consider the following equations:

$$\frac{dE_1}{dt} = \frac{1}{\tau}(-E_1 + S(K_1 - 3E_2))$$

$$\frac{dE_2}{dt} = \frac{1}{\tau}(-E_2 + S(K_2 - 3E_1)) \tag{6.18}$$

$$S(x) = \begin{cases} \dfrac{100(x)^2}{120^2 + (x)^2} & x \geq 0 \\ 0 & x < 0 \end{cases}$$

K_1 and K_2 here are the stimuli to the two neurons in the network, and $S(x)$ is again the Naka–Rushton function from (2.11). Assume $\tau = 20$ ms. Each neuron inhibits the other subtractively with a synaptic strength of -3. Explore the responses of this network by running **WTA2.m** using various combinations of excitatory inputs K_1 and K_2. Above a minimum level of excitation (about 50) and assuming initial conditions with all variables zero, the system always switches to an equilibrium point at which the more strongly stimulated neuron is active and the other neuron has been shut off by inhibition. This is the simplest example of a **winner-take-all** (WTA) network. This name has been used to describe such networks, because the neuron receiving the strongest stimulus will win the inhibitory competition with the other neurons and in turn suppress all of its competitors.

Let us analyze (6.18) in the case K_1 and $K_2 = 120$. Due to the competitive inhibition, one steady state is $E_1 = S(K) = 50$, and $E_2 = S(K - 3 \times 50) = S(-30) = 0$. Similarly, the reader can easily verify that $E_1 = 0$, and $E_2 = 50$ is also an equilibrium point. If you run **WTA2.m**, you will see that the isoclines intersect at a third equilibrium point in addition to the two above. From symmetry considerations you might expect this to occur where $E_1 = E_2$, and this is correct. If one sets $E_1 = E_2$ in either of the isocline equations in (6.18), the MatLab **roots** function gives the solution $E_1 = E_2 = 20$.

As in previous examples, the stability of each steady state must next be determined. As $(50, 0)$ and $(0, 50)$ will be the same, let us just examine the Jacobian matrix at the former state:

$$\overleftrightarrow{A} = \begin{pmatrix} -\dfrac{1}{\tau} & 0 \\ 0 & -\dfrac{1}{\tau} \end{pmatrix} \tag{6.19}$$

The eigenvalues here are obviously both identical: $\lambda = -1/\tau$, so $(50, 0)$ and $(0, 50)$ are both asymptotically stable nodes that are critically damped. At $(20, 20)$ we can use (6.10) to evaluate the Jacobian, with the result:

$$\overleftrightarrow{A} = \begin{pmatrix} -\dfrac{1}{\tau} & -\dfrac{8}{5\tau} \\ -\dfrac{8}{5\tau} & -\dfrac{1}{\tau} \end{pmatrix} \tag{6.20}$$

Setting $\tau = 20$ ms and using **Linearorder2.m**, the eigenvalues are: $\lambda = -0.13$ and $\lambda = +0.03$, so $(20, 20)$ is an unstable saddle point. Thus, trajectories will leave the vicinity of the saddle point at $(20, 20)$ and approach the nearer of the two asymptotically stable nodes. It is worth noting, however, that this decision network also exhibits hysteresis: once one neuron has suppressed the other, there must be a large increase in the stimulus to the second neuron before it can become active. Thus, there is resistance to change in this network once a decision has been made, which is also a characteristic of much human cognition!

We have now completed an analysis of several systems composed of two interacting neurons. Such systems can have multiple equilibrium points, which permit them to perform short-term memory tasks and to make decisions. Although two neuron systems are far too small to provide plausible explanations for complex mental phenomena, the next chapter will show that the principles learned above generalize to much larger and more interesting neural networks.

6.8   Exercises

1. Analyze the following divisive gain control circuit consisting of a bipolar cell B and an amacrine cell A. Let $\tau_A = 20$ ms and $\tau_B = 50$ ms. (a) Obtain the general formula for the equilibrium point in the first quadrant as a function of the light level L. (b) For $L = 100$ and for $L = 10\,000$ determine the stability, being suré to indicate the Jacobian. (c) Simulate these equations for 800 ms using **RungeKutta4.m** and plot $B(t)$ and $A(t)$ for each of the two L values above. As initial conditions let $A = 0$ and $B = 0$.

$$\frac{dB}{dt} = \frac{1}{\tau_B}\left(-B + \frac{L}{1+4A}\right)$$

$$\frac{dA}{dt} = \frac{1}{\tau_A}(-A + B)$$

2. The divisive gain control network below is more complex, because A depends on B^2. Analyze this network and its response when $L = 10\,000$, $\tau_A = 15$ ms and $\tau_B = 30$ ms. (You will have to use the MatLab **roots** function to solve for the steady states.) After analyzing the stability of any physiologically relevant steady states, simulate 500 ms of the response for $A(0) = 0$, $B(0) = 0$. Graph your results.

$$\frac{dB}{dt} = \frac{1}{\tau_B}\left(-B + \frac{L}{1+5A}\right)$$

$$\frac{dA}{dt} = \frac{1}{\tau_A}(-A + B^2)$$

3. Consider the following two-neuron network for short-term memory:

$$\frac{dE_1}{dt} = \frac{1}{\tau}\left(-E_1 + \frac{100E_2^2}{30^2 + E_2^2}\right)$$

$$\frac{dE_2}{dt} = \frac{1}{\tau}\left(-E_2 + \frac{100E_1^2}{30^2 + E_1^2}\right)$$

Assuming $\tau = 25$ ms, solve for all equilibrium points (simplify your task using symmetry), and determine the stability of each. Simulate the response of the network for each of the following initial conditions: (a) $E_1 = 4, E_2 = 7$; (b) $E_1 = 8, E_2 = 20$.

4. Suppose that the short-term memory network in eqn (6.8) involves excitatory connections between two neurons in different brain structures such as the frontal cortex and a subcortical nucleus. In this case, one would expect some delay in the transmission of excitation between the neurons. Modify the equations to include delay equations with time constants δ in the two excitatory paths using the approach developed in Chapter 4. Is there a critical delay δ_c at which the equilibrium point at $(80, 80)$ becomes unstable? Simulate the system with $\delta = 10$ ms and plot the time course of the responses. Assume that the stimulus has intensity 200 and lasts for 250 ms, and run your simulation for a total of 2000 ms. Compare your results with the data in Fig. 6.3.

5. The following network is a three-neuron generalization of the WTA network in (6.18):

$$\frac{dE_1}{dt} = \frac{1}{10}(-E_1 + S(K_1 - 5E_2 - 5E_3))$$

$$\frac{dE_2}{dt} = \frac{1}{10}(-E_2 + S(K_2 - 5E_1 - 5E_3))$$

$$\frac{dE_3}{dt} = \frac{1}{10}(-E_3 + S(K_3 - 5E_1 - 5E_2))$$

$$S(x) = \begin{cases} \dfrac{100(x)^2}{40^2 + (x)^2} & x \geq 0 \\ 0 & x < 0 \end{cases}$$

Assuming that $K_1 = K_2 = K_3 = 80$, find and analyze the stability of all steady states. Simulate the network for the values of K above and the following initial conditions: (a) $E_1 = 2, E_2 = 1, E_3 = 0$; and (b) $E_1 = 1, E_2 = 1, E_3 = 2$. What is the final state of the system in each case?

Computation by excitatory and inhibitory networks

With the techniques developed in the previous chapter it is now possible to analyze many rather complex networks with multiple steady states. Accordingly, let us examine larger neural networks that use differing balances of excitation and inhibition to perform a variety of sophisticated computations of functional significance. Several of these embellish the basic 'winner-take-all' (WTA) network, which was developed in the last chapter. First, we shall examine a WTA network that chooses the largest stimulus among a wide number of alternatives. Such networks make neural decisions based on the strength of the sensory evidence. Next, a WTA network with appropriate combinations of inputs will be shown to perform vector summation for any number of input vectors. Vector summation is important in motor control, somatosensory perception, and motion perception. An embellishment of this vector summation network will also predict whether motion will be seen as transparent or rigid, an example of perceptual categorization. Finally, we shall analyze the Wilson–Cowan (1973) equations for excitatory and inhibitory interactions in the neocortex. These equations give rise to a variety of dynamical phenomena for different ranges of parameter values and form the basis for an explanation of visual hallucinations (Ermentrout and Cowan, 1979).

7.1 Winner-take-all networks

Neural networks in which each neuron inhibits all other neurons except itself are known as **winner-take-all networks.** As noted in the previous chapter, mutual inhibition is a form of competition, and the neuron with the largest external input generally wins this competition by suppressing the other neurons below their thresholds. If the external input to each neuron is regarded as data supporting the response which that neuron would signal, then winner-take-all (WTA) networks make decisions based upon the preponderance of supporting evidence.

Let us consider an important experimental paradigm known as visual search. Visual search is a task that we are all confronted with frequently: pick out the face of a friend in a crowd, or find the utensil you need in a cluttered drawer. In the laboratory visual search is studied by requiring subjects to find a unique item in a collection of distracting items. Experiments typically measure the reaction time (time required to locate the unique item) as a function of the number of distractors. Two typical displays, each containing a unique target item and eight distractors are depicted in Fig. 7.1. When you look at the figure, the target item is so obvious in the top panel that it seems to pop-out of the background of

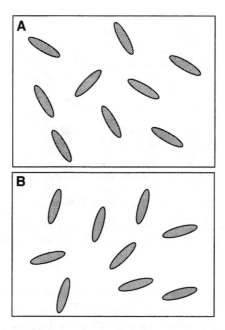

Fig. 7.1 Typical stimuli for a visual search experiment. The unique element in A 'pops-out' of the field of distractors, and reaction times are accordingly fast. The unique element in B is much harder to detect, and reaction times are much longer.

distractors. In the bottom panel, however, the task is considerably more difficult, and this is reflected in the longer reaction time required to locate the unique item. Treisman (1982) discovered that when the distractors are very different from the target item, the reaction time to detect the target is short and almost independent of the number of distractors, as is illustrated by the curve with open circles in Fig. 7.2. When the distractors are very similar to the target, however, reaction times are much slower and increase dramatically with the number of distractors. This is shown by the solid symbols in Fig. 7.2. As reaction times in this case increase by a similar number of milliseconds per distractor, many have inferred that this data pattern reflects serial processing in the brain. When a target pops out of the distractors so that reaction time is independent of the number of distractors, it has been argued conversely that the brain is employing a parallel processing strategy (Treisman, 1982; Treisman and Gelade, 1980; Bergen and Julesz, 1983).

A winner-take-all network be used to provide a novel interpretation of these visual search data. Using the Naka–Rushton function (2.11) to describe spike rates, the network equations are:

$$\tau \frac{dT}{dt} = -T + \frac{100(E_T - kND)_+^2}{\sigma^2 + (E_T - kND)_+^2}$$

$$\tau \frac{dD}{dt} = -D + \frac{100(E_D - k(N-1)D - kT)_+^2}{\sigma^2 + (E_D - k(N-1)D - kT)_+^2}$$

(7.1)

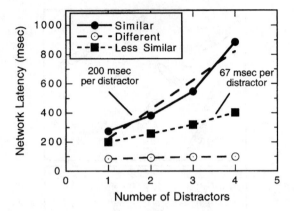

Fig. 7.2 Latencies of (7.1) as a function of the number of activated distractor neurons (D). Latency is the time for the target (T) neuron to reach 95% of its maximum rate while suppressing D responses.

where T is the response of whichever neuron receives information about the target and D is the response of each of the N distractor neurons. The constant k determines the strength of the inhibitory feedback, while the $()_+$ indicates that the parenthetical expression evaluates to zero for negative arguments. Two points about (7.1) are worthy of note. First, T and D are distinguished only by the level of their inputs, with $E_T > E_D$. Second, we have reduced an $(N+1)$-dimensional system (target plus N distractors) to a two-dimensional network by making use of symmetry concepts. On the assumption that initial conditions and inputs E_D to all distractors are identical, (7.1) accurately describes the dynamics of the entire $(N+1)$-dimensional system. This use of symmetry to reduce the dimensionality of a system will be termed **symmetry subsampling** and will be encountered again in Chapters 8 and 13. Note also that if some of the distractor inputs are zero, we may simply reduce N accordingly, as these neurons will have absolutely no influence on the resultant system dynamics.

A WTA network with five neurons, time constant $\tau = 20$ ms, and an inhibitory constant $k = 3$, is implemented in the MatLab program **WTA5Neurons.m**. The script has four distractors with neural inputs $E_D = 79.8$ and one target with $E_T = 80$. Running the script will reveal that the target neuron wins the competition with distractors and suppresses the distractor activity below threshold after a latency of about 880 ms. If two of the distractor inputs in the script are now reduced to zero so that there are only two remaining distractors in the WTA competition, the target neuron again wins but with a much shorter latency of 380 ms. Varying which neuron receives the stronger or target input will show that the winning neuron is always the one responding to the target.

As shown in Fig. 7.2, this WTA neural network makes decisions by choosing the maximum input and suppressing the others, but the dynamics cause the response latency (time to 95% of peak response) to increase with the number of distractors. Furthermore, the latency decreases as the distractor input E_D becomes less similar to the target input E_D (run the network with $E_D = 78$ for example). These results, plotted in Fig. 7.2, show that a WTA network will produce response latencies similar to those of humans (humans require some time for the motor response) in visual search tasks such as depicted in Fig. 7.1. As the network incorporates $N + 1$ neurons competing in parallel, the visual search data may be explained by a purely parallel process with inhibitory interactions, rather than by a postulated shift from parallel to serial processing as the task becomes more difficult. In support of this, Verghese and Nakayama (1994) have obtained data that argue against separate parallel and serial processes in visual search.

What is the dynamical basis for variable latencies in the WTA network (7.1)? The analysis of two inhibitory neurons in the previous chapter suggests that (7.1) will possess a saddle point separating various asymptotically stable steady states, with only one winning neuron active in each. When the input to the target neuron, E_T is only infinitesimally greater than that to distractor neurons, we can solve approximately for the equilibrium state by setting $T = D$ in (7.1). Taking $E_T = 80$, $\sigma = 120$, and $k = 3$, the first equation in (7.1) reduces to the following expression for a steady state in which all responses are equal:

$$T - \frac{100(80 - 3NT)_+^2}{120^2 + (80 - 3NT)_+^2} = 0 \tag{7.2}$$

This can be solved for $N = 1, 2, 4$, etc. using the **fzero** function in MatLab with the results: $T = 11.9$ for $N = 1$; $T = 7.6$ for $N = 2$; and $T = 4.5$ for $N = 8$. Indeed, reflection on the form of (7.2) indicates that the value of T at equilibrium must decrease as N increases in order for the input to just balance the recurrent inhibition.

Linearization of (7.1) around any of the steady states where $T \approx D$ gives:

$$\frac{d}{dt}\begin{pmatrix} T \\ D \end{pmatrix} = \frac{1}{\tau}\begin{pmatrix} -1 & -S'Nk \\ -S'Nk & -1 - (N-1)S'k \end{pmatrix}\begin{pmatrix} T \\ D \end{pmatrix} \tag{7.3}$$

where S' is the derivative of the Naka–Rushton function evaluated at the steady state. The eigenvalues of (7.3) are $\lambda = S'k - 1$ and $\lambda = -NS'k - 1$. As N, S', and k are all > 0, the equilibrium will be a saddle point so long as $S'k > 1$. The equilibrium state of (7.1) in which the T neuron fires at rate T_{eq} and all D neurons are suppressed will be asymptotically stable so long as $kT_{eq} > E_D$, a condition guaranteeing that all distractor responses will drop to zero. Thus, the WTA network will implement neural decisions as long as the inhibitory strength k is sufficiently large and the inputs are strong enough so that S' is large enough to produce a saddle point.

Given conditions producing a saddle point in the state space, it is now possible to see why WTA network latency increases with the number of distractors N as shown in Fig. 7.2 and in visual search data. Recall that the eigenvalues of (7.3) are $\lambda = S'k - 1$ and $\lambda = -NS'k - 1$ at the saddle point. In addition, solution of (7.2) showed that the values of T and D at the saddle point decrease as N increases. This in turn causes S' to decrease once values of T and D at the saddle point fall below the point of maximum slope of the Naka–Rushton function (0.58σ in this case). Accordingly, the positive exponent in the vicinity of the saddle point will *decrease* as N increases, and this will cause trajectories to diverge from the saddle point more slowly. Thus, increasing latencies in WTA networks result from a saddle point that occurs closer to the origin and therefore has a reduced positive exponent as the number of distractors N increases.

One final point about WTA networks: there are generally N additional equilibria that are asymptotically stable: ones in which one D neuron is active and all other activity is suppressed. (This assumes distractor input is not much weaker than target input.) If the WTA network is initially in the resting state where $T = D = 0$, however, the only state that can be reached under conditions producing a saddle point is the state in which the most strongly stimulated neuron is active and all others are inhibited. Once a winner has suppressed other activity, however, the network exhibits hysteresis, and emergence of a new winner requires a greater than normal input.

These properties of decisions implemented by WTA networks reflect properties of higher level cognitive decisions. It is common experience that decisions are more difficult and take longer when the number of appealing alternatives increases. Once a decision is definitively made, however, humans invariably exhibit more or less resistance to a change of mind and generally require more evidence to do so: we all exhibit hysteresis in our cognitive processes!

7.2   Vector summation

Although we rarely engage in vector summation consciously unless forced to do so in a math course, vertebrate brains are confronted with vector summation problems all the time. For example, it is crucial for your somatosensory system to have information about the location of your hands and feet at all times, yet the information sent to the brain from receptors in the muscles and joints effectively indicates only the angles of the joints (Rothwell, 1994). The only way in which the position of your hand or foot can be determined is for the brain to combine joint angle and bone length information by vector summation. If your goal is to reach out and grasp a wine glass in front of you, then your brain must also include visual direction and distance in the vector computation.

The phenomenon known as **path integration** represents a second striking example of vector summation in the mammalian brain. Figure 7.3 illustrates the basic phenomenon using information extracted from photographs in a review article by Whishaw *et al.* (1997). A rat emerges from its cage, which has been placed randomly below one of eight holes in a circular platform, and begins searching for a food pellet. The search takes a circuitous path until the food is located, at which point the rat returns directly to its cage before eating. As rats will return directly to their cage even when blindfolded, the return direction to the cage cannot be visually cued. Rather, the rat apparently encodes the direction and distance of its motion along each segment of its search path and then performs a vector computation to determine the location of its cage relative to its position at the end of the search. Evidence from a variety of sources suggests that information about the length and direction of each path component is stored in the hippocampus (McNaughton *et al.*, 1996; Whishaw *et al.*, 1997).

Motor control and motion perception are other areas where vector summation is important. In motor control, Georgopoulos and colleagues (1986, 1993) have shown that

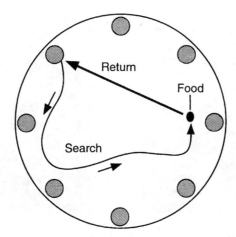

Fig. 7.3 Example of path integration by a rat searching for a food pellet. Despite a circuitous search path from the nest, hidden below one of eight holes in a platform, the rat returns directly to the nest with the located pellet. Path integration occurs even when the rat is blindfolded.

the direction of arm movements by monkeys is encoded by the vector sum of neural responses in motor cortex. Furthermore, individual motor neurons each fire maximally for a particular direction of arm movement Ω, and their responses fall off as $\cos(\Omega - \theta)$, where θ is the direction of actual movement.

In motion perception, neurons in primary visual cortex only respond to the direction of motion perpendicular to local lines and edges, and this information must be combined to determine the direction of object motion. Furthermore, there are circumstances in which local motion information cannot be integrated into a rigid motion percept, and transparent or sliding motion is seen (Kim and Wilson, 1993). The MatLab script **Motion_Demo.m** presents several motion stimuli so that these effects may be experienced. As shown in Fig. 7.4, the program produces a set of 12 circular apertures through which moving light and dark bars (cosine gratings) may be seen. If the program is run with directions of $\pm 22°$, half of the apertures display bars moving at an angle of $+22°$ and half at an angle of $-22°$ relative to vertically upward. In response to this stimulus in the laboratory, subjects perceive a rigid sheet of bars moving vertically upwards, which is the vector sum direction. Re-running the program with directions of $\pm 68°$ yields a very different percept: two sets of bars are seen sliding across one another. This transition from rigid to transparent motion reliably occurs for all subjects when the motion directions are approximately $\pm 45°$ (Kim and Wilson, 1993).

Let us examine a modified WTA network that can both compute the vector sum of input motion directions and simultaneously decide whether the motion is rigid or transparent. This network represents a simplified version of a detailed neural model for motion perception based on the physiology of a higher cortical motion area (Wilson *et al.*, 1992; Wilson and Kim, 1994). The model, with appropriate modifications, may be applied to motor control, path integration, and somatosensory computations. As shown in Fig. 7.5, 24 neurons tuned to directions of motion varying in 15° increments comprise the

Spikes, decisions, and actions

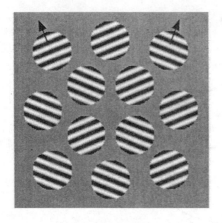 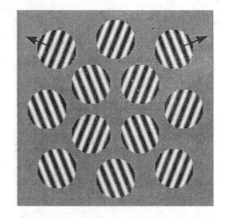

Rigid Motion   Transparent Motion

Fig. 7.4 Examples of the two motion displays produced by **Motion_Demo.m**. In the left-hand pattern, the motions are integrated into a percept of rigid upward motion, while in the right pattern motion transparency is seen.

network. Each unit in the network has a different preferred direction Ω and receives an excitatory input E_Ω which is a cosine-weighted sum of N input vectors:

$$E_\Omega = \sum_{\theta=1}^{N} L_\theta \cos(\Omega - \theta) \tag{7.4}$$

where L_θ is the length of the vector and θ is its direction. In neural terms, L_θ represents the firing rate of an input neuron optimally sensitive to motion in direction θ, and the cosine represents the synaptic weighting function. As mentioned above, cosine weighting is common in primate motor neurons (Georgopoulos *et al.*, 1986, 1993), and it is even found in the cricket (Theunissen and Miller, 1991) and leech (Lewis and Kristan, 1998) nervous systems. Thus, one may conjecture that cosine weighting is an early evolutionary adaptation to the ubiquitous requirement for animals to perform vector computations on their sensory inputs.

In addition to receiving a cosine-weighted input given by (7.4), the network neurons engage in a winner-take-all competition via recurrent inhibition. This, however, is a modified WTA scenario in two respects. As shown in Fig. 7.5, each neuron vigorously inhibits the range of neurons enclosed by gray regions, but it neither inhibits its two nearest neighbors, $N_{\pm 1}$ and $N_{\pm 2}$, nor neurons signaling roughly diametrically opposite directions of motion. (All inhibitory interactions are symmetrical; for clarity only the inhibitory connections of one unit are depicted.) This network is therefore a WTA network with a restricted range of competition. As you might conjecture from the connectivity pattern, there are two consequences of this restricted competition. First, near neighbors of the winner will survive the competition. Second, there can be two winners if they receive almost diametrically opposed inputs. We shall see in a moment that this is

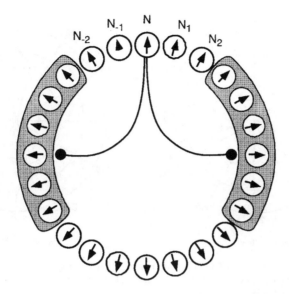

Fig. 7.5 Vector summation network for motion perception. Each of the 24 neurons signals a different direction of motion in 15° increments. Each unit also strongly inhibits all other units tuned to the relative directions depicted in the gray regions, excluding its two nearest neighbors $N_{\pm 1}$ and $N_{\pm 2}$ along with units close to the opposite direction of motion. Each unit has the same relative set of inhibitory connections centered on its preferred direction, but only one set is shown for clarity.

correct. The equations describing the network are:

$$\tau \frac{dR_\Omega}{dt} = -R_\Omega + \left(\sum_{\theta=\Omega-90°}^{\Omega+90°} L_\theta \cos(\Omega - \theta) - k \sum_{\alpha=\Omega\pm45°}^{\Omega\pm120°} R_\alpha \right)_+ \qquad (7.5)$$

where R_Ω is the response of the neuron with preferred direction Ω in 15° increments. The brackets with the $+$ subscript indicate that this expression is identically zero when the argument within the brackets is less than zero. This is an extremely simple nonlinearity with a threshold at zero and a linear response above threshold. However, this is sufficient for the network to perform vector computations. The strength of the inhibitory feedback k is not critical as long as it is sufficiently strong to suppress other neural activity.

The network in eqn (7.5) is implemented in MatLab script **VectorWTA.m**, where $k = 3$. If you run the simulation with vector lengths of 20 and angles of $\pm 40°$, you will see that the network indicates a perceived direction of 0°, which is the experimental observation depicted in Fig. 7.4. The response of the network in this case is a set of three neighboring active neurons signaling directions −15°, 0°, and 15° as shown in Fig. 7.6 (left), while all other neurons have been suppressed by the WTA inhibition. This triad of neighboring winners in the network results from the absence of recurrent inhibition onto the two neighbors $N_{\pm 1}$ and $N_{\pm 2}$ on either side of each unit. Activity in N_1 will inhibit the N_{-2} unit,

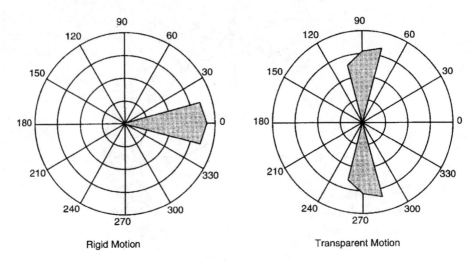

Fig. 7.6   Responses of (7.5) to $\pm 40°$ vectors (left) and $\pm 75°$ vectors (right) plotted in polar coordinates. The set of three responding neurons on the left signals rigid rightward motion, while the two sets of three active units on the right signal transparent motion.

and N_{-1} will inhibit N_2; so only three units will survive the inhibitory competition. By thus preserving the activity of the winning unit's nearest neighbors in the steady state, the network provides a sparse population code for vector direction that is much more accurate than the 15° interval between preferred directions might lead one to suspect. A **sparse population code** is a neural code in which activity of a small number of neurons represents a response much more accurately than any single neuron would be capable of by itself. In essence, a population code permits interpolation of intermediate responses between the optimal values for individual units. To see that the network in (7.5) will accurately encode intermediate values of direction, rerun the program with a variety of vector lengths and angles within a $\pm 45°$ range. For example, input vectors of length 20 at $-40°$ and length 8 at $+30°$ yield a predicted direction of $-22.4°$ as compared to a true vector sum direction of $-21.7°$. This error of 0.7° is as large as the network will produce, so the use of a three-unit sparse population code results in an improvement in accuracy by more than a factor of 10 ($\pm 7.5°$ versus $\pm 0.7°$). (Conversion of the three neural responses to a single number is accomplished by parabolic interpolation, see Wilson *et al.* (1992) for details.)

Exclusion of N_1 and N_2 from recurrent inhibition results in responses composed of a three-neuron sparse population code in the equilibrium state. As shown in Fig. 7.5, neurons with preferred directions differing by $\pm 135°$ or more also do not compete. Consequences of this absence of inhibition between nearly opposite directions will be apparent if you rerun **VectorWTA.m** with two vectors of length 20 and directions of $\pm 75°$. The result, depicted on the right of Fig. 7.6, is a pair of three-neuron responses in nearly opposite directions. The network in (7.5) thus switches from signaling a single direction to signaling two directions when the input vectors are sufficiently different. In the example of motion perception discussed in conjunction with Fig. 7.4, this corresponds to a switch from rigid to transparent motion as has been experimentally observed (Kim and Wilson, 1993). Such a switch in response is termed a **categorical decision**, because a continuous range of stimulus variation (in this case defined by the angle between local motion directions) is categorized into discrete responses: rigid or transparent. Furthermore, the response to two equal length vectors at $\pm75°$ is predicted to be a pair of motion vectors at $\pm82°$. Such repulsion between motion directions when transparency is perceived has been observed in several experimental studies using a variety of stimuli (Kim and Wilson, 1996; Marshak and Sekuler, 1979; Mather and Moulden, 1980). Direction repulsion is a direct consequence of the recurrent inhibition illustrated in Fig. 7.5. Although the units optimally responsive to directions of $\pm75°$ are not mutually inhibitory, their neighbors signaling $\pm60°$ are, so the population codes are biased away from one another by the neural dynamics.

How can a WTA network produce vector summation? The key is a mathematical proof that the vector sum direction Ω_V is given by the value of Ω for which the cosine-weighted sum in (7.4) is a maximum (Wilson *et al.*, 1992):

$$\Omega_V = \max_{\Omega}(E_{\Omega}) \tag{7.6}$$

Although true for any number of vectors, let us prove here that the sum of two vectors points in the direction is given by (7.6). Let the two vectors have lengths L_1 and L_2 and point in directions θ_1 and θ_2. Thus, the x and y coordinates of the first vector will be $L_1 \cos(\theta_1)$ and $L_1 \sin(\theta_1)$ respectively with similar terms for the second vector. Vector summation is accomplished by adding the x and y coordinates according to the formula:

$$V = \begin{pmatrix} L_1 \cos(\theta_1) \\ L_1 \sin(\theta_1) \end{pmatrix} + \begin{pmatrix} L_2 \cos(\theta_2) \\ L_2 \sin(\theta_2) \end{pmatrix} = \begin{pmatrix} L_1 \cos(\theta_1) + L_2 \cos(\theta_2) \\ L_1 \sin(\theta_1) + L_2 \sin(\theta_2) \end{pmatrix} \tag{7.7}$$

The direction in which V points, Ω_V, the vector sum direction, is just $\arctan(y/x)$ or:

$$\Omega_V = \arctan\left(\frac{L_1 \sin(\theta_1) + L_2 \sin(\theta_2)}{L_1 \cos(\theta_1) + L_2 \cos(\theta_2)}\right) \tag{7.8}$$

Now consider the cosine-weighted sum in (7.4):

$$E_{\Omega} = L_1 \cos(\Omega - \theta_1) + L_2 \cos(\Omega - \theta_2) \tag{7.9}$$

To find the maximum, set the derivative with respect to Ω equal to zero:

$$\frac{dE_{\Omega}}{d\Omega} = -L_1 \sin(\Omega - \theta_1) - L_2 \sin(\Omega - \theta_2) = 0 \tag{7.10}$$

Using the trigonometric substitution:

$$\sin(\Omega - \theta_1) = \cos(\theta_1)\sin(\Omega) - \cos(\Omega)\sin(\theta_1) \tag{7.11}$$

and a similar one for the second term in (7.10), we arrive at:

$$\{L_1 \cos(\theta_1) + L_2 \cos(\theta_2)\} \sin(\Omega) - \{L_1 \sin(\theta_1) + L_2 \sin(\theta_2)\} \cos(\Omega) = 0 \qquad (7.12)$$

This is easily solved for Ω to give the result in (7.8). It is also easy to show that the maximum value of E_Ω is the length of the resultant vector. This completes a proof of (7.6).

The mechanism permitting WTA networks to perform vector summation should now be apparent. In a WTA network, the neuron that wins the competition is the one receiving the maximum stimulus. So, a WTA network with cosine-weighted inputs given by (7.4) will compute the maximum and solve (7.6) for the vector sum, and it will do so using parallel processing for any number of input vectors. Sparse population coding makes it unnecessary for the winning neuron to signal the exact vector sum direction. The program **VectorWTA.m** can be used for any number of input vectors simply by activating (i.e. removing the %) the line asking the number of input vectors.

Vector summation provides an important example of a WTA network that can accomplish complex and important calculations. All that is required is cosine weighting of the network inputs according to eqn (7.4) plus competitive inhibition among the network outputs. Cosine weighting appears to be ubiquitous in the nervous system (Georgopoulos *et al.*, 1986, 1993; Theunissen and Miller, 1991; Lewis and Kristan, 1998). With respect to motion, we predicted that a WTA competition should occur in MT (middle temporal) cortex, a primate and human motion analysis area (Wilson *et al.*, 1992). Definitive evidence for a WTA computation in primate MT was subsequently provided by Salzman and Newsome (1994). One caveat: the vector summation explanation of motion perception requires computation of additional nonlinear or 'second order' motion signals from the stimulus, an operation for which there is extensive psychophysical and physiological evidence (Wilson *et al.*, 1992; Wilson, 1994a, 1994b; Smith, 1994; Albright, 1992; Mareschal and Baker, 1998).

In addition to our work (Wilson *et al.*, 1992; Wilson and Kim, 1994) on motion perception, both Touretzky *et al.* (1993) and Abbott (1994) subsequently explored neural network models for vector summation using cosine-weighting functions.

7.3   Retinal light adaptation

The retina represents one of the most heavily studied neural networks of the brain. (The retina develops as an outgrowth of the embryonic neural tube, so it is indeed part of the brain: the only part visible to the naked eye!) Although many details still remain to be learned about the retina, there is already a sufficient wealth of material to permit us to develop a fairly sophisticated understanding of its function. We know, for example, the major neural cell types and the anatomy of their interconnections. In addition, we know how the retinal ganglion cells, the cells that provide output to the brain, respond to various light patterns projected onto the retina. The model to be presented here incorporates key elements of a model proposed recently to explain retinal function in both light adaptation and afterimage formation (Wilson, 1997). Excellent summaries of the anatomical and physiological literature may be found in several reviews (Dowling, 1987;

Light

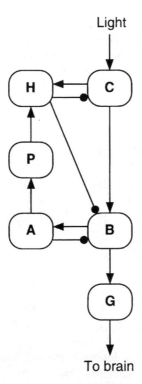

To brain

Fig. 7.7 Schematic diagram of anatomical connections among principal retinal cell types: cones (C), horizontal cells (H), bipolar cells (B), amacrine cells (A), ganglion cells (G), and interplexiform cells (P). Excitatory synapses are shown as arrowheads and inhibitory synapses as solid circles.

Wässle and Boycott, 1991) and a comprehensive recent book by Rodiek (1998). A summary of fundamental data on light adaptation may be found in Hood (1998).

Key aspects of retinal anatomy are summarized in Fig. 7.7. Light impinges on the cone photoreceptors (C) in the retina, and these respond by stimulating both the bipolar cells (B) and the horizontal cells (H). As we saw in Chapter 3, there is evidence that the horizontal cells provide subtractive inhibitory feedback onto the cones and also feedforward inhibition onto the bipolar cells. The bipolar cells activate both the amacrine cells (A) and the ganglion cells (G), which provide the retinal output by sending their axons to various brain centers. Amacrine cells provide negative feedback onto the bipolar cells, and this is hypothesized to be shunting or divisive in nature (Wilson, 1997). In addition, the amacrine cells contact interplexiform cells (P), which provide feedback onto the horizontal cells. Extensive evidence indicates that this amacrine to interplexiform to horizontal cell feedback circuit is neuromodulatory in nature and uses the neurotransmitter dopamine (Dowling, 1991; Witkovsky and Schütte, 1991). The interplexiform feedback circuit operates very slowly by neural standards, with a time constant on the order of several seconds rather than tens of milliseconds.

The simplified retinal model in Fig. 7.7 is comprised of six cell types at each spatial location. In actuality, there are multiple types of bipolar, amacrine, and ganglion cells, but restriction to one of each will suffice to explore the dynamical principles involved in retinal light adaptation. Our model ignores the fact that there are on the order to 10^6 neurons of each type in the retina, but this is equivalent to restricting visual stimuli to large uniform fields of light that adapt all parts of the retina equally (see Wilson (1997) for a model including spatial distributions of each cell type). Let us analyze the six cell circuit in Fig. 7.7 to examine light adaptation. The circuit equations are:

$$\frac{dC}{dt} = \frac{1}{10}(-C - PH + L)$$

$$\frac{dH}{dt} = \frac{1}{100}(-H + C)$$

$$\frac{dB}{dt} = \frac{1}{10}\left(-B + \frac{6C - 5H}{1 + 9A}\right)$$

$$\frac{dA}{dt} = \frac{1}{80}(-A + B)$$

$$\frac{dP}{dt} = \frac{1}{4000}(-P + 0.1A)$$

$$\frac{dG}{dt} = \frac{1}{10}\left(-G + \frac{50B}{13 + B}\right)$$

(7.13)

Several aspects of these equations deserve comment. First, the spike rate of the ganglion cell G is described by a Naka–Rushton function from (2.11) with $N = 1$. Indeed, this function was first applied in neurobiology to describe the spike rate of retinal ganglion cells, and $N = 1$ was found to give a good fit (Naka and Rushton, 1966). The dC/dt equation incorporates subtractive feedback from H cells. Note, however, that the strength of this is modulated by the P cell feedback loop. Thus, the H cell feedback will be strong at high light levels and weak in dim light. Feedback from A cells is described by a divisive term in the dB/dt equation. Finally, note that the time constants are 100 ms or faster, with the exception of $\tau_P = 4000$ ms. This means that the dP/dt equation will change extremely slowly relative to the other retinal dynamics, which agrees with the physiology cited above.

To analyze (7.13) we shall, as always, first examine the equilibrium state. At equilibrium:

$$C = -PH + L$$

$$H = C$$

$$B = \frac{6C - 5H}{1 + 9A}$$

$$A = B$$

$$P = 0.1A$$

$$G = \frac{50B}{13 + B}$$

(7.14)

As the retinal output G depends only on B, the remaining equations may be solved for B. The three simple identities at equilibrium $H = C$, $A = B$, and $P = 0.1A$, may be used to simplify the equations for C and B:

$$C = \frac{10L}{10 + B}$$

$$B = \frac{C}{1 + 9B}$$

(7.15)

Substituting the first equation in (7.15) into the second gives:

$$B(10 + B)(1 + 9B) = 9B^3 + 91B^2 + 10B = 10L$$

(7.16)

This cubic equation can be solved using the MatLab **roots** function for any given light level L. As a useful analytical approximation, however, let us consider two limiting cases, L very small and L very large. In the former case the B term will predominate, so $B \approx 10L$, and the network produces no stimulus attenuation. When $L \gg 1$, the B^3 term will be largest, so $B \approx L^{1/3}$. Thus, the network will produce dramatic stimulus compression in response to very intense light.

To appreciate the importance of light adaptation, let us first consider what would happen if the ganglion cell received a direct input from the cones that bypassed the adaptation circuitry. A fundamental task for the visual system is to discriminate between two stimuli to determine which has the higher light intensity. In the absence of light adaptation, two stimuli L and $L + \Delta$ will produce the following ganglion cell response difference δG in the steady state:

$$\delta G = \frac{50(L + \Delta)}{13 + (L + \Delta)} - \frac{50L}{13 + L}.$$

(7.17)

Suppose that at threshold human subjects can just discriminate between two light intensities when $\delta G = 1$, i.e. when the ganglion cell spike rate differs by a small criterion. On the assumption that Δ is small at threshold, (7.17) can be expanded as a first order Taylor series in Δ with the result:

$$\Delta = \frac{(13 + L)^2}{50 \times 13}$$

(7.18)

This equation, first derived by Shapley and Enroth-Cugell (1984), shows that the smallest discriminable difference in light intensity between two stimuli will increase with L^2 as L becomes large. Equation (7.18) is plotted as a function of light intensity L (in units of trolands) by the dashed line in Fig. 7.8. The solid circles plot data from a study by Finkelstein, Harrison, and Hood (1990). These authors observed that retinal light adaptation takes some time to develop, so briefly flashed stimuli can largely bypass the light adaptation process, and the solid data points were gathered under these conditions. In consequence, discrimination thresholds get very large (i.e. very poor), and discrimination

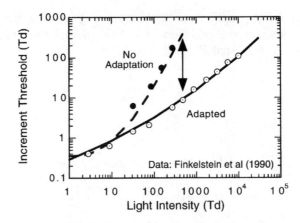

Fig. 7.8 Thresholds for detecting a brief light flash as a function of background light intensity. Solid circles plot thresholds under conditions minimizing retinal light adaptation, while open circles show analogous results for the fully adapted retina (Finkelstein *et al.*, 1990). Solid and dashed lines are the predictions of (7.13) under adapted and unadapted conditions, respectively.

becomes impossible for light intensities to the right of the double-headed arrow. The reason for this is saturation of the ganglion cell firing rate G.

One role of retinal light adaptation is to compress the neural input to the ganglion cells so as to avoid response saturation. The MatLab program **Retina.m** implements a fourth order Runge–Kutta simulation of (7.13). The program first requests a background light adaptation level L to which the model will be adapted by solving (7.16) for the steady state and then using this value to solve for the equilibrium values of all other variables in (7.14). The program then requests a change in luminance Δ and this is abruptly added to the background 100 ms into the simulation. If you run the simulation with $L = 100$ and $\Delta = 5$, you will see that there is a transient burst in ganglion cell (G) response to the increment that reaches a value of 1.53 above the adapted level and then drops back to a new level of adaptation. Given our previous assumption that the threshold value of Δ occurs when the G cell increment is 1.0 above its former baseline, this represents a suprathreshold response. By simulating responses to a number of smaller Δ values it is possible to home in on the threshold quite accurately (for $L = 100$ the threshold is between 3.0 and 3.5). Threshold values thus obtained are plotted as a solid curve in Fig. 7.8. The comparison data, again from Finkelstein *et al.* (1990), were obtained by first letting the retina adapt to a uniform background of intensity L and then flashing the incremental test stimulus, just as in the simulation of (7.13). As shown by the double-headed arrow at $L = 400$, light adaptation improves thresholds here by about 40 times. Furthermore, light adaptation increases the intensity range over which discrimination is possible from about $L = 400$ (no adaptation) to about $L = 40\,000$ (adapted state). Thus, retinal light adaptation greatly extends the range over which effective visual discrimination is possible.

One consequence of retinal light adaptation is the production of retinal afterimages. Afterimages occur when adjacent patches of retina are adapted to different light intensities. To experience a retinal afterimage, fixate the cross on the white and black bar

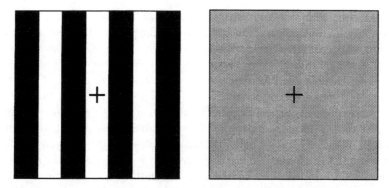

Fig. 7.9   Fixation of the cross on the left for about 15 s followed by a shift of gaze to the right cross will produce a negative retinal afterimage of the bars against the gray background.

pattern in Fig. 7.9 for about 15 s, then switch your gaze to the cross in the uniform gray field. You will perceive a set of low contrast bars with a darker bar located behind the cross (the bars will shift back and forth a bit due to small eye movements on your part). As a dark bar underlies the cross, this is a negative afterimage: light adaptation in your retina has caused it to signal that uniform gray areas are darker following adaptation to white but lighter following adaptation to black bars.

The MatLab script **Retina.m** can also be used to simulate this phenomenon. Suppose that the white bars in Fig. 7.9 have an intensity of 2000, the black bars an intensity of 1, and the gray field an average intensity of 1000. Running **Retina.m** with $L = 1$ and $\Delta = 999$ reveals that patches of retina adapted to dark bars will produce a steady state ganglion cell firing level $G = 22.2$ when subsequently stimulated by the gray pattern. To simulate the afterimage to white bars, run the program with $L = 2000$ and $\Delta = -1000$. The response to gray is now $L = 18.0$, so the uniform gray pattern will appear darker where the retina had adapted to light bars. Thus, the formation of negative afterimages can be explained by the dynamics of retinal light adaptation. The slow time course of afterimage decay is a further phenomenon explained by the dynamics of (7.13). Experimental measurements have revealed that negative afterimages decay exponentially with a time constant of 3–5 s (Burbeck, 1986; Kelly and Martinez-Uriegas, 1993). In (7.13) only the time constant for neuromodulation by the interplexiform cell P is sufficiently long to account for this. Thus, the slow build-up and decay of retinal neuromodulation results in the slow decay of negative afterimages. Other aspects of afterimage formation can also be explained by this mechanism (Wilson, 1997).

7.4   Wilson–Cowan cortical dynamics

Networks in the cortex can produce much more varied and complex dynamics than those exhibited by the retina. Indeed, cortical circuitry is sufficiently complex that we are nowhere close to understanding it in any detail. In consequence, it has been fruitful to analyze simplified models of cortical circuitry and dynamics. It has been known for at least half a century that cortical neurons may be subdivided into two classes: excitatory (E) neurons which are usually pyramidal cells providing cortical output, and inhibitory (I) interneurons with axons that generally remain within a given cortical area. Furthermore, anatomical evidence (Sholl, 1956; Szentagothai, 1967; Colonnier, 1968; Scheibel and Scheibel, 1970; Douglas and Martin, 1998) suggests that all forms of interconnection among these cell types regularly occur: $E \rightarrow E$, $E \rightarrow I$, $I \rightarrow E$, and $I \rightarrow I$. Given evidence for these common aspects of cortical architecture, we proposed that cortical dynamics might profitably be studied through analysis of the simple basic network illustrated in Fig. 7.10 (Wilson and Cowan, 1973). This network incorporates two cell types distributed in space and interconnected in all possible ways. Consistent with anatomy, the recurrent excitation remains relatively localized spatially, while inhibitory connections extend over a broader range. Similarly balanced interactions between E and I neurons have been proposed recently to explain a diverse range of cortical phenomena (Douglas and Martin, 1991; Adini *et al.*, 1997; DeBellis *et al.*, 1998; Somers *et al.*, 1998).

The original Wilson–Cowan (1973) equations were written in the form:

$$\tau \frac{dE(x)}{dt} = -E(x) + (1 - kE(x))S_E \left(\sum_x w_{EE}E(x) - \sum_x w_{IE}I(x) + P(x) \right)$$

$$\tau \frac{dI(x)}{dt} = -I(x) + (1 - kI(x))S_I \left(\sum_x w_{EI}E(x) - \sum_x w_{II}I(x) + Q(x) \right)$$

$$(7.19)$$

where $E(x)$ and $I(x)$ are the mean firing rates of neurons at position x, and P, Q are the external inputs to the network. The four connectivity functions w_{ij} represent the spatial spread of synaptic interconnections in the network. Based on the work of Sholl (1956) these functions were chosen to be decaying exponential functions of distance:

$$w_{ij}(x - x') = b_{ij} \exp(-|x - x'|/\sigma_{ij}) \qquad (7.20)$$

Cortical Position

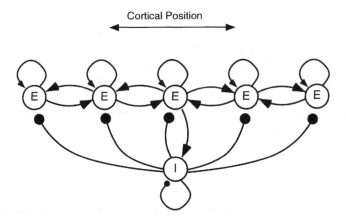

Fig. 7.10 Network of excitatory (arrows) and inhibitory (solid circles) interactions among E and I neurons in cerebral cortex. Excitatory interactions among the E cells have a short spatial extent, while inhibitory feedback from I cells is longer range. For clarity, only one I neuron is shown.

where b_{ij} gives the maximum synaptic strength, σ_{ij} is the space constant controlling the spread of connectivity, and $i, j = E, I$. One might just as easily have chosen a Gaussian function to describe the synaptic connectivity, but any monotonically decaying distance function will give rise to the same types of dynamical behavior. The sigmoidal nonlinear functions S were originally chosen to be hyperbolic tangents, but here we shall use the Naka–Rushton function from (2.11) with $N = 2$ instead:

$$S(P) = \frac{100P^2}{\theta^2 + P^2} \tag{7.21}$$

Finally, (7.19) can be transformed by setting $k = 0$ to obtain equations in which S describes the neural spike rate. The resulting network equations are:

$$\tau \frac{dE(x)}{dt} = -E(x) + S_E\left(\sum_x w_{EE}E(x) - \sum_x w_{IE}I(x) + P(x)\right)$$

$$\tau \frac{dI(x)}{dt} = -I(x) + S_I\left(\sum_x w_{EI}E(x) - \sum_x w_{II}I(x) + Q(x)\right) \tag{7.22}$$

where w_{ij} and S_i are defined in (7.20) and (7.21) respectively. These equations exhibit the same dynamics as (7.19).

In a complex neural network like (7.22) one is immediately confronted with the issue of choosing reasonable values for the various parameters, and this must generally be done in the absence of experimental evidence adequate to specify them all precisely. However, it is generally possible to use appropriate physiological constraints to place bounds on reasonable parameter values. Three such constraints are particularly important in studying spatially extended neural networks. First, the resting state $E = 0$, $I = 0$ should be asymptotically stable so that the network will not respond to small random inputs. Given the form of S_i, this condition is automatically satisfied, since (7.22) linearizes to $\tau \, dE/dt = -E, \tau \, dI/dt = -I$ in the neighborhood of the resting state. Second, the spatial spread of recurrent inhibition is generally greater than that of recurrent excitation. This condition will be satisfied if:

$$\sigma_{EI} = \sigma_{IE} > \sigma_{EE} \tag{7.23}$$

The final condition is that there should be no spatially uniform steady states other than the resting state in the absence of external stimulation. States of uniform excitation throughout a network have little interest, as they do not occur physiologically (except possibly during seizures). In a uniform state $E(x, t) = E(t)$ and $I(x, t) = I(t)$. Let us therefore approximate the spatial sums in (7.22) weighted by the connectivity functions w_{ij} as integrals over an effectively infinite distance (i.e. a distance large with respect to the space constants σ_{ij}):

$$\int_{-\infty}^{\infty} b_{ij} \exp(-|x|/\sigma_{ij}) \, dx = 2b_{ij}\sigma_{ij} \tag{7.24}$$

In other words, if neural responses are uniform across the network, then the spatial connectivity functions become constant products of their parameter values.

As the maximum possible values for E and I are both 100 given the form of S_i, the following two inequalities will be sufficient to guarantee that no spatially uniform steady state can exist in the network:

$$100(2b_{EE}\sigma_{EE} - 2b_{IE}\sigma_{IE}) < \theta_E$$
$$100(2b_{EI}\sigma_{EI} - 2b_{II}\sigma_{II}) > \theta_I \tag{7.25}$$

where the constants θ_i are the semi-saturation constants of the S_i functions in (7.21). The first of these inequalities guarantees that the maximum amount of recurrent excitation in the network can be overcome by the strength of recurrent inhibition and driven below the excitatory semi-saturation value θ_E. The second inequality in (7.25) guarantees that the excitatory input to the inhibitory neurons will drive the inhibitory neurons above their semi-saturation value θ_I. Thus, network inhibition will ultimately drive any uniformly excited state down to the resting state.

Despite choosing parameters that preclude the existence of any uniform activity states, the Wilson–Cowan equations exhibit a wide range of dynamical behaviors. To demonstrate this, let us first fix the spatial and temporal parameters at the following plausible values: $t = 10$ ms; $\sigma_{EE} = 40$ μm; $\sigma_{IE} = \sigma_{EI} = 60$ μm; $\sigma_{II} = 30$ μm. Notice that these values satisfy (7.23). In addition, the semi-saturation constants in (7.21) will be assigned the values $\theta_E = 20$ and $\theta_I = 40$. Finally, it will be convenient to reduce the number of remaining parameters by requiring $b_{IE} = b_{EI}$. The three remaining connection strengths, b_{EE}, b_{EI}, and b_{II} will always be chosen to satisfy inequalities (7.25).

As the simplest example to understand, let us first consider the short-term memory mode of (7.22), which can be produced by the parameter values $b_{EE} = 1.95, b_{EI} = 1.4$, and $b_{II} = 2.2$. The equations with these parameters are implemented in MatLab script **WCcortexSTM.m**. This program permits you to specify the width of a stimulus pulse which is then presented to the network at an intensity $P = 1.0$ for just 10 ms. Using a 100 μm-wide stimulus, for example, produces the final activity state plotted in the top panel of Fig. 7.11. Although the stimulus was only on for 10 ms, the network switches to an asymptotically stable steady state that is approximately the same width as the stimulus. Rerunning the program with a 1000-μm-wide stimulus triggers the network into the final state shown in the bottom panel of Fig. 7.11. Following brief stimulation, therefore, the network switches into a state that retains information about stimulus location and width encoded as a self-sustaining neural activity pattern. This represents a more realistic generalization of the prefrontal short-term neural memory discussed in the previous chapter, and it foreshadows aspects of hippocampal memory networks to be discussed in Chapter 14.

Although parameters have been chosen to satisfy (7.25) so that no spatially uniform steady states can exist, the simulations clearly demonstrate that spatially inhomogeneous states with asymptotic stability do exist. The qualitative reason for this is that short-range recurrent excitation can stabilize a narrow pulse of neural activity while the longer range inhibition prevents the excitatory activity from spreading. The mathematical principles

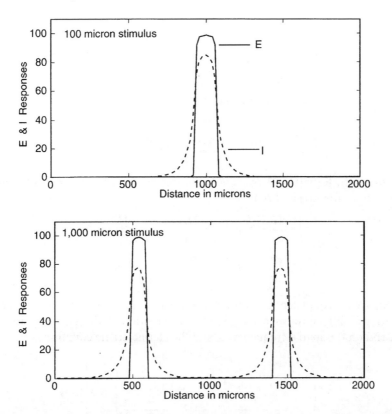

Fig. 7.11 Responses of Wilson–Cowan equations (7.22) to 10 ms stimulation by a 100 µm spatial stimulus (above) or a 1000 µm spatial stimulus (below). These are asymptotically stable short-term memory patterns maintained by recurrent excitation and inhibition within the network after stimulation has terminated.

supporting this behavior may be revealed by considering a spatial system of linear equations with all the forms of interconnection present in Fig. 7.10:

$$
\begin{aligned}
\frac{\mathrm{d}E}{\mathrm{d}t} &= -E + \frac{1}{4}\int_{-\infty}^{\infty} \exp(-|x-x'|/4)E\,\mathrm{d}x' - \frac{1}{4}\int_{-\infty}^{\infty} \exp(-|x-x'|/8)I\,\mathrm{d}x' \\
\frac{\mathrm{d}I}{\mathrm{d}t} &= -I + \frac{1}{8}\int_{-\infty}^{\infty} \exp(-|x-x'|/8)E\,\mathrm{d}x' - \frac{2}{3}\int_{-\infty}^{\infty} \exp(-|x-x'|/3)I\,\mathrm{d}x'
\end{aligned}
\tag{7.26}
$$

This is the typical form produced by linearization of (7.22) around a spatially uniform steady state produced by constant input values of P and Q. The spatial summations in (7.22) have been replaced by integrals here for analytical simplicity. Note, however, that the space constant for the E–E connections (4) is smaller than that for the E–I and I–E connections (8) in agreement with requirement (7.23). Equation (7.26) is an integro-differential equation that is differential in time but integral in space, and its stability characteristics may be examined by letting E and I be cosine functions of spatial position with spatial frequency w:

$$E(x, t) = E(t)\cos(wx) \quad \text{and} \quad I(x, t) = I(t)\cos(wx) \tag{7.27}$$

If these expressions are substituted into (7.26) it is now possible to evaluate all of the spatial integrals because (Gradshteyn and Ryzhik, 1980):

$$\int_{-\infty}^{\infty} \exp(-|x - x'|/\sigma)\cos(wx')\,dx = \frac{2\sigma\cos(wx)}{1 + (\sigma w)^2} \tag{7.28}$$

Given this result, Equation (7.26) is now reduced to the following pair of linear differential equations for $E(t)$ and $I(t)$:

$$\begin{aligned}
\frac{dE}{dt} &= -E + \frac{2E}{1 + (4w)^2} - \frac{4I}{1 + (8w)^2} \\
\frac{dI}{dt} &= -I + \frac{2E}{1 + (8w)^2} - \frac{4I}{1 + (3w)^2}
\end{aligned} \tag{7.29}$$

(This approach is equivalent to Fourier transformation of the equations with respect to x.) Equation (7.29) is now just a linear second order differential equation with coefficients that depend on the spatial frequency w, and the characteristic equation is:

$$\left| \begin{pmatrix} -1 + \dfrac{2}{1 + (4w)^2} & -\dfrac{4}{1 + (8w)^2} \\ \dfrac{2}{1 + (8w)^2} & -1 - \dfrac{4}{1 + (3w)^2} \end{pmatrix} - \lambda \right| = 0 \tag{7.30}$$

The eigenvalues of (7.30) display surprising behavior as a function of the spatial frequency w. For $w = 0$, (7.27) is spatially uniform, and (7.30) gives $\lambda = -1, -3$, so the uniform state is asymptotically stable. Similarly, as $w \to \infty$ the eigenvalues approach $\lambda = -1, -1$, so any rapidly varying spatial state also decays asymptotically to the resting state $E = 0$, $I = 0$. However, the behavior is very different over a range of intermediate spatial frequencies: for $0.085 < w < 0.22$, one value of λ becomes positive so that the resting state becomes an unstable saddle point. For $w = 0.15$, for example, $\lambda = +0.17, -4.03$. Thus, (7.26) is unstable with respect to perturbations of intermediate spatial frequency. (The human visual system is also most sensitive to intermediate spatial frequencies.)

The connection between (7.26) and the Wilson–Cowan (1973) equations (7.22) can now be made. Had we linearized (7.22) about a spatially homogeneous equilibrium state, the resulting linearization would have had exactly the form of (7.26), albeit with different coefficients (see Appendix to Wilson and Cowan (1973) for details). This leads to the prediction that certain spatial frequencies of stimulation, even of very small amplitude, will cause the system to spontaneously switch from a state of uniform neural activity to a spatially patterned state. You can verify this by running **WCstability.m**, which implements (7.22) but with constant spatial inputs $P = 31.5$ and $Q = 32.3$ that cause the

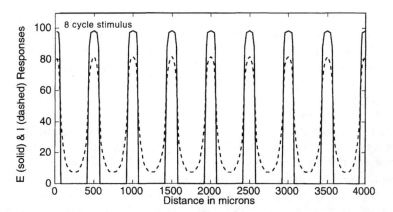

Fig. 7.12 Spatially inhomogeneous steady state of (7.22) triggered by a brief stimulus $0.01 \cos(2\pi\omega x)$ with $\omega = 8$. Very low or high values of the spatial frequency ω do not activate such states.

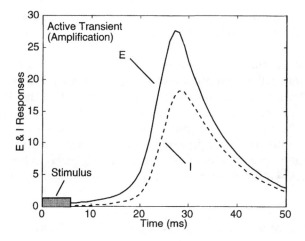

Fig. 7.13 Amplification (or active transient) produced by Wilson–Cowan equations (7.22) in response to a brief, weak stimulus. Recurrent excitation in the network causes $E(t)$ amplification, peaking about 20 ms after stimulus termination.

spatially uniform resting state to shift to $E = 0.25, I = 3.85$. Each time the script is run a 5.0 ms pulse of a tiny spatial perturbation, $0.01 \cos(2\pi\omega x)$, is added to P with the spatial frequency ω (an integer) specified by the user. For low spatial frequencies, $\omega = 0, 1, 2$, the small perturbation has no effect, and the system decays back to its uniform equilibrium state. Similarly, for $\omega \geq 40$, the uniform state is again asymptotically stable. However, for spatial frequencies in the approximate range $5 \leq \omega \leq 12$, the tiny perturbation causes the system to explode into a spatially inhomogeneous, asymptotically stable steady state, such that plotted in Fig. 7.12 for $\omega = 8$. Note that the program simulates 300 ms of time, so the 5.0 ms perturbation giving rise to the spatially structured state is long gone before the instability erupts. It is this spatially structured instability that gives rise to the short-term memory capabilities of (7.22) as illustrated in Fig. 7.11. These spatial instabilities result because the linearization of (7.22) takes the general form of (7.26).

In addition to a short-term memory mode, the Wilson–Cowan (1973) equations support several qualitatively different dynamical behaviors. One of these, which was termed the 'active transient mode', may be observed by running the script **WCcortexAT.m**, which uses the parameter values $b_{EE} = 1.5$, $b_{EI} = 1.3$, and $b_{II} = 1.5$ (all other parameters remaining the same). As illustrated in Fig. 7.13, a brief 5 ms spatial stimulus presented to a restricted area of the network gives rise to a delayed, but very large, amplification of the network response. In the case depicted, the E response grows from about 1.0 at the end of the stimulus period to about 28.0 some 20 ms later. The explanation for this dramatic response amplification in the active transient mode is that local recurrent excitation in the network causes excitatory activity to flare up, but this in turn triggers a delayed inhibitory pulse that subsequently extinguishes the network activity. As will be seen in the next few chapters, there are mathematical similarities between this network amplification and action potential generation. Such transient signal amplification was initially proposed to be appropriate for detection of weak stimuli by visual and other primary sensory cortical areas (Wilson and Cowan, 1973). Transient amplification generated by a balance of recurrent excitation and inhibition has been rediscovered and extended by Douglas *et al.* (1995).

A third dynamical mode of the Wilson–Cowan (1973) equations produces spatially localized oscillatory activity. Appropriate connectivity values, $b_{EE} = 1.9$, $b_{EI} = 1.5$, and $b_{II} = 1.5$, have been incorporated into the script **WCcortexOSC.m**. Running this script with a constant spatial stimulus of width 100 μm will produce a spatially localized oscillation with the $E(t)$ waveform plotted in Fig. 7.14. Each peak of this oscillation corresponds to a burst of action potentials at a peak rate of about 50 Hz. Even more complex oscillations can be produced by stimuli of greater widths. For example, a constant stimulus 400 μm wide produces a spatially adjacent pair of synchronized

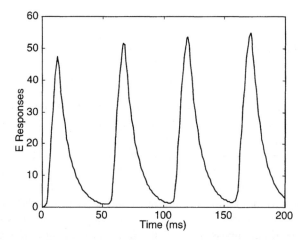

Fig. 7.14 Spatially localized oscillation produced by (7.22) in a different parameter range.

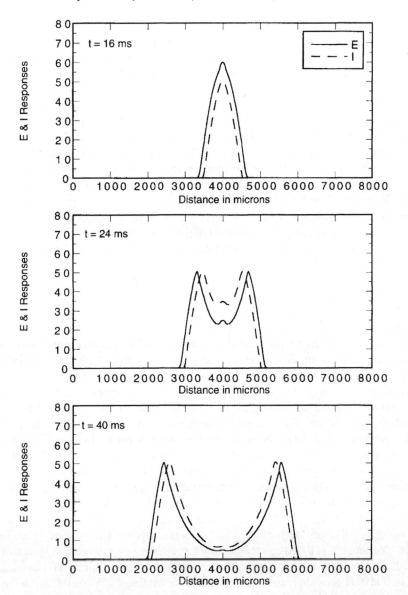

Fig. 7.15 Traveling neural activity waves produced by (7.22) when inhibition is sufficiently weak. From top to bottom, panels show spatial activity patterns at three successive times (16, 24, and 40 ms) after termination of a 5 ms stimulus pulse to the center of the network. Such waves may be a neural component of abnormal states like epilepsy.

oscillations. A stimulus 800 μm in width produces an even more complex oscillation with a pair of synchronized flanking regions oscillating out of phase with a narrow central oscillation. The dynamical basis for both nonlinear oscillations and synchrony between interacting oscillators will be covered in the next several chapters.

Finally, the Wilson–Cowan (1973) equations can produce traveling waves of neural activity in response to brief local stimulation in an appropriate parameter range. Sufficient conditions are that the parameters be the same as in the oscillatory mode just described but that the inhibition be severely reduced in effectiveness. This can easily be accomplished by providing a constant inhibitory input to the I cells. The MatLab script **WCcortexWAVES.m** incorporates the same parameters as the oscillatory mode but incorporates reduced inhibitory effectiveness by setting $Q = -90$. Running this script will generate the pair of traveling neural activity waves illustrated in Fig. 7.15 in response to a 5 ms stimulus to the center of the network. These waves originate because the reduced inhibitory activity is not sufficient to contain the activity generated by recurrent activation. These traveling waves are followed by a refractory area where the inhibition has built up. Because of this trailing refractory zone, two waves that meet will annihilate one another. This phenomenon can be observed by triggering waves at two points in the network simply by changing the commenting (lines beginning with %) as indicated in the program. Traveling waves such as these may occur in epileptic seizures and other forms of abnormal neural activity.

Waves of neural activity are also known to be important in retinal development (Feller *et al.*, 1996). Before the photoreceptors become active, a network of amacrine cells interconnected by excitatory synapses begins to generate traveling waves of neural bursting. These traveling waves are important in the development of normal connection patterns from the retina to the lateral geniculate nucleus (LGN), presumably causing coordinated synaptic modification in the latter structure. Feller *et al.* (1997) have developed a neural model to predict these wave phenomena. Short-range excitatory interconnections among amacrine cells cause waves to propagate across the tissue. The role of inhibition is played by neural adaptation or a refractory period in the amacrine cells following bursting. Finally, the amacrine activity is spatially pooled to produce ganglion cell outputs to the LGN. Whereas the Wilson–Cowan (1973) equations generate waves via recurrent excitation and delayed inhibition, the Feller *et al.* (1997) model employs only one cell type with recurrent excitation but with a delayed refractory period following activity. Thus, common dynamical principles (but differing physiological realizations) underlie these two examples of neural traveling waves. More complex methods for producing neural traveling waves are discussed in Chapters 12 and 13.

As virtually all areas of the brain incorporate both excitation and inhibition, the Wilson–Cowan (1972, 1973) equations may be regarded as a canonical model of a wide range of complex dynamics emerging from such interactions. Indeed, Hoppensteadt and Izhikevich (1997) provide a current and extensive catalog of Wilson–Cowan dynamics. Furthermore, the Wilson–Cowan equations have recently provided explanations for psychophysical data (Adini *et al.*, 1997), EEG waveforms during epileptic seizures (DeBellis *et al.*, 1998), and other aspects of neural dynamics (Jirsa and Haken, 1997).

7.5   Visual hallucinations

A variety of drugs can induce hallucinations involving stereotypical geometric patterns: concentric circles, radial spokes, spirals, or checkerboards with expanding check sizes away from central vision (Siegel, 1977). An elegant study by Ermentrout and Cowan (1979) demonstrated that these geometric hallucinations can be predicted from spatially inhomogeneous steady states of (7.22) if the equations are extended to two spatial dimensions by making each summation a function of both x and y. In addition to extending (7.22) to two dimensions, two additional factors were found to be necessary to explain visual hallucinations. First, the visual cortex must become sufficiently excitable so that spatially inhomogeneous steady states like the one-dimensional example in Fig. 7.12 become asymptotically stable. (Such states are appropriate for short-term memory but *not* for normal visual cortex.) In drug-induced hallucinations, the drugs presumably potentiate the $E \rightarrow E$ connections in visual cortex, thus producing hyperexcitability.

The second explanatory factor in hallucinations is the mapping from retina to visual cortex. For reasons discussed elsewhere (Wilson *et al.*, 1990), there is an enormous magnification of the central retinal representation in cortex relative to that of peripheral retina. Furthermore, the left and right halves of the retinas are each separately mapped onto the opposite visual cortex. Thus, the map of the right half of each retina onto the left visual cortex in humans is distorted approximately as shown in Fig. 7.16, where the vertical meridian maps onto the boundary of the cortical representation (Horton and Hoyt, 1991). As observed by Schwartz (1980) and Fischer (1973), the polar coordinate representation of the retina (R, θ) is mapped approximately into cortical coordinates ln($1+R$) and θ, the transformation used to produce the maps in Fig. 7.16. This retino-cortical mapping transforms images falling on the retina in highly characteristic ways. As shown in Fig. 7.16, concentric circles A–D on the retina would be mapped into the (almost) parallel vertical lines A–D in the cortical representation. Likewise, the radial spoke pattern E–G maps into (almost) parallel horizontal lines E–G on the cortex. Finally, spirals on the retina will map into oblique, parallel cortical lines.

The important insight of Ermentrout and Cowan (1979) was that asymptotically stable steady states of the cortex where neural activity is organized into roughly parallel stripes would of necessity be perceived by the subject as patterns of concentric circles, radial spokes, or spirals depending on what retinal stimulus would have produced them normally. The final point in explaining hallucinations was to prove that such parallel stripes of cortical activity do indeed represent asymptotically stable steady states in a two-dimensional generalization of (7.22). We have already seen that periodic activity patterns can be asymptotically stable in the simulation reproduced in Fig. 7.12, and the analytical explanation was presented in eqns (7.26) to (7.30). The two-dimensional generalization of (7.30) will contain products of terms in the spatial frequencies related to the x and y coordinates. Thus, two-dimensional steady states can be periodic in both x and y dimensions at once. The various alternatives may be observed by running the MatLab script **Hallucinations.m**, which introduces brief, noisy stimuli to trigger various parallel activation patterns in a small (64×64) generalization of (7.22). These would cause perception of concentric circles, radial spokes, and spirals. The program also demonstrates that the network can be triggered into a steady state consisting of roughly equally spaced circular patches of neural activity. Such patterns are the cortical correlate of a hallucinated checkerboard. (The similarity of these neural patterns to tiger stripes and leopard spots is not accidental: Murray (1989) has shown that similar dynamical principles

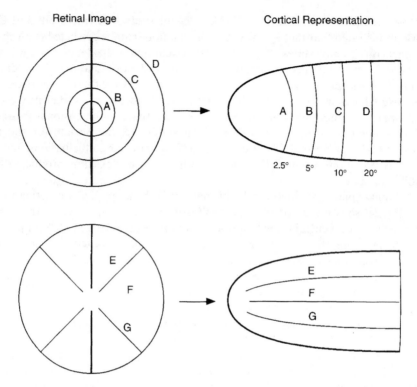

Fig. 7.16 Retino-cortical mapping in the human visual system (after Schwartz, 1980 and Horton and Hoyt, 1991). The right half of each retinal image maps to the right visual cortex as shown. Thus, concentric circles A–D map into almost parallel vertical lines, while radii E–G map into almost parallel horizontal lines. Parallel lines of cortical activity during hallucinations, simulated in **Hallucinations.m**, would thus be perceived as circles or radial spokes, depending on orientation.

mediated by diffusion produce the patterns on animal coats.) Ermentrout (1998) provides a more detailed summary of theoretical work on visual hallucinations.

This chapter has explored some of the richness and complexity of neural networks involving various combinations of recurrent excitation and recurrent inhibition. Such networks can accomplish vector summation, light adaptation, make perceptual decisions, and provide plausible models of short-term memory. The general theme of such networks is that short-range recurrent excitation must be balanced by longer range recurrent inhibition. Furthermore, networks that are imbalanced for their appropriate task, typically in the direction of excessive excitability (or equivalently reduced inhibition), provide important models of such neurological states as epilepsy and drug-induced hallucinations. Current research trends are to further embellish excitatory–inhibitory networks by introducing multiple subpopulations of E and I neurons. As one example, Somers *et al.* (1998) have developed a network model of orientation tuning in visual cortex that incorporates selective long-range excitatory connections among neurons with similar orientation tuning in addition to the local excitatory and longer range inhibitory interconnections embodied in the Wilson–Cowan (1973) equations. In the following chapters, interactions between positive and negative feedback, whether from excitatory and inhibitory neurons or from different ionic currents, will be shown to produce a complex array of nonlinear neural oscillations and traveling waves.

7.6 Exercises

1. Write down the 5×5 Jacobian (omit the equation for dG/dt) for the retinal model in (7.13). Solve for the steady state and obtain the eigenvalues using MatLab for $L = 1$, 10, 100, 1000. Does the nature of the singularity (spiral, node, saddle, etc.) change with light intensity?

2. In the text it is proved that (7.6) with E_Ω given by (7.4) is identical to the vector sum direction. Prove that $\max(E_\Omega)$ is the length of the resultant vector.

3. The vector summation network (7.5) can produce either one or two responses to signal coherent or transparent motion. Use the program **VectorWTA.m** to determine the largest angle (in whole degrees) between two vectors, each of length 10, that will produce a single response peak. Call this angle θ_{max}. Now remove the comment in the program so that you can stimulate with three vectors, all of length 10. Let the angles of two vectors be $\pm 1.2\theta_{max}$, so that these two vectors alone would produce two peaks in the response. What is the response if the third vector is at $0°$? Can you formulate a general rule determining whether one or two responses will occur?

4. To see that light adaptation postpones the onset of ganglion cell saturation, perform the following simulations using **Retina.m**. First, let the background $L = 10$ and determine incremental responses for $\Delta = 1$, 10, 100, 1000, and 10^5. Now run a second series for background adaptation to $L = 1000$ and increments $\Delta = 10$, 100, 1000, and 10^5. Plot your results on log–log coordinates. How does the ganglion cell response change as the level of background light adaptation increases?

5. The following equations might be obtained as a linearized approximation to nonlinear dynamical equations of the form (7.22):

$$\frac{dE}{dt} = -E + 1.5 \int_{-\infty}^{\infty} \exp(-|x - x'|) E \, dx' - \frac{1}{4} \int_{-\infty}^{\infty} \exp(-|x - x'|/4) I \, dx'$$

$$\frac{dI}{dt} = 2I + \frac{5}{8} \int_{-\infty}^{\infty} \exp(-|x - x'|/4) E \, dx' - \frac{1}{4} \int_{-\infty}^{\infty} \exp(-|x - x'|/4) I \, dx'$$

Analyze the stability of these equations with respect to spatial perturbations of frequency ω by utilizing (7.27) and (7.28). What are the eigenvalues for $\omega = 0$ and $\omega = \infty$? Using the MatLab **eig()** function, determine the range of spatial frequencies ω for which the equations are unstable.

8   *Nonlinear oscillations*

The study of nonlinear dynamical systems in the last two chapters has enabled us to analyze rather complex neural networks in terms of the stability of their equilibrium states. However, we have yet to consider one of the most exciting and important topics in all of dynamical systems theory: nonlinear oscillations. Indeed, nonlinear oscillations, or rhythms, are ubiquitous in living organisms. Circadian rhythms, cardiac rhythms, hormonal cycles, the rhythms of breathing and locomotion (walking, running, swimming): all are of the essence of life. Not only are many of these rhythms generated by neural networks, but we shall shortly see that even the generation of action potentials in single neurons is the result of inherently nonlinear oscillations.

In linear systems the only possible oscillations involve sines and cosines. These linear oscillations form closed circular or elliptical trajectories around an equilibrium point that is a center. Furthermore, if the initial conditions are changed even slightly, the result is a neighboring oscillatory solution with the same form in the phase plane but a different amplitude. No other type of oscillation is possible in a linear system, regardless of whether it has two or two thousand dimensions. In the nervous system and other biological systems, there is always some degree of noise resulting from physiological or environmental fluctuations. Such noise would continuously alter the amplitude of a linear oscillation, causing it to wander around the state space rather aimlessly. Clearly, such a sloppy linear oscillation could not control one's breathing or heartbeat very effectively. As we shall see, nonlinear oscillations are largely immune to this noise problem. For this reason alone, it is safe to conclude that biological rhythms evolved to be inherently nonlinear. To even begin to understand the rhythms of life and the nervous system, it is thus essential to study nonlinear oscillations.

8.1   Limit cycles

Let us start discussion of nonlinear oscillations with the definition of an oscillation itself. A trajectory $X(t)$ of a dynamical system with any number of dimensions is an **oscillation** if:

$$\vec{X}(T + t) = \vec{X}(t) \quad \text{for some unique } T > 0 \text{ and all } t. \tag{8.1}$$

This states that the system will always return to exactly the same state after time T, and T is therefore called the **period** of the oscillation. Note that if (8.1) is true for T, it will also be true for NT where N is any integer > 0, so the period is defined to be the minimum T for which (8.1) is holds. Also, the requirement that T be unique excludes equilibrium points, for which (8.1) would otherwise hold. The reciprocal of T, $1/T$, is termed the **frequency** of the oscillation. Note that (8.1) applies to linear systems as well as nonlinear.

If a linear system is periodic, there are infinitely many periodic solutions within any small neighborhood of a given oscillation: as the solution is a sum of sines and cosines, an oscillation of any amplitude whatsoever is a solution. (The amplitude is, of course, determined by the initial conditions.) Nonlinear systems can also produce analogous oscillations, but this is rare in biological systems. Of vastly greater significance is the fact that nonlinear systems can generate isolated oscillations that are surrounded by open, non-oscillatory trajectories that either spiral towards or else away from the oscillation over time. Let us make this precise with a definition:

Definition: An **oscillatory** trajectory in the state space of a nonlinear system is a **limit cycle** if all trajectories in a sufficiently small region enclosing the trajectory are spirals. If these neighboring trajectories spiral towards the limit cycle as $t \rightarrow \infty$, then the limit cycle is **asymptotically stable**. If, however, neighboring trajectories spiral away from the limit cycle as $t \rightarrow \infty$, the limit cycle is **unstable**.

Figure 8.1 shows schematic illustrations of both asymptotically stable and unstable limit cycles in the phase plane of a two-dimensional system. Notice that the definition of a limit cycle only requires that trajectories which are sufficiently close be open spirals. The reason for this restriction is that many nonlinear systems contain several limit cycles separated from one another.

Before exploring limit cycles in neuroscience, it will be necessary to develop some analytical tools to predict their existence. Let us first restrict our consideration to two-dimensional systems, as the relevant theorems are both more numerous and more intuitive in this case. A very useful theorem due to Poincaré, the discoverer of limit cycles, states that a limit cycle must surround one or more equilibrium points:

Theorem 9: If a limit cycle exists in an autonomous two-dimensional system, it must necessarily surround at least one equilibrium point. If it encloses only one, that one must be a node, spiral point, or center, but not a saddle point. If it surrounds more than one equilibrium point, then the following equation must be satisfied:

$$N - S = 1$$

where N is the number of (nodes + spiral points + centers), and S is the number of saddle points.

The requirement that the system be autonomous means that all coefficients must be constant. This guarantees that no trajectory can cross itself. The reason is simple: the differential equations describing the system define a unique direction at every point in

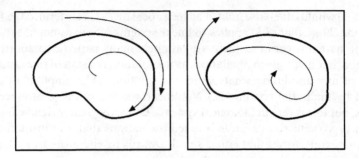

Fig. 8.1 Closed curves depict an asymptotically stable limit cycle (left) and an unstable limit cycle (right). Neighboring trajectories are plotted by arrows.

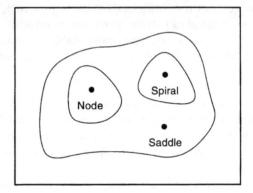

Fig. 8.2 Phase plane of a two-dimensional dynamical system with the three steady states indicated. Theorem 9 indicates that any possible limit cycle must surround one of the sets of steady states indicated by the three closed curves.

state space. If a trajectory were to cross itself, there would have to be two different directions specified by the equations at some point, but this is impossible unless the coefficients of the equations change with time, which has been prohibited. An intuitive grasp of Theorem 9 can therefore be gained from the observation that trajectories originating inside a limit cycle in the phase plane can never cross the limit cycle because it is a closed oscillatory trajectory. Therefore, these trajectories must either originate or terminate somewhere, and that must be a steady state (or another limit cycle which itself surrounds a steady state, etc.)

Figure 8.2 illustrates the possible locations for limit cycles in a two-dimensional system with three steady states: a node, a spiral point, and a saddle point. In this example, Theorem 9 precludes a limit cycle around any pair of the steady states. Note that Theorem 9 tells us nothing about the exact location or size or even the existence of the limit cycle but only about the set of steady states it would have to enclose if it existed.

Theorem 9 is a necessary but by no means sufficient condition for the existence of limit cycles in a nonlinear system. In Chapter 6, for example, we encountered several nonlinear systems with multiple steady states, such as two nodes and a saddle point, and yet there were no limit cycles to be found in those cases. What we need is a theorem that specifies conditions under which a system must have a limit cycle. Fortunately, such a theorem exists for two-dimensional systems. Let us first state the **Poincaré–Bendixon Theorem** and then sketch a proof using diagrams.

Theorem 10 (Poincaré–Bendixon): Suppose there is an annular region in an autonomous (i.e. constant coefficient) two-dimensional system that satisfies two conditions: (1) the annulus contains no equilibrium points; and (2) all trajectories that cross the boundaries of that annulus enter it. Then the annulus must contain at least one asymptotically stable limit cycle.

Theorem 10 is easily understood by examining Fig. 8.3. This figure shows an annular region (gray) that satisfies the conditions of the theorem. To be consistent with Theorem 9, the annulus must surround a node, spiral point, or center, which is plotted as a dot. Note, however, that this steady state, although surrounded by the annulus, is not within the annular region itself, so it does not violate the conditions of the theorem. Arrows in Fig. 8.3 show representative trajectories entering the annulus over both its inner and outer boundaries as required by Theorem 10. Once these trajectories enter the annulus, the conditions of the theorem guarantee that they can never leave. Also, they can never come to rest, because there are no equilibrium points in the annulus. Finally, because the system is autonomous, no two entering trajectories can ever cross one another. As trajectories entering from region A and region B move closer together, therefore, there must be at least one closed trajectory that they approach asymptotically. Thus, there must be at least one asymptotically stable limit cycle enclosed within the annulus. This completes an intuitive proof of the Poincaré–Bendixon theorem.

It is important to recognize that Theorem 10 specifies that the annulus must contain at least one asymptotically stable limit cycle, but the theorem also permits there to be an odd number of limit cycles. In this case, the outer and inner limit cycles would have to be asymptotically stable, because trajectories entering the annulus across the outer and inner

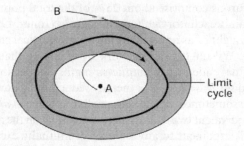

Fig. 8.3 Annulus (gray region) fulfilling the requirements of Theorem 10. Region A contains a steady state (unstable), and trajectories enter the annulus from both regions A and B. As indicated, a limit cycle must exist within the annulus.

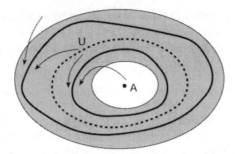

Fig. 8.4 Schematic of an annulus (gray region) satisfying Theorem 10 but containing three limit cycles. Two are asymptotically stable (solid curves), but the intervening one (dashed curve) must be unstable. A is an unstable node or spiral point. Representative trajectory directions are shown by the arrows.

boundaries must all approach limit cycles (not necessarily the same one). If there is more than one limit cycle, asymptotically stable limit cycles must alternate with unstable limit cycles. You can convince yourself of this by imagining what would happen to trajectories originating between two nested, asymptotically stable limit cycles: they would have to be separated by an unstable limit cycle, which is illustrated schematically in Fig. 8.4. Although the existence of alternate asymptotically stable and unstable limit cycles may seem to be an unlikely occurrence, they are actually predicted by the Hodgkin–Huxley equations, and their existence has been experimentally verified! Armed with Theorems 9 and 10, we are now ready to study limit cycles in two-dimensional neural systems.

8.2 Wilson–Cowan network oscillator

As a first application of these criteria to neural oscillations, let us consider a localized (i.e. non-spatial) version of the Wilson–Cowan (1972) equations. The equations presented here are the simplest example of these equations that possesses a limit cycle. Consider a four-neuron network consisting of three mutually excitatory E neurons which in turn stimulate one inhibitory I neuron that provides negative feedback onto the three E cells as depicted on the left in Fig. 8.5. Neural circuits like this are typical of the cortex, where inhibitory GABA neurons comprise about 25% of the total population of cortical cells with the rest being mainly excitatory glutamate neurons (Jones, 1995). Thus, the network in Fig. 8.5 may be thought of as approximating a local cortical circuit module.

Let us simplify the Wilson–Cowan network by assuming that all E neurons receive identical stimuli and have identical synaptic strengths. Under these conditions we can invoke symmetry and set $E_1 = E_2 = E_3$, thereby reducing the number of neurons in the network, a procedure sometimes termed **subsampling** (Wallén *et al.*, 1992). This results in the mathematically equivalent two-neuron network shown on the right in Fig. 8.5. In fact, we can generalize this argument to any number of mutually excitatory and inhibitory neurons with identical interconnections, so the key concept is that of recurrent excitation coupled with recurrent inhibition. Note that by reducing the network to two neurons (or two neural populations), the recurrent excitation is transformed into equivalent

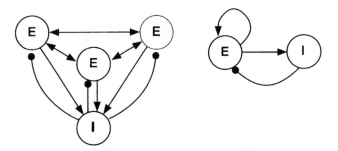

Fig. 8.5 Neural circuit of a network oscillator (Wilson and Cowan, 1972). Excitatory connections are shown by arrows and inhibitory by solid circles. The simplified network on the right is mathematically identical to that on the left by symmetry if all $E \rightarrow E$ connections have the same strength, etc.

self-excitation by the E neuron. The equations for the spike rates are:

$$\frac{dE}{dt} = \frac{1}{5}(-E + S(1.6E - I + K))$$
$$\frac{dI}{dt} = \frac{1}{10}(-I + S(1.5E)) \tag{8.2}$$

The function S in (8.2) is the Naka–Rushton function from (2.11) with $N = 2$, $M = 100$, and $\sigma = 30$. These equations indicate that the E neuron receives recurrent excitation with synaptic weight 1.6 and also receives subtractive inhibition from the I neuron. The external input that drives the network is K, which is assumed constant. The I neuron receives excitatory input from the E neuron with weight 1.5, and the time constants for E and I are 5 ms and 10 ms respectively. When $K = 0$, it is easy to verify that $E = 0, I = 0$ is the only equilibrium point and is asymptotically stable. In an intermediate range of K values the dynamics change, however, and limit cycle oscillations result.

Let us examine the state space of (8.2) in order to prove the existence of limit cycles for $K = 20$. The isocline equations are:

$$E = S(1.6E - I + K)$$
$$I = S(1.5E) \tag{8.3}$$

The second of these equations is easily plotted in its current form and is shown by the dashed line in Fig. 8.6. To plot the first isocline, however, we must employ the inverse of $S(x)$, which is obtained as follows:

$$S(x) = \frac{Mx^2}{\sigma^2 + x^2}$$

so $y = S(x)$ has the inverse:

$$x = \sigma\sqrt{\frac{y}{M - y}} \quad \text{for } 0 \leq y \leq M \tag{8.4}$$

Spikes, decisions, and actions

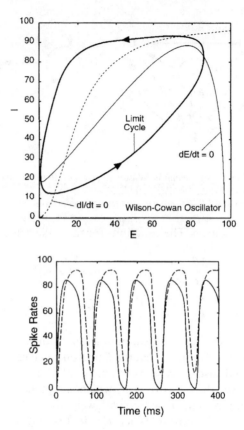

Fig. 8.6 Limit cycle of the Wilson–Cowan (1972) equations (8.2). Results are plotted in the phase plane (above) along with the two isoclines from (8.3) and (8.5). The lower panel plots $E(t)$ (solid line) and $R(t)$ (dashed line) as functions of time.

Therefore, the first isocline in (8.3) becomes:

$$I = 1.6E + K - \sigma \sqrt{\frac{E}{M - E}} \tag{8.5}$$

As $M = 100$ and $\sigma = 30$, the resulting isoclines for $K = 20$ are plotted in the E–I state space in Fig. 8.6. Note that there is a unique equilibrium point, which is the solution of (8.3) with the first equation transformed into form (8.5). To solve for the equilibrium, we simply substitute the second equation in (8.3) into (8.5) to get:

$$\frac{M(1.5E)^2}{\sigma^2 + (1.5E)^2} - 1.6E - 20 + \sigma \sqrt{\frac{E}{M - E}} = 0 \tag{8.6}$$

MatLab provides an easy method for solving (8.6): write a function script for the left-hand side of (8.6) (called **WCequilib.m** on the disk) and use the command **fzero('WCequilib', guess)** where 'guess' is a first approximation to the answer. This is all implemented in MatLab script **Equilibrium WC.m**, which finds that $E = 12.77$ at equilibrium, so $I = 28.96$ from (8.3).

Given the values of E and I at equilibrium, we can now calculate the Jacobian of (8.2). Using the formula for dS/dx in (6.10):

$$\overset{\leftrightarrow}{A} = \begin{pmatrix} 0.42 & -0.39 \\ 0.32 & -0.1 \end{pmatrix}; \quad \lambda = 0.16 \pm 0.24i \tag{8.7}$$

Thus, the only equilibrium point of the system is an unstable spiral point. We can now use the Poincaré–Bendixon theorem to prove that (8.2) must have at least one asymptotically stable limit cycle. Given the fact that the neural response function $0 \le S \le 100$, it follows that trajectories can never leave the state space box bounded by 0 and 100. This can be proven by considering the values of both derivatives in (8.2) on the boundaries of this box: $dE/dt \ge 0$ when $E = 0$; $dE/dt \le 0$ when $E = 100$; $dI/dt \ge 0$ when $I = 0$; and $dI/dt \le 0$ when $I = 100$. This represents an enormous simplification when dealing with nonlinear dynamics of neurons: spike rates are always bounded by zero below and a maximum value determined by the absolute refractory period. Thus, all trajectories that enter the box $0 \le E \le 100$, $0 \le I \le 100$, must stay within it, and all trajectories must also leave some small neighborhood of the unstable equilibrium point. Therefore, we have created an annulus containing no interior steady states, so by Poincaré–Bendixon Theorem 10, an asymptotically stable limit cycle must exist. If you run the MatLab simulation **WCoscillator.m**, you will see that an asymptotically stable limit cycle does indeed exist, and it is plotted in Fig. 8.6 for $K = 20$. Experimentation with a wide range of initial conditions shows that all trajectories do indeed approach the limit cycle asymptotically. In addition, it is interesting to experiment with other values of K in (8.2) to determine the stimulus range producing limit cycles.

8.3 FitzHugh–Nagumo equations

The simplest equations that have been proposed for spike generation are the FitzHugh–Nagumo equations. Like the Hodgkin–Huxley equations (see Chapter 9), these equations have a threshold for generating limit cycles and thus provide a qualitative approximation to spike generation thresholds. FitzHugh was well aware that his equations did not provide a detailed model for action potentials but emphasized: 'For some purposes it is useful to have a model of an excitable membrane that is mathematically as simple as possible, even if experimental results are reproduced less accurately.' (FitzHugh, 1969). This simplicity will aid in studying limit cycle oscillations.

The FitzHugh (1961) and Nagumo *et al.* (1962) equations describe the interaction between the voltage V across the axon membrane, which is driven by an input current I_{input} and a recovery variable R. R may be thought of as mainly reflecting the outward K^+ current that results in hyperpolarization of the axon after each spike. As these variables are only qualitatively related to the underlying ion currents, different authors choose slightly different parameter values in these equations. Here they will be written as:

$$\frac{dV}{dt} = 10\left(V - \frac{V^3}{3} - R + I_{input}\right)$$

$$\frac{dR}{dt} = 0.8(-R + 1.25V + 1.5) \tag{8.8}$$

Note that the time constant for V is 12.5 times faster than that for R (0.1 ms versus 1.25 ms), which reflects the fact that activation processes in the axon are much more rapid than the recovery processes. The isocline equations are:

$$R = V - \frac{V^3}{3} + I_{input}$$

$$R = 1.25V + 1.5 \tag{8.9}$$

For $I_{input} = 0$, MatLab procedure **roots** shows that the only equilibrium point is:

$$V = -1.5$$

$$R = -\frac{3}{8} \tag{8.10}$$

The Jacobian is:

$$\overset{\leftrightarrow}{A} = \begin{pmatrix} 10 - 10V^2 & -10 \\ 1 & -0.8 \end{pmatrix} \tag{8.11}$$

At equilibrium, therefore:

$$\overset{\leftrightarrow}{A} = \begin{pmatrix} -12.5 & -10 \\ 1 & -0.8 \end{pmatrix} \quad \text{so} \quad \lambda = -11.6, -1.7 \tag{8.12}$$

Thus, in the absence of any input, the equilibrium is an asymptotically stable node. For hyperpolarizing or inhibitory (negative) inputs I, the equilibrium remains an asymptotically stable node, as the reader can easily verify.

For depolarizing (positive) I, the situation changes, however. As an example, let $I = 1.5$. For this input, the only steady state is found from **roots** to be $V = 0$, $R = 1.5$, and (8.11) shows that it is an unstable node. So, trajectories must leave some small region surrounding the steady state. If we can now construct an outer closed boundary surrounding this node through which all trajectories enter, Theorem 10 can be used to prove the existence of an asymptotically stable limit cycle. To construct this boundary, begin at point 1 on the $dR/dt = 0$ isocline in Fig. 8.7, which need only be chosen at an adequately large V value ($V = 3$ in this instance). Next, draw a horizontal line until it intersects the $dV/dt = 0$ isocline at point 2. From (8.8), all trajectories must cross this line moving downward and to the left, as $dV/dt < 0$ and $dR/dt < 0$ in this region of state space. Now draw

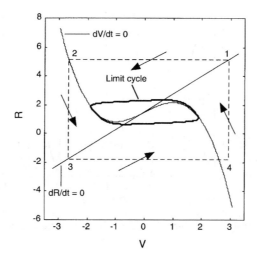

Fig. 8.7 Phase plane for FitzHugh–Nagumo equations (8.8) with isoclines (8.9) for $I_{input} = 1.5$. Dashed rectangle 1234 is the outer boundary of an annulus satisfying the Poincaré–Bendixon theorem. As illustrated by the arrows, all trajectories crossing the boundary of this rectangle enter it.

a vertical line down to the intersection with the $dR/dt = 0$ isocline at point 3. Again (8.8) shows that all trajectories must cross this boundary moving diagonally downward to the right, because $dV/dt > 0$ and $dR/dt < 0$ in this region. Similar reasoning shows that the horizontal line 3–4, drawn so that point 4 lies directly below point 1, will be crossed by trajectories moving diagonally upward into the rectangle. One repeats this procedure with the vertical line from 4 to 1 and finds that all trajectories cross moving toward the upper left. This completes our proof of the existence of an asymptotically stable limit cycle in (8.8), as all trajectories enter the region defined by the rectangle 1–2–3–4–1, and all leave the neighborhood of the unstable steady state.

To explore the FitzHugh–Nagumo limit cycle further, it is necessary to simulate (8.8) using the Runge–Kutta method. Running the script **FitzHugh.m** with $I_{input} = 1.5$ produces the series of simulated action potentials shown in Fig. 8.8. You can also verify that $I = 0.5$ is subthreshold, as no action potentials are generated. Comparing Fig. 8.8 with responses from the Hodgkin–Huxley equations depicted in Fig. 9.1 and 9.3 of the next chapter shows that (8.8) does not provide a very accurate description of action potential shapes. Nevertheless, the FitzHugh–Nagumo equations provide mathematical insight into the nature of neuronal excitability, namely that spikes are generated when I_{input} becomes strong enough to destabilize the equilibrium state (see Exercises).

8.4 Hopf bifurcations

Theorems 9 and 10 provide us with powerful means to determine whether limit cycles exist in a nonlinear system. Unfortunately, however, both theorems are limited to two-dimensional systems and do not apply when more than two neurons or two ion flows

Spikes, decisions, and actions

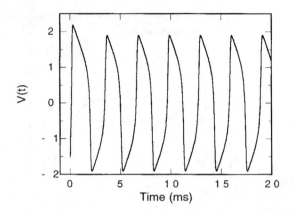

Fig. 8.8 Action potentials $V(t)$ produced by the FitzHugh–Nagumo equations (8.8) for $I_{input} = 1.5$.

are involved. Theorem 9 does not generalize to higher dimensions, because a closed trajectory defining a limit cycle oscillation cannot be said to enclose equilibrium points in three or more dimensions in any meaningful sense. One might think at first that the Poincaré–Bendixon theorem (Theorem 10) would generalize to higher dimensions if instead of specifying an annulus into which trajectories flow, one specified a solid 'doughnut' shape in three dimensions or a 'hyper-doughnut' in higher dimensional systems. However, such a ploy also fails to guarantee the existence of a limit cycle, because trajectories need not be closed to avoid crossing in higher dimensions. This means that a trajectory might remain within the doughnut but be chaotic (see Chapter 11) rather than being a limit cycle oscillation. Fortunately, there is one powerful theorem that applies to a system with any number of dimensions from two up, the Hopf bifurcation theorem.

Theorem 11 (Hopf bifurcation theorem): Consider a nonlinear dynamical system in $N \geq 2$ dimensions that depends on a parameter β:

$$\frac{d\vec{X}}{dt} = \vec{F}(\vec{X}, \beta)$$

Let X_0 be an isolated equilibrium point of this system. Assume that there is a critical value $\beta = \alpha$ with the following properties determined from the Jacobian, $A(\beta)$: (1) X_0 is asymptotically stable for some finite range of values $\beta < \alpha$. (2) When $\beta = \alpha$ the system has one pair of pure imaginary eigenvalues $\lambda = \pm i\omega$ while all other eigenvalues have negative real parts. (3) X_0 is unstable for some range of values $\beta > \alpha$. Then either the system possesses an asymptotically stable limit cycle over a range $\beta > \alpha$ or else it possesses an unstable limit cycle over some range $\beta < \alpha$. Near $\beta = \alpha$ the frequency of this oscillation will be approximately $\omega/2\pi$, and the oscillation will emerge with infinitesimal amplitude sufficiently close to α.

The Hopf bifurcation theorem is rather complex, and the proof is even more so. Qualitatively, however, it is possible to gain a good intuitive grasp of the theorem. As the theorem applies to $N \geq 2$ dimensions, let us consider a classic two-dimensional example: the van der Pol equation, which provided the first model of heart rhythms. Its normal form is:

$$\frac{dx}{dt} = y$$
$$\frac{dy}{dt} = -\omega^2 x + y(\beta - x^2)$$

(8.13)

where β is the parameter specified in Theorem 11 and ω will be shown to be the frequency. Inspection shows that $(0, 0)$ is the only equilibrium point of this system. At $(0, 0)$ the Jacobian of the associated linear equations is:

$$\overset{\leftrightarrow}{A} = \begin{pmatrix} 0 & 1 \\ -\omega^2 & \beta \end{pmatrix}$$

(8.14)

so the eigenvalues are:

$$\lambda = \frac{1}{2}(\beta \pm \sqrt{\beta^2 - 4\omega^2})$$

(8.15)

The three requirements on the eigenvalues listed in Theorem 11 are satisfied when $\beta = 0$, at which $\lambda = \pm i\omega$. For $\beta < 0$ the origin is an asymptotically stable spiral point, and for $\beta > 0$ the origin is an unstable spiral. Thus, the Hopf bifurcation theorem indicates that near the critical value $\beta = 0$ there will be a limit cycle solution to (8.13) with a frequency near $\omega/2\pi$.

The Hopf bifurcation theorem does not tell us whether this is an unstable limit cycle occurring when $\beta < 0$ or an asymptotically stable limit cycle for $\beta > 0$. Furthermore, Theorem 11 does not reveal how close β must be to zero for the limit cycle to exist. However, Theorem 11 has done most of our work for us: we know where to look in the parameter space to find limit cycle behavior. To explore this further, run the simulation **VanDerPol.m** with values of β near zero and with $\omega = 5$, so the predicted limit cycle frequency will be 0.796. Using initial conditions $x = 1, y = 0$ with $\beta = \pm 0.5$, you will readily discover that an asymptotically stable limit cycle exists for $\beta > 0$. Furthermore, you will see that even for $\beta = 0.5$, which is not all that close to $\beta = 0$, the frequency is remarkably close to 0.8.

This example helps to provide an intuitive feel for the Hopf bifurcation theorem. When $\beta < 0$ in (8.13), the origin is asymptotically stable. When $\beta = 0$ the linearized system defined by the Jacobian in (8.14) is stable but not asymptotically stable, and hence linearization does not reveal whether the origin of (8.13) is asymptotically stable or unstable. This is the special case not covered by Theorem 8 in which stability is determined entirely by the nonlinear terms. It is instructive to run **VanDerPol.m** with $\beta = 0$ in order to view the trajectory directions indicated by the arrows in the phase plane.

The effects of the nonlinear terms in (8.13) on stability when $\beta = 0$ can be determined through a clever calculation. Multiply the first equation in (8.13) by $\omega^2 x$ and the second equation by y and add the two to obtain:

$$\omega^2 x \frac{dx}{dt} = \omega^2 xy, \qquad y\frac{dy}{dt} = -\omega^2 xy - x^2 y^2$$

so

$$\omega^2 x \frac{dx}{dt} + y\frac{dy}{dt} = \frac{1}{2}\frac{d}{dt}(\omega^2 x^2 + y^2) = -x^2 y^2 \tag{8.16}$$

The intermediate step in the last equation contains the expression $(\omega^2 x + y^2)$, which is just the formula for an ellipse. So we have shown that any such elliptical contour will change in time by getting smaller, as the time derivative is negative (or zero on the axes). Therefore, the nonlinear terms in (8.13) guarantee that trajectories will decay towards the origin for $\beta = 0$. When $\beta > 0$, the origin becomes an unstable spiral point, so trajectories must leave its immediate vicinity. However, the nonlinear terms will still dominate far from the origin, so an asymptotically stable limit cycle must exist at intermediate distances. The technique used in (8.16) to determine the stability effect of nonlinear terms is a simple application of Lyapunov functions, which will be developed more fully in Chapter 14.

The final point concerning Hopf bifurcations is the requirement that all roots of the characteristic equation in more than two dimensions must have negative real parts at $\beta = 0$, except for the one pure imaginary pair. This means that trajectories of the multi-dimensional system will decay onto a two-dimensional subspace on which the remaining requirements of Theorem 11 guarantee a limit cycle. This is easy to understand if (8.13) is modified to make it into a very simple three-dimensional system:

$$\frac{dx}{dt} = y$$

$$\frac{dy}{dt} = -\omega^2 x + y(\beta - x^2) \tag{8.17}$$

$$\frac{dz}{dt} = kz$$

The additional variable z obeys an equation independent of x and y for simplicity. If $k < 0$, the third eigenvalue of (8.17) is negative, and z will just decay to zero, so an asymptotically stable limit cycle will still exist in the x–y plane. However, if $k > 0$, then trajectories will move off to infinity in the z direction, so a limit cycle cannot exist.

In Chapter 6 we studied bifurcations where pairs of equilibrium points, one asymptotically stable and one unstable (see Fig. 6.5), were created or destroyed. Let us create a similar diagram for (8.13) to better understand the sense in which the Hopf theorem relates to **bifurcations**. Because this is the bifurcation of a limit cycle from an equilibrium point, we must plot not only the locus of the equilibrium point as β changes but also a measure of the amplitude of the limit cycle. For the van der Pol equation (8.13) the steady state is always at the origin and simply changes from asymptotically stable to unstable as β passes through zero. The amplitude of the limit cycle as a function of β must be obtained from simulations using **VanDerPol.m**, and several conventions can be used to plot it in the bifurcation diagram. The convention I shall adopt is to choose one of the variables, in this case x, and plot both the maximum and minimum values that occur on the limit cycle.

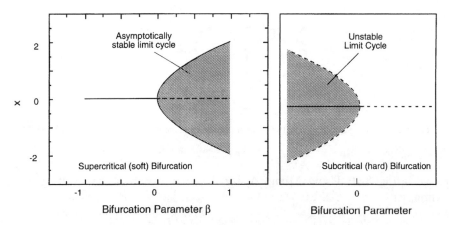

Fig. 8.9 Hopf bifurcations. A supercritical or soft bifurcation is depicted on the left, while a subcritical or hard bifurcation is plotted on the right. These are also called pitchfork bifurcations because of their shape. Limit cycles are indicated by a gray region connecting the maximum and minimum values of a variable $x(t)$ on the cycle. Solid lines indicate asymptotically stable states, and dashed lines indicate unstable states.

To emphasize that these are just two points on the same trajectory, the area between them will be shaded light gray. The results for (8.13) are plotted on the left in Fig. 8.9. The other convention for bifurcation diagrams is to plot asymptotically stable solutions with solid lines and unstable solutions with dashed lines. As can be seen, the asymptotically stable limit cycle emerges with infinitesimal amplitude as β passes through zero and the origin becomes unstable. This is a **Hopf bifurcation** in which the steady state loses its stability to the emerging, asymptotically stable limit cycle. A Hopf bifurcation at which an asymptotically stable limit cycle emerges as the steady state becomes unstable is called a **supercritical bifurcation** or a **soft bifurcation**. The particular one in Fig. 8.9 is sometimes also called a 'pitchfork bifurcation' because of its shape. The alternative form of bifurcation in which an unstable limit cycle merges with an asymptotically stable equilibrium state, making it unstable, is depicted on the right of Fig. 8.9. This is termed a **subcritical bifurcation** or a **hard bifurcation**. It will be seen in the next chapter that the Hodgkin–Huxley equations exhibit a subcritical bifurcation. Note that for both supercritical and subcritical bifurcations, the limit cycle amplitude is infinitesimally small arbitrarily close to $\beta = \alpha$.

8.5 Delayed negative feedback

The Hopf bifurcation theorem provides an enormously powerful tool for demonstrating the existence of limit cycles in dynamical systems in higher dimensions. Let us now apply it to a four-dimensional example of a negative feedback circuit with delays. The delays will be approximated by additional differential equations as developed in Chapter 4 and

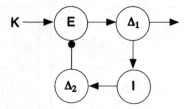

Fig. 8.10 Negative feedback network with two delay stages, Δ_1 and Δ_2 described by (8.18).

depicted in Fig. 8.10. Using Δ_1 and Δ_2 for the delay variables, the resulting system of equations is:

$$\frac{dE}{dt} = \frac{1}{20}(-E + S(K - \Delta_2))$$

$$\frac{d\Delta_1}{dt} = \frac{1}{\tau}(-\Delta_1 + E)$$

$$\frac{dI}{dt} = \frac{1}{50}(-I + 6\Delta_1) \tag{8.18}$$

$$\frac{d\Delta_2}{dt} = \frac{1}{\tau}(-\Delta_2 + I)$$

$$S(P) = \frac{100P^2}{50^2 + P^2} \quad \text{for } P \geq 0; \quad \text{otherwise } S(P) = 0$$

The time constants here are in milliseconds. For stimulus $K = 350$, the steady state can be shown to be: $E = 50$, $I = 300$. Let us now linearize (8.18) around the steady state and attempt to find a value of the delay time τ at which a Hopf bifurcation occurs. The Jacobian is:

$$\overset{\leftrightarrow}{A} = \begin{pmatrix} -1/20 & 0 & 0 & -1/20 \\ 1/\tau & -1/\tau & 0 & 0 \\ 0 & 6/50 & -1/50 & 0 \\ 0 & 0 & 1/\tau & -1/\tau \end{pmatrix} \tag{8.19}$$

The Routh–Hurwitz criterion can be used to determine the critical value of τ for which (8.19) has a pair of pure imaginary roots. If you type this matrix into MatLab function **Hopf.m** (this is why it was called Hopf!), save it, and then run **Routh_Hurwitz.m**, you will find that a Hopf bifurcation occurs when $\tau = 10.74$ ms and that the imaginary eigenvalues are $\lambda = \pm 0.0556i$, while the remaining two eigenvalues have negative real parts. Thus, a Hopf bifurcation will produce a limit cycle with initial frequency $1000 \times 0.0556/(2\pi) = 8.85$ Hz (the factor of 1000 converts from cycles/ms to cycles/s or Hz). To determine the range of τ values for which (8.18) has a limit cycle and to determine whether it is asymptotically stable or unstable, run **FBdelay.m** with values of τ near 10.74. You will discover that the limit cycle is asymptotically stable for $\tau > 10.74$.

This is a very important example, because it demonstrates that this simple neural feedback system cannot oscillate if the transmission delays are very short. If, however, the delays become long enough, limit cycle oscillations will occur. Whether these oscillations are desirable or dysfunctional depends on the role that the feedback circuit plays in the nervous system.

8.6 Adaptation and perceptual reversals

In Chapter 6 it was shown that spike rate adaptation in a short-term memory network (6.14) could eventually cause the network activity to shut off; thereby losing the information in short-term memory. Will there be a similar effect if adaptation is introduced into the two-neuron winner-take-all (WTA) network described by (6.18)? The network equations with adaptation variables A_1 and A_2, which operate by increasing the semi-saturation constants σ in (2.11), are:

$$\frac{dE_1}{dt} = \frac{1}{\tau}\left(-E_1 + \frac{100[K_1 - 3.2E_2]_+^2}{(120 + A_1)^2 + [K_1 - 3.2E_2]_+^2} \right)$$

$$\frac{dE_2}{dt} = \frac{1}{\tau}\left(-E_2 + \frac{100[K_2 - 3.2E_1]_+^2}{(120 + A_2)^2 + [K_2 - 3.2E_1]_+^2} \right) \tag{8.20}$$

$$\frac{dA_1}{dt} = \frac{1}{\tau_A}(-A_1 + \beta E_1)$$

$$\frac{dA_2}{dt} = \frac{1}{\tau_A}(-A_2 + \beta E_2)$$

where the small plus sign on the bottom right of the brackets indicates that the entire bracket evaluates to zero whenever the quantity within is <0. Let $\tau = 20$ ms, and $\tau_A = 600$ ms, thus again recognizing that adaptation is a much slower process that neural activation. Let $K_1 = K_2 = 150$ be the inputs for E_1 and E_2, and introduce a slight asymmetry in the initial conditions by setting $E_1 = 1$, $E_2 = 0$, $A_1 = 0$, and $A_2 = 0$. If you execute MatLab script **WTAadapt.m**, with adaptation parameter $\beta = 1.5$, you will find that the network oscillates as depicted in Fig. 8.11. Note that as E_1 adapts, E_2 is disinhibited to the point where it finally switches on, and E_1 is then inhibited. This cycle repeats itself and generates a limit cycle oscillation. This adaptation driven oscillation has a period of about 3 s, even though neural responses have a 20 ms time constant.

When a system has more than two dimensions, it is impossible to plot the entire state space or the isoclines, which are now intersecting surfaces rather than lines. We can, however, frequently gain insight into the dynamics by examining various two-dimensional projections of the state space. Given the four variables in (8.20), there are six pairs we might choose to plot. Examination of the various combinations reveals that useful information is obtained by plotting E_1 versus A_1 (or, equivalently, E_2 versus A_2). As shown in Fig. 8.12, this state space projection clearly reveals the closed limit cycle trajectory.

This limit cycle is more complex than those examined so far. We already know from Chapter 6 that when there is no adaptation, the asymptotically stable steady states involve

Spikes, decisions, and actions

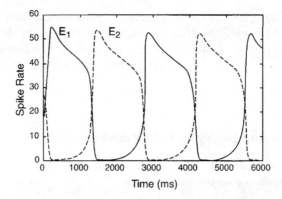

Fig. 8.11 Limit cycle produced by neural adaptation in a WTA network (8.20).

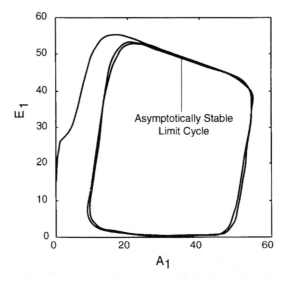

Fig. 8.12 Two-dimensional projection of the four-dimensional state space of (8.20) onto the E_1, A_1 plane. The limit cycle is clearly revealed in this projection.

activity in one neuron and suppression of the other. Weak adaptation does not affect this scenario and therefore cannot produce a limit cycle. When the adaptation becomes sufficiently strong, the firing rate of the active E neuron adapts to the point where it disinhibits the second E neuron, which then becomes active and suppresses the first cell, which is the mechanism for limit cycle genesis here. The Hopf theorem can be applied in this case, but the computation becomes rather complex. You are encouraged to experiment using the **WTAadapt.m** program. For example, try varying β in (8.20) to determine the values at which limit cycles arise and vanish (see Exercises).

There are several well-known perceptual phenomena that are believed to result from adaptation in competitive neural networks. One phenomenon is the perceptual reversal of

Fig. 8.13 Three examples of perceptual oscillations: the Necker cube (top), binocular rivalry (center), and the Marroquin figure (bottom). The cube A oscillates between the two three-dimensional interpretations in B and C. If one eye views the horizontal bars in the middle, while the other views the verticals, binocular rivalry ensues. The Marroquin figure generates the percept of scintillating circles.

ambiguous figures such as the Necker cube (Fig. 8.13, top). A second is binocular rivalry, which is produced by viewing vertical stripes with one eye and horizontal stripes with the other eye. One sees an alternation between horizontal and vertical stripes with a period of 2–4 s (Fox and Herrmann, 1967). Binocular rivalry may be experienced by running the MatLab program **BinocRivalry.m**. If you view the colored bar pattern produced by the program through red–green stereoscopic glasses (i.e. place green cellophane over one eye and red cellophane over the other), you should see quite dramatic perceptual alternations between horizontal and vertical bars. These perceptual reversals are believed to reflect adaptation of inhibitory circuits in the visual cortex (Blake, 1989). Finally, the Marroquin figure at the bottom of Fig. 8.13 produces a scintillation of circular shapes at various locations in the pattern, and these can be explained using a spatial network version of (8.20) (Wilson *et al.*, 1999).

These three perceptual alternations are by no means as precise as the limit cycle in Fig. 8.11, but this is doubtless due to the involvement of many more than just two neurons in the relevant networks and to the presence of physiological noise in the visual system. The important observation is that there are a variety of perceptual alternations that can be explained on the basis of limit cycles in competitive networks incorporating neural adaptation (Wilson *et al.*, 1999). In addition, limit cycles based on neural adaptation form a basis for lamprey swimming control (see Chapter 13).

8.7 Exercises

1. The following equations are a somewhat more complex version of (8.8) for describing action potential generation:

$$\frac{dV}{dt} = -V^3 + 3V^2 + 0.125 - R + I$$

$$\frac{dR}{dt} = -R + 5V^2 - 1$$

Plot the isoclines and analyze all of the equilibrium states for $I = 0$. Based on this analysis, draw closed curves indicating all possible locations in the phase plane at which limit cycles might occur. You need not simulate these equations, as we will be discussing more sophisticated versions of them in Chapter 9.

2. Consider the following pair of equations, which represent a somewhat different form of (8.8):

$$\frac{dV}{dt} = 10\left(\beta V - \left(\frac{V^5}{3}\right) - R\right)$$

$$\frac{dR}{dt} = 0.8(-R + V)$$

After showing that $V = 0, R = 0$ is an equilibrium point, prove the existence of a limit cycle as the Hopf parameter β passes through a critical value α. Determine the value of α analytically, and estimate the frequency of the limit cycle that will emerge at that point. Using methods analogous to those in (8.16), prove that the nonlinear terms cause all trajectories to decay to the origin when $\beta = \alpha$. Finally, simulate these equations for a value of β that produces a small amplitude limit cycle and compare the frequency of the simulation with your analytic result.

3. Consider the FitzHugh–Nagumo equations in (8.8), which have many of the qualitative properties of the Hodgkin–Huxley equations. Using I_{input} as the relevant parameter, prove that these equations undergo two Hopf bifurcations, and indicate whether each is subcritical or supercritical. Using your proof and Runge–Kutta simulations, produce a bifurcation diagram analogous to that in Fig. 8.9 that includes both bifurcations. (Hint: solve for V at the bifurcations using the Jacobian in (8.11), then use this information to solve for I_{input} at each bifurcation.)

4. The Wilson–Cowan (1972) oscillator originally incorporated a different mathematical form of the nonlinear firing rate function: the logistic function instead of the Naka–Rushton function. To see that this other form of sigmoidal (i.e. S-shaped) nonlinearity does not change the general features of solutions, consider the following modification of (8.2):

$$\frac{dE}{dt} = \frac{1}{5}\left(-E + \frac{100}{1 + \exp(-0.1(1.6E - I + k - 40))} \right)$$

$$\frac{dI}{dt} = \frac{1}{10}\left(-I + \frac{100}{1 + \exp(-0.1(1.5E - 40))} \right)$$

(Note that the connection strengths are identical to 82.). Conduct the following analysis of limit cycles in this system: (a) determine the stability characteristics of the steady state when $K = 0$; (b) prove that there is a Hopf bifurcation near $K = 10$; and (c) simulate and plot results (along with a phase plane plot) for $K = 15$.

5. Using **WTAadapt.m** determine the values of β (to the nearest 0.1) in (8.20) at which limit cycles arise and then vanish (you will find two values of β). Draw the bifurcation diagram near the higher β value and indicate what type of bifurcation this appears to be.

6. The following equations provide a much more accurate description of action potentials in the giant axon of the squid than do the FitzHugh–Nagumo equations:

$$0.05\frac{dV}{dt} = -\{1.35 + 3.67V + 2.5V^2\}(V - 0.55) - 2R(V + 0.92) + I$$

$$\frac{dR}{dt} = \frac{1}{\tau_R}(-R + 1.1V + 0.82)$$

$$\tau_R = 1.3\,\text{ms}$$

where the time constants are in ms and I is the input current. (a) Give a complete analysis of all steady states for $I = 0$. (b) Give a complete analysis of all steady states for $I = 0.1$. In this case plot the isoclines and prove that the system must have at least one asymptotically stable limit cycle. (c) Simulate these equations for 20 ms using input current $I = 0.1$ and plot $V(t)$. For initial conditions, start with $V(0)$ and $R(0)$ at their resting values for the case where $I = 0$. Indicate what step size you have used to obtain an error estimate that is < 0.001, and indicate the actual error estimate you have obtained.

9   *Action potentials and limit cycles*

In the previous chapter we developed criteria for the existence of limit cycles in nonlinear dynamical systems, namely the Poincaré–Bendixon theorem and the Hopf bifurcation theorem. As one example, we examined the FitzHugh–Nagumo equations, the simplest approximation to the dynamics of action potentials. However, these equations are not closely related to physiology, as they fail to include ionic currents and equilibrium potentials.

We are now poised to study the dynamics of ionic currents underlying the generation of action potentials in the Hodgkin–Huxley equations, where it will be shown that a periodic spike train is in fact a limit cycle. Following a brief review of the concepts behind the Hodgkin–Huxley equations, we shall study a set of equations that are simple to analyze mathematically but that provide a remarkably accurate description of action potentials. Subsequent topics examine hysteresis in spike generation, a dynamical categorization of neuron types, and various nonlinear resonance phenomena, including stochastic resonance.

9.1   Hodgkin–Huxley equations

The Hodgkin–Huxley (1952) equations describe the change in membrane potential or voltage V as a function of the sodium (I_{Na}), potassium (I_K), leakage (I_{leak}), and stimulating (I_{input}) currents across the membrane as well as membrane capacitance C. The most general form of the Hodgkin–Huxley equations is:

$$C \frac{dV}{dt} = -I_{Na} - I_K - I_{leak} + I_{input} \tag{9.1}$$

As each current obeys Ohm's law, the current $I = g(V - E)$, where g is the electrical conductance (reciprocal of the resistance), V is the voltage across the membrane, and E is the equilibrium potential of the ion in question computed from the Nernst equation (2.16). The capacitance C (in micro-Farads/cm^2, $\mu F/cm^2$) arises from the fact that the lipid bilayer of the axon membrane forms a thin insulating sheet that serves to store electrical charge in the same way as an electrical capacitor (cf. Hille, 1992; Johnston and Wu, 1995; Delcomyn, 1998). Hodgkin and Huxley discovered empirically that the conductances were not constant but rather were functions of the membrane potential V, and this voltage dependence is the key to understanding action potentials. Therefore, (9.1) was rewritten as:

$$C \frac{dV}{dt} = -g_{Na}m^3h(V - E_{Na}) - g_K n^4(V - E_K) - g_{leak}(V - E_{leak}) + I_{input}$$

$$\frac{dm}{dt} = \frac{1}{\tau_m(V)}(-m + M(V))$$

$$\frac{dh}{dt} = \frac{1}{\tau_h(V)}(-h + H(V)) \tag{9.2}$$

$$\frac{dn}{dt} = \frac{1}{\tau_n(V)}(-n + N(V))$$

In the first equation, E_{NA}, E_K, and E_{leak} are the equilibrium potentials at which each of the three currents is balanced by ionic concentration differences across the membrane. Evidently, the Hodgkin–Huxley equations are a fourth order system of nonlinear differential equations. The additional variables m, h, and n represent the rates of Na conductance channel activation, Na channel inactivation, and K channel activation respectively. Nonlinearity results from the fact that the equilibrium values of these variables, $M(V)$, $H(V)$, and $N(V)$ are all functions of the membrane potential V, as are the time constants τ_m, τ_h, and τ_n. Explicit mathematical forms for all these functions may be found in Hodgkin and Huxley (1952), Cronin (1987), and Johnston and Wu (1995).

The scientific content of the Hodgkin–Huxley equations comes from two sources. First is the observance of Ohm's law for the individual currents. The second is the hypothesis that the Na, K and leakage currents are all independent and therefore sum in (9.1). This hypothesis was tested by solving the mathematical model (9.2) that resulted (on a mechanical desk calculator!) and showing that it reproduced the experimentally observed shape and duration of the action potential in the squid giant axon. The mathematical forms chosen for the functions τ_m, τ_h, τ_n, $M(V)$, $H(V)$, and $N(V)$, were biologically motivated curve fits to the data. Hodgkin and Huxley's (1952) work represents the first and perhaps most dramatic success of nonlinear dynamics in predicting neurophysiological data.

The mathematical forms chosen by Hodgkin and Huxley for functions τ_m, τ_h, τ_n, $M(V)$, $H(V)$, and $N(V)$ are all transcendental functions. Both this and the presence of four equations in (9.2) make the Hodgkin–Huxley equations difficult to analyze mathematically. Fortunately, detailed study of these equations has led to several insightful simplifications. Rinzel (1985) noticed that τ_m is so small for all values of V that the variable m rapidly approaches its equilibrium value, $M(V)$. As a good approximation, therefore, the second equation in (9.2) can be eliminated and $m = M(V)$ substituted into the equation for dV/dt. Second, Rinzel noted that the equations for h and n were similar in time course and in their equilibrium values $H(V)$ and $N(V)$. In fact, an accurate approximation is obtained by setting $h = 1 - n$. What this means in ionic terms is that Na^+ channel closing, h, occurs at the same rate but in the opposite direction to K^+ channel opening, n. This relationship permits one to eliminate the equation for h, thereby reducing (9.2) to a two-dimension dynamical system. Under these assumptions, (9.2) assumes the form (Rinzel, 1985):

$$C\frac{dV}{dt} = -g_{Na}M(V)^3(1-R)(V-E_{Na}) - g_K R^4(V-E_K) - g_{leak}(V-E_{leak}) + I$$

$$\frac{dR}{dt} = \frac{1}{\tau_R(V)}(-R + G(V))$$

$$\tau_R(V) = 1 + 5\exp\left(\frac{-(V+60)^2}{55^2}\right)$$

(9.3)

To emphasize the simplifications and changes of variables, R has been used to describe the K^+ channel opening and Na^+ channel closing, which together constitute the recovery variable (hence the appellation R). The explicit expression for $\tau_R(V)$, the recovery time constant, has been included to indicate the general transcendental nature of these functions.

Using the forms of $H(V)$ and $G(V)$ derived from the original Hodgkin–Huxley equations as described above, let us examine the action potentials and the state space of the system. (Mathematical forms for $M(V)$ and $G(V)$ are contained in the MatLab scripts **MM.m** and **GG.m**, but they are too complex to provide much analytical insight.) The script **RinzelHH.m** implements (9.3), and action potentials are plotted in Fig. 9.1 for $I_{input} = 10\,\mu A$. The state space for (9.3) is plotted on the left of Fig. 9.2. Note in particular that the $dR/dt = 0$ isocline is straight over most of its range, while the dV/dt isocline is approximately cubic, although it does not agree with the exact cubic shape assumed in the FitzHugh–Nagumo equations (8.8). Changes in the spike rate and isocline shapes can be explored by running **RinzelHH.m** for different values of the input current. You can also estimate the threshold current necessary for spike generation: it lies in the range $0 < I_{input} < 10$.

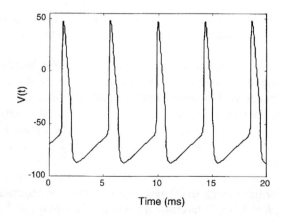

Fig. 9.1 Action potentials generated by (9.3), the Rinzel (1985) approximation to the Hodgkin–Huxley (1952) equations. In this instance $I_{input} = 10\,\mu A$, and the resultant spike rate is 250 Hz.

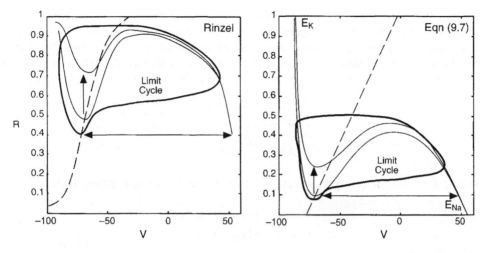

Fig. 9.2 Phase planes for (9.3) on left and (9.7) on right. For both equations the primary effect of increasing I_{input} is to shift the lower left lobe of the dV/dt isocline (solid curves) upwards as shown by the vertical arrows. Spike height (horizontal double-headed arrows) is primarily determined by the distance between the equilibrium point on the dR/dt isocline (dashed) and the right branch of the dV/dt isocline.

9.2 Essential dynamics of the Hodgkin–Huxley equations

The two-dimensional Rinzel approximation to the Hodgkin–Huxley equations in (9.3) can be simplified still further to reveal the essential dynamical principles underlying action potential generation. It is evident from the phase plane diagram in Fig. 9.2 (left) that the dV/dt isocline is roughly cubic in shape, while the dR/dt isocline is linear over most of its range. These observations were exploited by FitzHugh (1961) in developing the simplified FitzHugh–Nagumo equations discussed in the previous chapter. In the interests of mathematical simplicity, however, the FitzHugh–Nagumo equations ignored most physiological aspects of the Hodgkin-Huxley equations, such as adherence to Ohm's law and explicit reference to the Na^+ and K^+ equilibrium potentials. Let us develop a more accurate approximation to the Hodgkin–Huxley equations that rectifies the short-comings of the FitzHugh–Nagumo equations while retaining their mathematical tract-ability. To maintain biophysical significance, Ohm's law and the dependence on Na^+ and K^+ equilibrium potentials must be made explicit. This approach, which exposes the biological significance of the isoclines, leads to equations of the form:

$$C\frac{dV}{dt} = -g_{\text{Na}}(V)(V - E_{\text{Na}}) - R(V - E_{\text{K}}) + I$$
$$\frac{dR}{dt} = \frac{1}{\tau_{\text{R}}}(-R + G(V))$$

(9.4)

These equations have the same form as (9.3), namely an equation for dV/dt that is the sum of Na^+ and K^+ currents plus the stimulating current I, and a second equation for the recovery variable R. (The passive leakage current in (9.3), which plays no role in spike generation, has been absorbed into the Na^+ current for convenience.) For mathematical tractability, the first equation can be restricted to a cubic in V (based on isocline shapes discussed above), so $g_{Na}(V)$ must be restricted to a quadratic polynomial. Similarly, $G(V)$ can only be quadratic if the term $R(V - E_K)$ in the first equation is to remain no higher than cubic. Given these constraints, let us examine the isoclines of (9.4), which are:

$$R = \frac{-g_{Na}(V)(V - E_{Na}) + I}{(V - E_K)} \quad \text{for} \quad \frac{dV}{dt} = 0$$

$$R = G(V) \quad \text{for} \quad \frac{dR}{dt} = 0 \tag{9.5}$$

Setting $I = 0$ for the moment, it is evident from the first isocline equation that:

$$R = 0 \quad \text{when} \quad V = E_{Na}$$

$$R = \infty \quad \text{when} \quad V = E_K \tag{9.6}$$

These points are marked on the right-hand phase plane in Fig. 9.2. Thus, simply writing the dynamics in a form obeying Ohm's law leads to a $dV/dt = 0$ isocline with a natural biophysical interpretation in terms of E_{Na} and E_K!

A fit of (9.5) to the isoclines on the left of Fig. 9.2 leads to the following differential equations:

$$C\frac{dV}{dt} = -(17.81 + 47.71V + 32.63V^2)(V - 0.55) - 26.0R(V + 0.92) + I$$

$$\frac{dR}{dt} = \frac{1}{\tau}(-R + 1.35V + 1.03) \tag{9.7}$$

where the capacitance $C = 0.8\,\mu F/cm^2$ and $\tau_R = 1.9\,ms$. In generating these equations, the voltages were divided by 100 to keep the parameter values near unity. Thus, the equilibrium potentials are $E_{Na} = 0.55$ (or $+55\,mV$), and $E_K = -0.92$ (or $-92\,mV$), which are the same values that were used in (9.3). Thus, (9.7) expresses potential in deci-volts, and the input current I is in $\mu A/100$. For all comparisons with Hodgkin–Huxley results the solutions of (9.7) will therefore be multiplied by 100.

Having derived these equations, let us first determine whether they produce spikes that are accurate reflections of the Hodgkin–Huxley solutions. Figure 9.3 compares spike shapes at two different spike frequencies and also plots spike rates for both (9.7) and the full Hodgkin–Huxley formulation over the entire physiological range of input currents I. Equation (9.7) produces a good approximation to action potential shape, the correlation between the two spike shapes being > 0.96, and it also reproduces the reduction in spike amplitude with increasing spike rate observed in the Hodgkin–Huxley equations. The reader can explore the dependence of spike rate and spike height on input current I by running MatLab script **HHWeqn.m** and varying input current over the range $0 < I < 2.0$. Note that the numerical values of I are also 100 times smaller than those for the Hodgkin–Huxley equations, so the threshold value is in the range $0.0 < I < 0.090$. Figure 9.2 also shows that the dV/dt isocline in (9.5) deforms primarily on the lower left side as I is

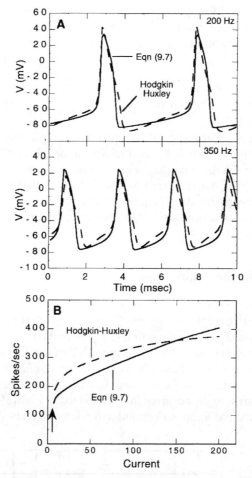

Fig. 9.3 Comparison of spike trains generated by (9.7) and the Hodgkin–Huxley equations (9.2). (A) Spike trains at 200 and 350 Hz for (9.7) (solid lines) and Hodgkin–Huxley (dashed lines). In addition to the similarity in shape, both equations produce a reduction in spike amplitude with increasing frequency. (B) Spike rate as a function of input current for Hodgkin–Huxley (dashed line) and (9.7) (solid line). Spike threshold is indicated by the arrow.

changed from threshold to 10 times threshold. Finally, Fig. 9.2 also shows that the limit cycles for (9.3) and (9.7) are quite similar in shape when plotted in state space. Thus, (9.7) provides a reasonably accurate approximation to the Hodgkin–Huxley equations given the simplifying assumptions made by Rinzel (1985) to obtain (9.3).

Let us now see how easy the analysis of (9.7) can be. The equilibrium state is given by the simultaneous solution of (9.5), which becomes, with parameters from (9.7):

$$-40.788V^3 - 81.079V^2 - 63.302V + 1.25I - 18.553 = 0 \qquad (9.8)$$

This can either be solved for specific values of I using the **roots** function in MatLab. For example, $I = 0$ yields only one real root: $V = -0.70$, or $-70\,\text{mV}$, the same as for the Hodgkin–Huxley simulations in Fig. 9.3, and $R = 0.088$ at rest. The Jacobian matrix for (9.7) is:

$$\overset{\leftrightarrow}{A} = \begin{pmatrix} -122.36V^2 - 118.28V - 22.937 & -32.5V - 29.9 \\ 0.71053 & -0.52632 \end{pmatrix} \tag{9.9}$$

where the equilibrium equation for R has been used to eliminate it from the matrix. Equations (9.8) and (9.9) can now be used to determine the stability of the equilibrium point for any value of I in the usual way. For example, if $I = 0.25$, $V = -0.67$ from (9.8), and the eigenvalues of the A matrix are $\lambda = 0.53 \pm 2.18i$, so the steady state is an unstable spiral point. The Poincaré–Bendixon Theorem can now be used to prove the existence of a limit cycle using a construction similar to that employed for the FitzHugh–Nagumo equation (see Exercise 1).

9.3 Hysteresis in the squid axon

One of the most striking aspects of dynamical modeling in neuroscience is the fact that nonlinear equations frequently predict novel phenomena that the creators of the equations had never imagined. As a case in point, Hodgkin and Huxley created their equations in 1952, yet many years elapsed before it was shown that the equations predicted a novel hysteresis effect that had never been observed (Cooley *et al.*, 1965; Rinzel, 1978; Best, 1979). Subsequently, Guttman, Lewis, and Rinzel (1980) tested this prediction and showed that hysteresis actually occurred in the squid axon. Figure 9.4 shows the results of their experiment. A squid axon was stimulated with a current I that began at 0 and

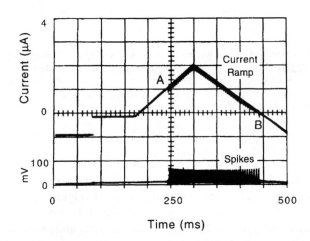

Fig. 9.4 Hysteresis in the giant axon of the squid (reproduced with permission, Guttman *et al.*, 1980). In response to a triangular current ramp, spiking activity begins at a high current at A but then continues to the much lower current at B, thus demonstrating hysteresis.

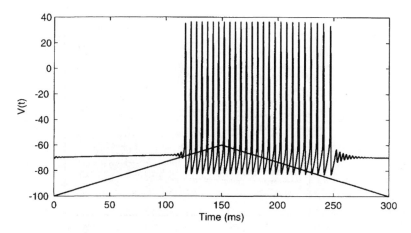

Fig. 9.5 Hysteresis produced by (9.7) in response to a triangular current ramp. Compare with data in Fig. 9.4.

increased linearly up to a maximum value. Following this it decreased linearly back to its original value. This triangular variation of I is indicated by the upper trace in Fig. 9.4. The spikes produced by this stimulation are plotted in the lower trace, where the solid black area indicates that the spike frequency was too high for the individual spikes to be resolved on the oscilloscope (time divisions are at 50 ms intervals). The striking observation is that the axon did not commence spiking until the stimulating current was about half way to its maximum value, but once spiking had begun, it continued almost all the way back down to zero! This is an example of neural hysteresis in a single axon!

This hysteresis experiment can be simulated using (9.7) by varying the input current I as a triangular function of time. If you run the MatLab program **HHWhysteresis.m** and choose a slow ramp and $I = 0.2$, you will obtain the result plotted in Fig. 9.5, which agrees well with the data in Fig. 9.4. You should try experimenting with other values of I, which defines the maximum value of the triangular current variation, to see how it affects the hysteresis.

Why does eqn (9.7) (or Hodgkin–Huxley) predict hysteresis in the squid axon? Let us apply the Hopf bifurcation theorem to (9.7) using the Jacobian in (9.9). Because the characteristic equation will only be quadratic in V, we can easily solve for the value of V at which the eigenvalues become pure imaginary, which occurs at $V = -0.688$. Now this value of V may be substituted into (9.8) to find the current that will produce this steady state. The result is $I = 0.0777$, which is thus the value of the current at which a Hopf bifurcation occurs. Therefore, one might guess that a slightly larger current, such as $I = 0.078$, would generate a low amplitude limit cycle in (9.7). If you try this by running **HHWeqn.m**, you will indeed see a limit cycle, but it will exhibit the full spike amplitude. This means that $I = 0.0777$ is actually a subcritical or hard Hopf bifurcation: (9.7) must have an unstable limit cycle for $I < 0.0777$ (refer back to Fig. 8.9). In the real world, unstable limit cycles can never be observed (due to their instability); only the

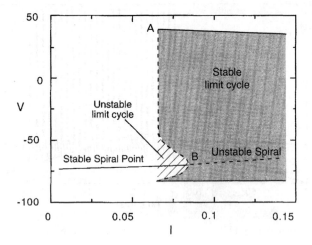

Fig. 9.6 Bifurcation diagram for (9.7) as a function of input current I. Over the range A–B an asymptotically stable limit cycle (gray region) and an unstable limit cycle (hatched region) coexist. The steady state becomes unstable at B in a subcritical or hard Hopf bifurcation.

consequences of their presence can be seen. In two-dimensional dynamical systems, however, we can 'observe' an unstable limit cycle by making time flow backwards (would that this were also possible in the real world sometimes!). Try running **HHWeqn.m** by making both time constants Tau and TauR negative in the script. Change the second initial condition to $X(2, 1) = 0.2$, and run the program with $I = 0.068$. You will find that there is now a small, asymptotically stable limit cycle surrounding the equilibrium point. Making time run backwards like this switches the stability of all limit cycles in the system, so the limit cycle that generates spikes is now unstable. (This is why the initial conditions were altered to be inside this second unstable cycle.) Note that making time run backwards in simulations does not work in higher dimensional systems, because it reverses the signs of all eigenvalues.

The situation may be clarified with a bifurcation diagram for (9.7). This is plotted in Fig. 9.6 where the I range over which both the asymptotically stable (gray) and the unstable (hatched) limit cycle exist is depicted. When the current I is changed very slowly from its resting value, the steady state remains asymptotically stable, and spikes will not be triggered until point B is reached. Beyond B spiking will continue until I finally drops below point A. Below A the asymptotically stable limit cycle vanishes, and the spike train stops. This is exactly what is seen in Fig. 9.5 and in the experimental squid axon data in Fig. 9.4. It was through such a mathematical analysis of the Hodgkin–Huxley equations that Rinzel (1978) predicted this hysteresis and the experimental conditions for observing it. A second equally fascinating consequence of the bifurcation diagram in Fig. 9.6 is the prediction that a *depolarizing* pulse at just the right time during a spike train should permanently extinguish it. This, too, was predicted by Rinzel (1978) and Best (1979) and experimentally demonstrated by Guttman *et al.* (1980). The phenomenon is explored in the Exercises.

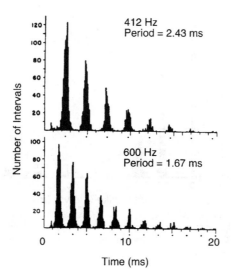

Fig. 9.7 Interspike interval histograms of auditory nerve axons in the monkey in response to pure tones of 412 Hz (top) and 600 Hz (bottom). Data from Rose *et al.* (1967) (reproduced with permission). The neuron shows a higher probability of firing at the same phase of each stimulus period.

9.4 Noisy neurons: improving auditory thresholds

All neural systems contain some physiological noise due to fluctuations in ionic concentrations, etc. Noise is normally regarded as being deleterious to the operation of a system and as something to be avoided or minimized. This is always true in linear systems, but there are circumstances in which noise can actually improve the performance of a nonlinear system. Let us examine one such example that has been explored by Longtin (1993, 1995).

Consider the data in Fig. 9.7, which were recorded from an auditory nerve fiber in a squirrel monkey (Rose *et al.*, 1967). This is an interspike interval (ISI) histogram in which the data refer to the intervals between successive spikes fired by a neuron in response to sinusoidal stimulation at 412 or 600 Hz. The stimulus period is listed in each plot, and the dots on the abscissa indicate integral multiples of that period. It is evident from the graph that the neuron did not fire during each period of the stimulus. Rather, firing was probabilistic: the first peak in the histogram indicates the frequency with which two spikes occurred one period apart; the second peak is the frequency with which the neuron missed one period so that the spikes occurred two periods apart, etc. It is obvious from the records that this neuron can miss five or more periods between action potentials. Note also that successive peaks decrease in amplitude approximately exponentially, which suggests a Poisson process.

How might this erratic neural behavior be explained? More interestingly, could there be any functional advantage to this noisy behavior? Both questions can be answered by simulating this experiment on the neuron described by (9.7). Suppose that the stimulating

Spikes, decisions, and actions

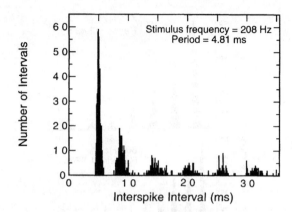

Fig. 9.8 Interspike interval histogram for (9.7) in response to a subthreshold 208 Hz sinusoidal input in the presence of Gaussian noise. Clustering of responses at integral multiples of the stimulus period is indicative of stochastic resonance (compare with Fig. 9.7).

current I to this neuron is a sinusoidal function of time with a frequency of 208 Hz. If you run the MatLab script **HHnoise.m** without noise (set SDnoise = 0), you will find that the threshold stimulus amplitude to generate a train of spikes is about 0.041. Reducing the amplitude to 0.025, about 60% of the threshold value, the equations generate only a weak subthreshold oscillation in V, and no spikes are produced. Now suppose some physiological noise is added to the input current. Let us assume the noise is Gaussian and has zero mean and standard deviation SDnoise = 0.12. Now the equations generate approximately 60 spikes/s. Figure 9.8 shows the ISI histogram for eqn (9.7) under these conditions, and it clearly reflects all of the major characteristics of the auditory neuron in Fig. 9.7. In particular, successive histogram peaks decrease in size and are centered near integral multiples of the 4.81 ms stimulus period.

This simulation makes it easy to understand the effects of noise on the neural response: the noise added to the stimulus increases the probability of firing predominantly when the stimulus sinusoid is near its peak phase. However, the noise alone has a low probability of generating spikes in the absence of the sinusoidal stimulus: on average it generates 4 spikes/s at random intervals. This example demonstrates that when there is a threshold for spike generation, subthreshold noise can *increase* the sensitivity to periodic sensory stimulation. In this example, the sensory threshold is improved (i.e. decreased) by about 40% through addition of noise. Such noise-induced enhancements of sensitivity are known as **stochastic resonance** (see Douglas *et al.*, 1993). Thus, noise can improve sensory performance via stochastic resonance because of neural thresholds. The price paid for stochastic resonance is a noisy or stochastic output, but the nature of the stimulus is encoded into the interspike intervals in the response.

Finally, this simulation provides an explanation for the Poisson character of the ISI histogram. Referring to Fig. 9.8, assume that the probability of firing a spike during any one stimulus period is p. That means that the ratio between the first and second peak amplitude should be about $(1 - p)$, which is the probability of not having fired in the previous interval. If we let $(1 - p) = 0.3$ and take the first peak height as about 60, successive peaks should therefore have heights of 18.0, 5.4, and 1.62. Figure 9.8 shows that this is about right, although there are too few spikes in this simulation (about a thousand) to accurately delineate the higher order peaks.

9.5   Human and mammalian cortical neurons

Equation (9.7) provides an accurate and tractable description of the Hodgkin–Huxley equations for action potentials in the squid axon. However, the squid axon is unusual in having only one Na^+ and one K^+ current. As a consequence, the squid axon cannot fire at rates below about 175 spikes/s and produces only a modest increase in spike rate with increasing input current (see Fig. 9.3). The vast majority of neurons in other animals also possess a rapid, transient K^+ current that permits the cell to fire at very low spike rates with a long latency to the first spike when the input current is low (Rogawski, 1985). This current, known as I_A, was first characterized and added to the Hodgkin–Huxley model by Connor, Walter, and McKown (1977). I_A currents are found in a wide range of neurons, including human and mammalian neocortical neurons (Avoli and Williamson, 1996; Gutnick and Crill, 1995). Because of its ubiquity and also because neurons with an I_A current exhibit a mathematically distinctive bifurcation from the resting to the spike generation state, let us extend our action potential model to incorporate it.

As with the Na^+ and K^+ currents in the squid axon, Connor *et al.* (1977) incorporated a mathematically complex, transcendental description of the I_A current. More recently, Rose and Hindmarsh (1989) demonstrated that many effects of the I_A current could be approximated by making the equation for the recovery variable R quadratic. Let us therefore examine the dynamical consequences of a quadratic dR/dt equation. The equations are:

$$\frac{dV}{dt} = -\{17.81 + 47.58V + 33.8V^2\}(V - 0.48) - 26R(V + 0.95) + I$$

$$\frac{dR}{dt} = \frac{1}{\tau_R}(-R + 1.29V + 0.79 + 0.33(V + 0.38)^2) \qquad (9.10)$$

$$\tau_R = 5.6\,\text{ms}$$

The capacitance $C = 1.0\,\mu F/cm^2$ and therefore has not been written explicitly. These equations have parameter values chosen to provide a good approximation to the action potentials produced by human neocortical neurons (Wilson, 1999). (Human neurons have been studied intracellularly in tissue, usually from the temporal lobe, removed during epilepsy surgery.) Note that the dR/dt equation is written as the sum of a linear and a quadratic voltage term, which may be interpreted qualitatively as the normal I_K and the transient I_A current contributions.

Figure 9.9 shows the state space for (A) $I = 0$ and (B) $I = 0.5\,\text{nA}$ from MatLab script **HumanNeuron.m**. For $I = 0$ there are three steady states at a, b and c. The resting state at a, where $V = -0.75$ ($-75\,\text{mV}$), is an asymptotically stable node, while b and c are an unstable saddle point and an unstable spiral point respectively. Threshold is reached for (9.10) when the stable node (a) and unstable saddle point (b) coalesce and vanish at a

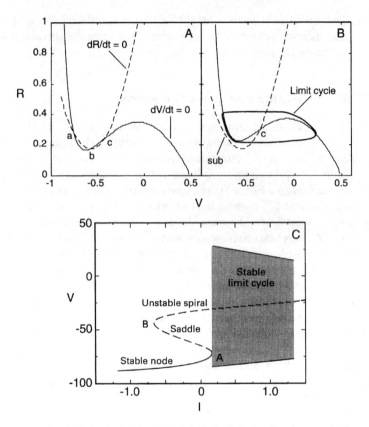

Fig. 9.9   Phase planes for (9.10) for $I = 0$ (A) and $I = 0.5$ (B). In A the isoclines intersect in three steady states: an asymptotically stable node (a), a saddle point (b), and an unstable spiral point (c). In B the node and saddle point have vanished at a saddle-node bifurcation, resulting in a limit cycle around spiral point c. C shows the bifurcation diagram for (9.10) as a function of I.

bifurcation, which is therefore very different from a Hopf bifurcation. This leaves only the unstable spiral point c, around which a limit cycle emerges. This is evident in the bifurcation diagram in Fig. 9.9C, where limit cycles surround the single unstable spiral point for $I > A$, while three steady states, including the resting state at the stable node, exist between A and B. Because a saddle point and a node coalesce and vanish at A, this is called a **saddle-node bifurcation**, and the limit cycle that emerges in such cases will have a finite amplitude at its inception. This is very different from a Hopf bifurcation where the limit cycle must originate with infinitesimal amplitude when a single steady state changes from stable to unstable. Not all saddle-node bifurcations produce limit cycles, and simulations are generally required to determine whether or not a limit cycle does occur.

   Figure 9.10 illustrates the spike train produced by eqn (9.10) for $I = 0.22$ nA. Note the two characteristic changes produced by the simulated I_A current: a long latency to the first spike (about 200 ms), and a much lower firing rate (about 5.0 spikes/s) than that produced by either (9.3) or (9.7) and shown in Fig. 9.3. In fact, just above threshold ($I = 0.2149$ nA),

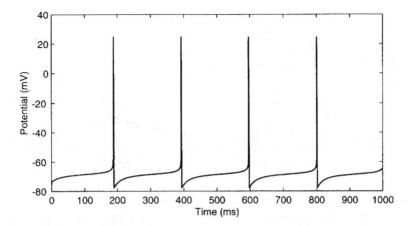

Fig. 9.10 Spike train from (9.10) for $I = 0.22$ nA. Note long latency and low spike rate compared to Fig. 9.3.

(9.10) produces latencies of more than 1 s and spike rates below 1 spike/s! Both the lower spike rates and the long latencies result from the fact that the $dR/dt = 0$ isocline lies very close to the limit cycle in the subthreshold region marked 'sub' in Fig. 9.9B. This means that dR/dt will be very small in this subthreshold region, so the interval between spikes is greatly increased relative to the case where the dR/dt isocline is far away (compare with Fig. 9.2). The greatly reduced spike rates produced by I_A have the functional significance of extending spike rate coding over a much greater dynamic range (Connor *et al.*, 1977; Rogawski, 1985).

Neurons that can be stimulated to fire at arbitrarily low frequency due to saddle-node bifurcations are termed **Class I** neurons, while those that can only begin firing at a relatively high frequency resulting from a subcritical Hopf bifurcation are called **Class II** neurons (Ermentrout, 1998). Almost all mammalian neurons are Class I, while the squid axon is of Class II.

The cortex of humans and other mammals is currently thought to contain four major types of neurons, each of which is characterized by its distinctive nonlinear dynamical properties. Spike shapes for two of these cell types, regular-spiking (RS) and fast-spiking (FS), are plotted in Fig. 9.11C (McCormick *et al.*, 1985). Action potentials of RS neurons have a rapid rise but a much slower decay phase. FS cells, on the other hand, have a similar rise but a much more rapid decay phase than regular-spiking cells, and in consequence, the width of the spikes is significantly narrower. In humans, spike widths at half amplitude average 0.95 ms for RS cells and 0.60 ms for FS cells (Wilson, 1999). The model in eqn (9.10) has already been optimized to approximate the size and shape of RS neuron action potentials obtained from human neocortical neurons, and a comparison between model results and human data (Foehring *et al.*, 1991) is shown in Fig. 9.11A. The model provides an accurate quantitative fit to the spike shape and, with the addition of a slow potential to be discussed in the next chapter, to the spike rates as well. Equation (9.10) can also provide an excellent quantitative approximation to the action potentials of FS cells if just one parameter is changed: the recovery time constant must be reduced to $\tau_R = 2.1$ ms.

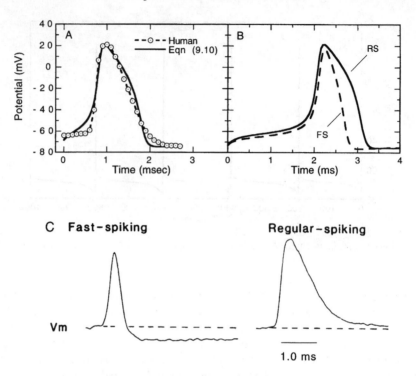

Fig. 9.11 Regular-spiking (RS) and fast-spiking (FS) neurons. (A) Spike shape of a human RS neuron (data from Foehring *et al.*, 1991) compared to spike from (9.10). (B) RS and FS spike shapes produced by changing τ_R in (9.10) from 5.6 ms (RS) to 2.1 ms (FS). (C) FS and RS spike shapes recorded from neocortical neurons (reproduced with permission, McCormick *et al.*, 1985).

This speed-up of the recovery phase by a factor of 2.7 produces the FS action potential plotted in Fig. 9.11B. You can verify this by changing TauR in the MatLab script, **HumanNeuron.m**. Note that the FS spike is not only narrower but also slightly reduced in height relative to the RS cell, and this is also evident in the data in Fig. 9.11C. This is a dynamical consequence of the faster R variable in the equations.

9.6 Subharmonic resonance and phase shifts

When the stimulus to a linear system is periodic with frequency ω the response is always periodic at the same frequency. In nonlinear systems, however, limit cycles can respond at subharmonics of the stimulus frequency, a phenomenon known as subharmonic resonance. Cochlear neurons in the auditory system exhibit subharmonic resonance in response to pure tones of appropriate frequencies (Kiang *et al.*, 1965). The phenomenon of subharmonic resonance may be illustrated by the response of (9.7) to the periodic stimulus $I = A \sin(2\pi\omega t)$. If you run the MatLab script **HHnoise.m** with $A = 0.25$, SDnoise $= 0$ (noise is not relevant here) and $\omega = 200$ Hz, the result at the top of Fig. 9.12

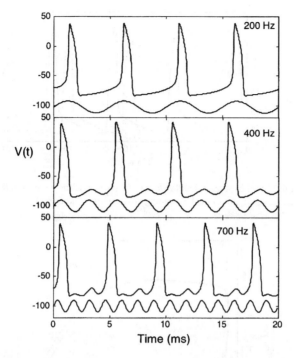

Fig. 9.12 Subharmonic resonance exhibited by (9.7) in response to sinusoidal stimulation at 200, 400, and 700 Hz.

will result. These conditions generate one spike locked to the rising phase of each stimulus cycle, hardly a surprising result. When the frequency is increased to $\omega = 400$ Hz, however, the more unusual result in the middle panel is obtained. Now the neuron only generates an action potential during each second cycle of the stimulus oscillation. This is the phenomenon of **subharmonic resonance**, also called **frequency division**. Strikingly, the spike rate is the same for 200 Hz and 400 Hz stimuli! Finally, the bottom panel shows that at 700 Hz a spike is only generated during each third stimulus cycle, a further example of subharmonic resonance. Experimentation with other stimulus values using **HHnoise.m** will yield interesting and sometimes very irregular results, some of which are chaotic (see Chapter 11).

Why does subharmonic resonance occur in neurons? The physiological answer is that when stimuli oscillate at sufficiently high rates, one or more cycles will occur during the absolute refractory period following the previous spike when no additional spikes can be generated. A mathematical interpretation of this may be gained by considering a simpler problem: the effect of a brief current pulse applied at various times during a spike train. The script **PhaseShift.m** applies a constant depolarizing current $I = 0.15$ to eqn (9.7) and allows one to specify both the amplitude and initiation time of a 0.1 ms pulse. If no pulse is applied (pulse amplitude of zero), the equations produce a spike train with a period of 5.1 ms, as shown by the solid line in Fig. 9.13. If a pulse with amplitude $= 1.0$ is applied at

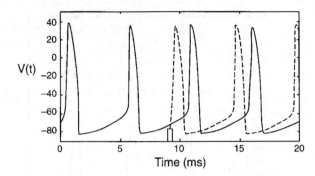

Fig. 9.13 Phase shift produced by a brief depolarizing pulse (rectangle) delivered at 9.0 ms to (9.7). The pulse causes a phase advance (dashed curve) in the ongoing spike train (solid curve).

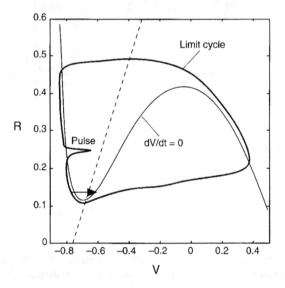

Fig. 9.14 Phase plane of (9.7) with a depolarizing pulse that produces the distortion of the limit cycle indicated. If the same pulse is delivered at the phase shown by the arrow, the trajectory is shifted across the bottom lobe of the dV/dt isocline, and an early spike is triggered.

$t = 9.0$ ms, however, a significant phase shift occurs: the next spike occurs much earlier than in the absence of the pulse (dashed line in Fig. 9.13). An earlier pulse with same amplitude triggered at 8.0 ms produces almost no phase shift. Experimentation with **PhaseShift.m** will reveal that pulses delivered within about 1.0 ms after completion of a spike have almost no effect.

Further understanding of these phase shifts can be gained by examination of the state space depicted in Fig. 9.14. The labeled pulse shows the deformation of the limit cycle trajectory produced by a 0.1 ms pulse delivered at a phase that produces almost no phase shift. The effect of the pulse is to shift the trajectory almost parallel to the V axis by a short distance. As this causes the trajectory to cross the $dV/dt = 0$ isocline to a region where

$dV/dt < 0$, the trajectory rapidly decays back to the original limit cycle following the pulse. If the same pulse were delivered at the slightly later phase indicated by the horizontal arrow, however, it would push the trajectory entirely across the bottom lobe of the $dV/dt = 0$ isocline to a region where $dV/dt > 0$, and the next spike would be triggered early. Thus, spikes are triggered only at those phases where the pulse causes the trajectory to jump across the bottom lobe of the dV/dt isocline. The refractory period is just the phase range where stimuli are not sufficiently strong to produce a large enough jump in the trajectory. These observations also provide an explanation for subharmonic resonance. A high-frequency sine wave stimulus may be regarded as similar to a periodic train of pulses. When the stimulus frequency is too high, one or more pulses will occur within the phase range after each spike where spike triggering cannot occur. Other pulses, however, may be expected to occur at phases where they trigger an early spike. The combination of these two factors can cause both synchronization and subharmonic resonance.

9.7 Morris–Lecar equations

In addition to eqns (9.3), (9.7), and (9.10), a number of additional two-equation models for action potential generation have been developed, including those by Hindmarsh and Rose (1982), Kepler *et al.* (1992), and Morris and Lecar (1981). Due to an excellent analysis of the Morris–Lecar equations by Rinzel and Ermentrout (1989), these equations have become popular among neural theorists. Accordingly, I shall present them briefly here and relegate an exploration of their properties to the Exercises. The normalized or dimensionless form of these equations presented here is obtained by dividing all voltages by the equilibrium potential for the excitatory ion (Ca^{2+} in the original model or Na^+ when applied to squid or cortical neurons) and scaling time appropriately (Rinzel and Ermentrout, 1989). The resulting Morris–Lecar equations are:

$$\frac{dV}{dt} = -g(V)(V - 1.0) - 2R(V + 0.7) - 0.5(V + 0.5) + I$$

$$\frac{dR}{dt} = \frac{0.2}{\tau_R(V)}(-R + G(V))$$

$$g(V) = \frac{1}{1 + \exp(-(V + 0.01)/0.075)} \qquad (9.11)$$

$$G(V) = \frac{1}{1 + \exp(-(V - 0.1)/0.07)}$$

$$\tau_R(V) = \frac{1}{\cosh((V - 0.1)/0.14)}$$

The first two equations have the same form as (9.4) except for the explicit addition of a leakage conductance term, $-0.5(V + 0.5)$. The transcendental expressions for $g(V)$ and $G(V)$ are logistic functions, which are just scaled and shifted hyperbolic tangents. Finally,

Spikes, decisions, and actions

the time constant τ_R varies with V, being smallest when $V = 0.1$. The Morris–Lecar equations are implemented in MatLab script **Morris_Lecar.m**, and their properties are explored in the Exercises. The dynamical behavior of (9.11) with the constants given is very similar to (9.10).

9.8 Exercises

1. Use the Poincaré–Bendixon theorem to prove the existence of limit cycles for eqn (9.7) with $I = 0.5$. Determine the stability of the steady state and then construct an appropriate closed curve in the phase plane through which all trajectories must pass to complete the proof. Show an accurate graph of your construction.

2. When the K^+ concentration outside a neuron gets too high, the equilibrium potential E_K becomes less negative. Prove that the resting state for $I = 0$ of eqn (9.7) can become unstable under these conditions for a range of E_K values. (Note that $E_K = -0.92$ in the equation; you must replace this with a parameter.) Simulate this situation to show that spike trains are generated. This is one possible mechanism for the inception of neural activity in visual cortex during migraine auras. What happens if E_K increases even more?

3. Prove the existence of a limit cycle for eqn (9.10) with $I = 0.2$. You should use the Poincaré–Bendixon approach. This case is a bit more complex than Exercise 1.

4. Determine the value of I for which eqn (9.10) exhibits a Hopf bifurcation. Is this a supercritical or a subcritical bifurcation?

5. The bifurcation diagram in Fig. 9.6 shows the presence of an unstable limit cycle for eqn (9.7) when $I = 0.068$. To explore further consequences of this, use program **PhaseShift.m** with ConstStim $= 0.068$ (it is 0.15 in the program). First verify that a pulse with amplitude $= 1.0$ presented at $t = 4.0$ ms, will terminate the spike train. You have just shut off all neural activity with a depolarizing pulse! Now vary the amplitude and timing of your pulse to estimate the range over which termination rather than phase shifting occurs. Also find a time and pulse strength at which a hyperpolarizing pulse (i.e. a pulse of negative amplitude) will terminate the spike train. To plot the unstable limit cycle in the phase plane, modify the script to make time run backwards by making the two time constants (Tau and TauR) negative. This will make the unstable limit cycle asymptotically stable. Also change the initial conditions so $R(0) = 0.25$, which falls within the domain of attraction of this limit cycle.

6. You might have noticed that eqn (9.3) has a factor of $(1 - R)$ in the Na^+ current to represent channel inactivation. Equation (9.7), however, does not contain this term, so it strictly represents a neuron in which Na^+ currents never inactivate. The following equations describe a neuron containing the $(1 - R)$ term:

$$0.8 \frac{dV}{dt} = -(31.6 + 85.3V + 58.0V^2)(1 - R)(V - 0.55) - 10R(V + 0.92) + I$$

$$\frac{dR}{dt} = \frac{1}{1.9}(-R + 2.2V + 1.75)$$

Give a complete analysis of these equations, including the resting state and all bifurcations produced by varying I. Be sure to indicate the nature of each bifurcation. Simulate the equations for several appropriate values of I and use this to sketch the bifurcation diagram for the system.

7. Give a complete analysis of the Morris–Lecar (1981) equations in (9.11). First determine the effects of voltage dependence of τ_R using **Morris_Lecar.m**. In particular, compare spikes with the voltage dependence in (9.11) with those produced when $\tau_R = 0.5$, i.e. constant. Use $I = 0.09$ as the stimulating current. For the remaining analysis set $\tau_R = 0.5$ for simplicity. Find and analyze the stability of all steady states for $I = 0$ (this will require the MatLab function **fzero**, see Appendix). What is the threshold value of I, and what type of bifurcation occurs at threshold?

8. Just as the Morris–Lecar equations in (9.11) have been normalized, so (9.10) describing a neocortical neuron can also be simplified to reveal its canonical (i.e. essential) dynamical structure. Consider the following equations, which obey Ohm's law with normalized equilibrium potentials of 1.0 for Na^+ and $-1/5$ for K^+:

$$\frac{dV}{dt} = -4\left(V^2 - \frac{V}{10}\right)(V - 1) - R\left(V + \frac{1}{5}\right) + I$$

$$\frac{dR}{dt} = \frac{1}{5}(-R + 3V^2)$$

These equations have been implemented in the script **CanonicalNeuron.m**. Give a complete dynamical analysis of these equations, finding all steady states and their stability when $I = 0$. Determine the value of I at the threshold for spike generation, and indicate the type of bifurcation that occurs at threshold. Now reduce the time constant of the R equation from 5 to 1.4. What happens if the equations are simulated with $I = 0$ (i.e. no external input) and initial conditions $V = 0.4$, $R = 0.3$? Why?

10   *Neural adaptation and bursting*

We have now explored the nonlinear dynamics of the Hodgkin–Huxley equations and seen how the balance between rapid depolarizing and slower hyperpolarizing currents produces limit cycles and action potentials. Furthermore, it has been shown that an additional current I_A, has the effect of greatly increasing the dynamic range of mammalian neurons by causing firing to begin at a rate of less than 1.0 Hz rather than a rate close to 200 Hz characteristic of Hodgkin–Huxley neurons. These two modes of spike initiation result from different underlying dynamics characterizing the bifurcations leading to repetitive firing: hard Hopf bifurcations in the case of Hodgkin–Huxley or Class II neurons, and a saddle-node bifurcation in the case of mammalian or Class I neurons. These are only the basics of single neuron dynamics, however, as many other currents have been identified in neurons from a wide range of species. Mammalian and human neocortical neurons, for example, are currently known to incorporate at least 12 ion currents, each of which has a somewhat different effect on action potential production (Gutnick and Crill, 1995). The detailed biophysics of these currents will not be developed here, as excellent treatments may be found elsewhere (Hille, 1992; Johnston and Wu, 1995). Rather, we shall examine the simplest dynamical models that can explain two rather complex neural phenomena: spike rate adaptation and bursting. This will require incorporation of one or two additional ionic currents into the basic models from the last chapter, each described by an additional differential equation. This will ultimately lead to a mathematical taxonomy of single-neuron dynamics as currently understood.

10.1   Spike frequency adaptation

Human neocortical neurons incorporate an I_A current, and we saw in the last chapter that eqn (9.10) provided an accurate description of the shape of action potentials of regular-spiking cells (see Fig. 9.11) and also of fast-spiking neurons (by reducing the time constant τ_R. Regular-spiking neurons have one other dramatic dynamical characteristic lacking in fast-spiking neurons: the spike rate rapidly decreases or adapts during continued stimulation (Connors and Gutnick, 1990; Gutnick and Crill, 1995). This results from activation of a very slow hyperpolarizing potassium current, which is generally mediated by an influx of calcium (Ca^{2+}) ions. This is frequently termed an afterhyperpolarizing potential or I_{AHP}. Data from a human regular spiking neuron (Lorenzon and Foehring, 1992) showing spike frequency adaptation are plotted in Fig. 10.1C. Adaptation similar to this was incorporated into models of short-term memory and neural competition in Chapters 6 and 8.

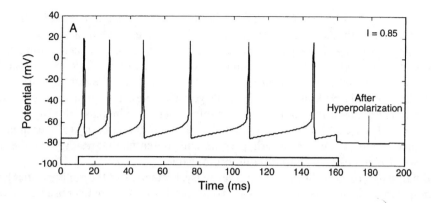

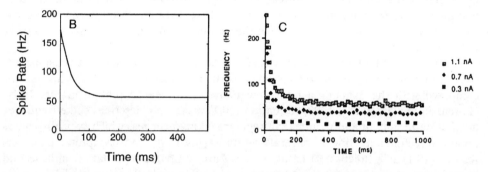

Fig. 10.1 Spike frequency adaptation in a regular-spiking (excitatory) cortical neuron. The spike train produced by (10.1) is plotted in A for $I = 0.85$. Spike rate as a function of time for (10.1) is plotted in B for $I = 1.8$. Comparable spike rate adaptation data for a human regular-spiking neuron (Lorenzon and Foehring, 1992) are plotted in C.

The dynamical effects of an I_{AHP} current can be produced by adding an additional K$^+$ current to eqn (9.10) that is governed by the conductance variable H:

$$\frac{dV}{dt} = -\{17.81 + 47.58V + 33.8V^2\}(V - 0.48)$$
$$- 26R(V + 0.95) - 13H(V + 0.95) + I$$
$$\frac{dR}{dt} = \frac{1}{5.6}\left(-R + 1.29V + 0.79 + 3.3(V + 0.38)^2\right) \qquad (10.1)$$
$$\frac{dH}{dt} = \frac{1}{99.0}(-H + 11(V + 0.754)(V + 0.69))$$

The most obvious point about this potential is that the time constant for H is 99 ms, almost 20 times slower than R and 100 times slower than V. Also note that the reversal potential for this current is $V = -0.95$ (i.e. -95 mV), the same reversal potential as for K$^+$ in the recovery current governed by R. Finally, we have already seen that $V = -0.754$ in the resting state of (9.10) when $I = 0$, and (10.1) has been constructed so that $H = 0$ in this resting state. Thus, the I_{AHP} conductance H only has an effect at suprathreshold levels, where it grows quadratically with V. Because the H variable has such a long time-constant, it has no effect on action potential shape but instead functions to slowly counteract part of the stimulating current I, thus reducing the spike rate. If you run the MatLab script **RegularSpiking.m** with $I = 0.85$, you will obtain the spike train depicted in Fig. 10.1A. The I_{AHP} current causes the interspike interval to increase almost threefold within 100 ms. Figure 10.1A also shows that after the stimulating current I is terminated, the neuron displays a prominent afterhyperpolarizing potential that results from the very slow decay of H in (10.1) back to its equilibrium value.

The analysis of (10.1) via linearization is very easy, because all of the nonlinearities have been limited to cubic order, so the Jacobian is only quadratic in V (see Exercise 1). Because $H = 0$ at threshold and its time constant is so long, action potential initiation in (10.1) again begins at a saddle-node bifurcation as depicted in Fig. 9.9. Because of its accelerating quadratic dependence on V, H is mainly driven by suprathreshold current levels, so it slowly reduces the effective value of the simulating current I, thus producing spike frequency adaptation.

The spike rate adaptation produced by H in (10.1) is plotted as instantaneous spike rate (reciprocals of successive interspike intervals) in Fig. 10.1B for $I = 1.8$. This may be compared with the data for a human regular-spiking neuron plotted in Fig. 10.1C (Lorenzon and Foehring, 1992). Equation (10.1) accurately describes spike frequency adaptation in human and mammalian regular-spiking neurons, which typically causes about a threefold reduction in spike rate within 100 ms. Initial and asymptotic spike rates for eqn (10.1) as a function of input current I are compared with data from a second human regular-spiking neuron (Avoli *et al.*, 1994) in Fig. 10.2. The model correctly predicts that both spike rates will be almost linear functions of I above a threshold value (Wilson, 1999), although additional factors (V dependence of τ_R) are necessary to explain saturation effects at high current levels.

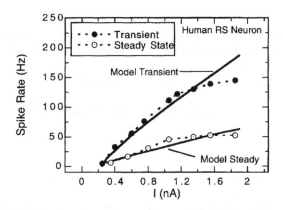

Fig. 10.2 Initial transient and steady state spike rates for a regular-spiking neuron as a function of stimulus current I. Solid and open circles show transient and sustained data for a human regular-spiking neuron respectively (reproduced with permission, Avoli *et al.*, 1994, copyright Springer-Verlag). Solid lines show transient and steady state responses of (10.1).

We have now developed and analyzed dynamical models of two types of human cortical neurons: the excitatory regular-spiking neurons in (10.1), and the inhibitory fast-spiking neurons in (9.10) with $\tau_R = 2.1$ ms. It is significant that only regular-spiking neurons and not fast-spiking exhibit spike frequency adaptation. Furthermore, the I_{AHP} current can be reduced or blocked in humans by acetylcholine, histamine, norepinephrine, or serotonin, all of which are modulatory neurotransmitters in the nervous system (McCormick and Williamson, 1989). Thus, the excitability of regular-spiking neurons can be controlled by varying the strength of the I_{AHP} current in eqn (10.1). Inhibitory, fast-spiking neurons, however, are always in a maximally excitable state, presumably so that they can prevent runaway excitation and seizures.

10.2 Neural bursting and hysteresis

Slow afterhyperpolarizing I_{AHP} potentials provide the basis for spike frequency adaptation, and they can also produce more dramatic patterns of neural activity. Chay and Keizer (1983) developed a simple model for the action potentials produced by β cells in the pancreas by adding a slow hyperpolarizing potential to the Hodgkin–Huxley equations. Although their parameters were adapted to the pancreas, a neural version of their model can be produced by including a hyperpolarizing potential in our Hodgkin–Huxley model:

$$0.8 \frac{dV}{dt} = -(17.81 + 47.71V + 32.63V^2)(V - 0.55)$$

$$- 26.0R(V + 0.92) - 0.54H(V + 0.92) + I$$

$$\frac{dR}{dt} = \frac{1}{1.9}(-R + 1.35V + 1.03) \tag{10.2}$$

$$\frac{dH}{dt} = \frac{1}{250}(-H + 9.3(V + 0.70))$$

This is just eqn (9.7) with the addition of an H-mediated hyperpolarizing current having a reversal potential equal to that of K^+ (−0.92 or −92 mV). Because $V = -0.70$ when $I = 0$ in the resting state, (10.2) has been designed so that $H = 0$ as well. The H variable is very slow here with a time constant of 250 ms. To see the effects of this slow I_{AHP} current on a Hodgkin–Huxley-type neuron, you can run the MatLab script **HHburster.m** with $I = 0.14$. The resulting spike train is shown in Fig. 10.3A: following a prolonged initial burst of spikes, (10.2) fires bursts of six spikes each about once every 200 ms. As shown in the expanded plot of a single burst (Fig. 10.3B), each burst is preceded and followed by a rapid subthreshold membrane potential oscillation, and the spikes within each burst occur at a constant spike rate of roughly 175 Hz. Some other very interesting bursting patterns occur within the range $0.12 \le I \le 0.25$, both very regular patterns like Fig. 10.3 and with some surprisingly irregular ones which, as we shall see in the next chapter, can be chaotic.

The astute reader may already have guessed the dynamical basis of the bursting produced by (10.2). Recall first that eqn (9.7), from which (10.2) was derived, exhibits hysteresis because there is a range of inputs I over which a spike-generating limit cycle coexists

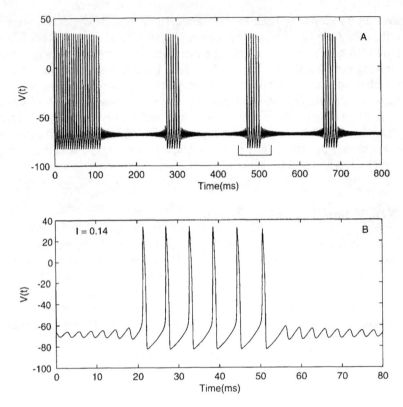

Fig. 10.3 Bursting produced by hysteresis in (10.2). The time scale in B is expanded to show details of the burst in A above the bracket at 500 ms.

with an asymptotically stable resting state. This is shown in the bifurcation diagram in Fig. 9.6. The effect of the very slow H equation in (10.2) is to sweep the effective stimulating current I back through the hysteresis region determined by the V and R equations. Thus, upon initial suprathreshold stimulation, spikes are generated by a limit cycle at a fairly high rate. The depolarizations produced by these spikes, in turn, cause a slow increase in the H-mediated hyperpolarizing current, which effectively reduces the net stimulus current until firing ceases. As the neuron is then in a resting state, H will slowly decay to near zero, thus removing the hyperpolarization. At some point, the hyperpolarizing current will be low enough so that the stimulus current will again initiate firing. This is manifested in a series of bursts separated by periods when the neuron is hyperpolarized below its threshold.

These periods between successive bursts vary enormously with I. For $I = 0.1$, bursts of four spikes occur at about 1.3 s intervals. For I sufficiently large, however, all bursting disappears, because the I_{AHP} current is no longer strong enough to hyperpolarize the neuron to a level below the point where spiking ceases. Thus, a slow hyperpolarizing current can produce neuronal bursting via a hysteresis loop when it is incorporated into a

Class II neuron model exhibiting a hard Hopf bifurcation as the resting state becomes unstable.

10.3 Calcium currents and parabolic bursting

The previous example of neural bursting is the simplest type to analyze mathematically, because it involves just one additional current that drives the spike-generating apparatus through a hysteresis loop. However, most bursting neurons are driven by a more complex combination of ionic currents. The most important of these are neurons that incorporate both a transient inward Ca^{2+} (calcium) current I_T, which depolarizes the neuron, and an I_{AHP} hyperpolarizing current that is mediated by Ca^{2+}. Several examples of neural bursting thought to be driven by the combination of I_T and I_{AHP} currents are illustrated in Fig. 10.4. The model presented here is a modified version of the Plant (1981) model for bursting in aplysia neurons (Fig. 10.4A) that has been subjected to detailed mathematical analysis by Rinzel and Lee (1987). Let us adapt eqn (9.10) by adding two additional currents controlled by the conductance variables X and C:

$$\frac{dV}{dt} = -\{17.81 + 47.58V + 33.8V^2\}(V - 0.48)$$
$$- 26R(V + 0.95) - 1.93X(1 - 0.5C)(V - 1.4) - 3.25C(V + 0.95)$$
$$\frac{dR}{dt} = \frac{1}{5.6}\left(-R + 1.29V + 0.79 + 3.3(V + 0.38)^2\right) \qquad (10.3)$$
$$\frac{dX}{dt} = \frac{1}{30}(-X + 7.33(V + 0.86)(V + 0.84))$$
$$\frac{dC}{dt} = \frac{1}{100}(-C + 3X)$$

The variable X represents the Ca^{2+} conductance, which is a quadratically accelerating function of V at subthreshold levels. Because the equilibrium potential for Ca^{2+} is about $+1.4$ (i.e. $+140$ mV), the X variable controls a depolarizing current that affects V through the term $1.93X(1 - 0.5C)(V - 1.4)$. The factor $(1 - 0.5C)$ in this expression, which is not present in the original Plant (1981) model, reflects evidence that I_T currents inactivate (Hille, 1992; Johnston and Wu, 1995). This method of approximating inactivation was developed by Rinzel (1985) to simplify the Hodgkin–Huxley equations by letting $h = 1 - n$ and was discussed in deriving (9.3) in the previous chapter. As will be seen, the current controlled by X triggers action potential bursts.

The second effect of X is to increase the concentration of internal Ca^{2+} represented by the variable C, by providing the input to the dC/dt equation. C in turn is the very slow (time constant of 100 ms) Ca^{2+}-modulated conductance of a separate K^+ channel, which is represented in the V equation by the term $-3.25C(V + 0.95)$, the equilibrium potential of K^+ being chosen as -0.95 (-95 mV). This is a more accurate description of an I_{AHP} current that is modulated by Ca^{2+}. The dC/dt equation in (10.3) lacks the V dependence of the comparable equation in the Plant (1981) model, because of evidence that most mammalian I_{AHP} currents are not voltage gated (Johnston and Wu, 1995).

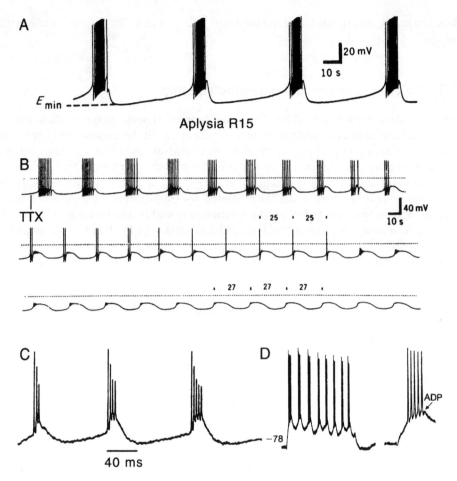

Fig. 10.4 Examples of bursting neurons. Endogenous bursting of aplysia neuron R15 is shown in A. In B a second aplysia bursting neuron had Na^+ channels blocked by tetrodotoxin (TTX) causing a reduction and then cessation of action potentials, but Ca^{2+}-mediated membrane potential oscillations persisted (A and B (reproduced with permission) from Mathieu and Roberge, 1971). Mammalian neocortical neurons that burst in response to constant stimulation are illustrated in C (mouse somatosensory cortex, Agmon and Connors, 1989) and D (chattering cell in cat visual cortex, Gray and McCormick, 1996). Each burst in these mammalian cells is typically comprised of 2–5 spikes.

Equation (10.3) is relatively difficult to analyze (see Rinzel and Lee's (1987) analysis of the Plant model), but a lot can still be learned by linearizing and using a Hopf bifurcation analysis as outlined here. Recall first that eqn (9.10), from which (10.3) was derived, has the three steady states depicted in Fig. 9.9 when $I = 0$, so spike generation in (9.10) begins at a saddle-node bifurcation. There is no input current in eqn (10.3), but the X-modulated current produces a slow depolarization of the membrane which ultimately leads to burst firing by driving the neuron through a bifurcation. Once bursting begins, the X-controlled C conductance causes a very slow hyperpolarizing current to build up and also inactivates the X current until the burst is terminated. This sequence of events then repeats itself as a complicated limit cycle in four dimensions.

Running MatLab script **PlantBurster.m** will produce the bursting pattern shown in Fig. 10.5A,B. This may be compared with the response of an aplysia (an invertebrate) R15 neuron, in Fig. 10.4A,B. These neurons can be identified in all aplysias, and the burst

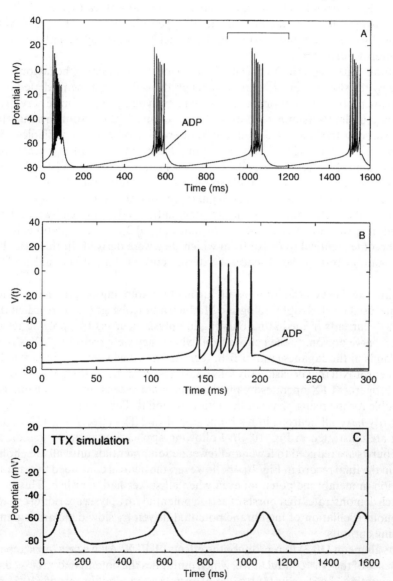

Fig. 10.5 Responses of endogenous bursting neuron described by (10.3). B shows an expansion of the burst near 1000 ms bracketed in A. Mathematical simulation of TTX application to (10.3) is shown in C. Compare to Fig. 10.4A,B.

pattern is common across individuals. (Note that model time constants have not been scaled to fit the very long time-course of the aplysia neuron.) Each burst is followed by an afterdepolarizing potential (ADP), which has been observed in bursting neurons of many species (see Fig. 10.4). Mathematically, depolarizing afterpotentials occur because each spike is followed by a rapid hyperpolarization caused by the R variable mediated K^+ current, yet the slow depolarizing current caused by X has not yet decayed away fully after the end of the R hyperpolarization. Thus, the ADP is not produced by a separate ionic current in the model but rather by a combination of fast R and slower X currents.

Equation (10.3) describes an **endogenous burster**, as bursting behavior is triggered entirely by the slow inward I_T current through the cell membrane mediated by X. Such neurons are particularly important as pacemakers in many invertebrate nervous systems. The program **PlantBurster.m** plots an X–C projection of the four-dimensional state space of (10.3) to show the limit cycle produced by interaction of these two variables. The role of X can be analyzed as follows. Solving for the steady states of (10.3) reveals only one in the physiological range at $V = -0.727$, $X = 0.11$, $R = 0.25$, and $C = 0.33$, and linear analysis shows that this is an unstable spiral point. This does not, of course, demonstrate the existence of a limit cycle, but it does guarantee that the system will not remain at rest. If, however, the coefficient 7.33 in the dX/dt equation were replaced by zero, both X and C would be identically zero at all steady states of (10.3), so the equations for V and R would become identical to (9.10) from which they were derived. In this case the system must remain at rest in the absence of external current, so (10.3) could no longer fire bursts.

We can take these considerations still further by employing a mathematical analog of a technique used by electrophysiologists. Mathieu and Roberge (1971) revealed the role of slow Ca^{2+} currents in mediating bursting in aplysia neurons by applying tetrodotoxin (TTX), a nerve poison, to the neuron from which they were recording. TTX is extracted from glands of the Japanese puffer fish, regarded as a sushi delicacy in Japan, but it is a deadly neurotoxin that occasionally kills gourmets when the sushi chef fails to remove the glands properly. TTX operates by blocking the voltage-sensitive Na^+ channels that are responsible for the rising phase of the action potential. Thus, someone who ingests TTX will shortly have all neurons in his brain stop firing! The effects of TTX on an aplysia neuron are illustrated in Fig. 10.4B. Following application at the time indicated, successive bursts are reduced to fewer and fewer action potentials until all are abolished. As shown in the final record in Fig. 10.4B, however, the neuron continued to generate a slow oscillation in membrane potential even when all spikes had vanished. This biophysical approach demonstrates that bursts of action potentials in aplysia are riding on crests of an endogenous oscillation of membrane potential driven by slow depolarizing and hyperpolarizing currents.

Let us alter eqn (10.3) to reproduce the effects of TTX. All we need do is replace the V dependence of the Na^+ conductance, which multiplies the potential term $(V - 0.48)$ in the dV/dt equation, by its constant value at rest. As the equilibrium of (10.3) occurs at $V = -0.727$, let us evaluate the Na^+ conductance at this point where its value is found to be 1.0836. This represents a passive leakage conductance at rest that was explicit in eqns (9.1) to (9.3) but was absorbed in subsequent analysis. This changes the first equation in (10.3) to:

$$\frac{dV}{dt} = -1.0836(V - 0.48) - 26R(V + 0.95)$$
$$- 1.93X(1 - 0.5C)(V - 1.4) - 3.25C(V + 0.95) \tag{10.4}$$

In this equation the active aspect of the Na^+ conductance has been blocked with a mathematical application of TTX, thus leaving only the passive properties of the membrane with respect to Na^+. The effects of incorporating (10.4) into (10.3) may be observed by running the script **PlantBursterTTX.m**. This produces the voltage oscillations in Fig. 10.5C, which may be compared with the aplysia TTX data in Fig. 10.4B. Clearly, our mathematical application of TTX has abolished all spiking activity. In doing so, however, it has revealed an intrinsic limit cycle driven by the X and C variables. The program shows this by plotting the X–C projection of the four-dimensional phase space of the system. One can now complement this simulation with a mathematical analysis of the TTX poisoned system in (10.4). If the coefficient 7.33 in the dX/dt equation in (10.3) is treated as a variable parameter, it cane shown by linearized analysis that the system undergoes a Hopf bifurcation to a limit cycle as this coefficient increases from zero. This demonstrates that the slow depolarizing Ca^{2+} current mediated by X is responsible both for the intrinsic membrane oscillations in the TTX simulation and for the bursting exhibited by (10.3).

Before leaving the Plant (1981) model for bursting in aplysia, let us consider one final aspect of the spike bursts. A close examination of Fig. 10.4A reveals that each burst begins and ends with interspike intervals that are longer than those in the middle of the burst. Thus, the instantaneous spike frequency in these bursts begins low, climbs to a peak, and then decays before the burst is finally extinguished by hyperpolarization. This is known as **parabolic bursting** because the spike rate approximates a parabola during each burst. Rinzel and Lee (1987) have shown that the Plant (1981) equations will produce parabolic bursting if the parameters are modified slightly. To see this, let us increase the time constants for X and C in (10.3) to 125 ms and 300 ms respectively. This will slow the endogenous membrane oscillation significantly, thus producing bursts with many more spikes each. With these modifications eqn (10.3) produces bursting like that shown in Fig. 10.6, which was obtained from a 2500 ms simulation using **ParabolicBurster.m**. The instantaneous spike frequency during the burst is plotted below and shows the characteristic parabolic shape of the burst. Thus, parabolic bursting represents a variation on the theme of interacting I_T and I_{AHP} currents.

10.4 Neocortical bursting

Although bursting neurons were first discovered in invertebrates, such as aplysia, similar neurons have more recently been discovered in mammalian neocortex. For example, Fig. 10.4C shows activity from a bursting neuron in mouse somatosensory cortex (Agmon and Connors, 1989), while Fig. 10.4D shows responses of a 'chattering' cell in cat visual cortex (Gray and McCormick, 1996). In both these cases, the bursting is not endogenous but instead must be driven by a constant stimulating current injected into the cell by the

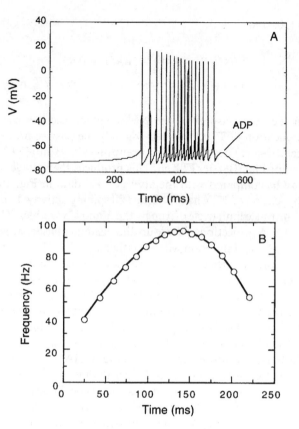

Fig. 10.6 Parabolic burst of spikes in A and spike rate in B (reciprocal of interspike intervals) from (10.3) with modified time constants described in the text.

experimenter. However, the bursting still reflects intrinsic membrane properties of these cells rather than network properties.

As an interplay of I_T and I_{AHP} currents also provides the basis for bursting in mammalian neurons, let us explore a model for them that is derived from (9.10) and is similar in form to eqn (10.3). The equations are:

$$\frac{dV}{dt} = -\{17.81 + 47.58V + 33.8V^2\}(V - 0.48) - 26R(V + 0.95)$$
$$- 1.7X(V - 1.4) - 13C(V + 0.95) + I$$
$$\frac{dR}{dt} = \frac{1}{2.1}\left(-R + 1.29V + 0.79 + 3.3(V + 0.38)^2\right) \qquad (10.5)$$
$$\frac{dX}{dt} = \frac{1}{15}(-X + 9(V + 0.754)(V + 0.7))$$
$$\frac{dC}{dt} = \frac{1}{56}(-C + 3X)$$

The R variable now has the faster 2.1 ms time constant characteristic of fast-spiking cells (see Fig. 9.11), because chattering cells have a much narrower spike than regular-spiking neurons (Gray and McCormick, 1996). The key difference between (10.5) and (10.3), however, is found in the dX/dt equation. As it was shown in the last chapter that $V = -0.754$ at equilibrium in (9.10), the factor $(V + 0.754)$ in the dX/dt equation guarantees that $X = 0$ and therefore also $C = 0$ in the absence of an external current I. Therefore, eqn (10.5) will remain in the resting state with $V = -0.754$ when $I = 0$, guaranteeing that (10.5) cannot generate endogenous bursts.

Let us study eqn (10.5) using a combination of analytical methods and simulations. The isocline equations are easily obtained, and substitution then leads to the following equation for the equilibrium points:

$$485.9V^3 + 1056.24V^2 + 768.736V + 187.424 - I = 0 \tag{10.6}$$

Six-figure precision in this equation is necessary in order to obtain accurate roots. Differentiation of (10.5) then leads to the following expression for the Jacobian, where V must be evaluated at the steady state:

$$\overleftrightarrow{A} = \begin{pmatrix} a_{11} & -26(V + 0.95) & -1.7(V - 1.4) & -13(V + 0.95) \\ 3.1429V + 1.8086 & -0.47619 & 0 & 0 \\ \dfrac{6}{5}V + 0.8724 & 0 & \dfrac{-1}{15} & 0 \\ 0 & 0 & \dfrac{3}{56} & \dfrac{-1}{56} \end{pmatrix}$$

$$a_{11} = -553.5V^2 - 694.06V - 221.23$$

$$\tag{10.7}$$

The first entry a_{11} has been written separately, as it will be altered later. It should be emphasized that while the steps involved in obtaining (10.6) and (10.7) only involve simple algebra and differentiation, this is a case where the reliability of a symbolic mathematics program such as MathView, Maple, or Mathematica is extremely helpful in avoiding errors.

If you solve (10.6) for $I = 0$ using the MatLab **roots** function and then analyze the stability of the only equilibrium point, which occurs at $V = -0.754$, you will find that it is an asymptotically stable node. As (10.5) is a four-dimensional system in which we seek spike-producing limit cycles, it is natural to look for a Hopf bifurcation, and the input current I is the natural variable to choose for the analysis. This is a case in which the problem can be solved in reverse: first V is treated as a parameter in (10.7) and the Routh–Hurwitz criterion is used to determine the value of V at which two roots are pure imaginary. Equation (10.7) for A has been entered into the function **ChatteringHopf.m**, and this can now be solved by running the script **R_Hchatter.m** which gives $V = -0.7017$ as the bifurcation value. (Note that there are two bifurcation points here, so the one obtained

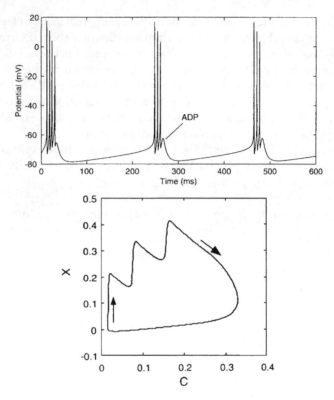

Fig. 10.7 Bursting behavior of (10.5) for $I = 0.3$. Spike bursts in A with prominent ADP are comparable to data in Fig. 10.4C,D. The X–C projection of the four-dimensional state space is plotted in B to illustrate how these variables drive the bursting limit cycle.

depends on the initial guess.) The program finds that the four eigenvalues are $\lambda = -7.14$, -0.16, and $\pm 0.0453i$. Thus, the system satisfies the Hopf Bifurcation Theorem 11, and the emergent limit cycle of (10.5) begins with frequency $(0.0453/2\pi) \times 1000 \text{ ms/s} = 7.21$ Hz. Substitution of the bifurcation value $V = -0.7017$ back into (10.6) now shows that $I = 0.197$ nA is the input current at which the bifurcation occurs.

   This mathematical analysis can be corroborated by running the script **Chattering.m** with a range of I values in the vicinity of $I = 0.197$. If you then run **Chattering.m** with $I = 0.3$, you will obtain the sequence of bursts plotted in Fig. 10.7. As you can see, this has the characteristics of spike trains from the chattering cell and the mouse somatosensory neuron shown in Fig. 10.4C,D. In particular, spike height decreases after the first spike in the burst, there are a small number of spikes per burst, and each burst is terminated with an afterdepolarization (ADP). It is interesting to explore the effect of different current intensities on both the burst frequency and the number of spikes per burst.

   This mathematical analysis of eqn (10.5) for a chattering cell reveals that the system will undergo a Hopf bifurcation with an initial frequency of 7.21 Hz when $I = 0.197$. To complete analysis of (10.5), it is necessary to demonstrate that these slow membrane potential oscillations above the Hopf bifurcation achieve sufficient amplitude to push the spike-generating apparatus in the fast V and R equations back and forth through their saddle-node bifurcation. Demonstrating this analytically is extremely difficult, so let us be content to show that it does indeed occur using the simulation of (10.5) in **Chattering.m**. The X–C projection of the four-dimensional state space plotted in Fig. 10.7B for $I = 0.3$ shows that these two variables do indeed produce a limit cycle that drives the V and R equations through their saddle-node bifurcation into a spiking regime and back again. Note that the X–C oscillation is coupled to spike generation via V so this is not a simple example of one autonomous limit cycle driving a second limit cycle. This is why the mathematical analysis is complex.

Our previous analysis of the Plant (1981) burster and the physiology of aplysia neurons showed that the Ca^{2+} oscillations in V could be revealed using a mathematical version of TTX application, and the same is true for the model in (10.5). In this case analysis of simulated TTX poisoning of the Na^+ conductance in (10.5) is developed in Exercise 2, where you will see that slow membrane oscillations are indeed revealed.

10.5 A dynamical taxonomy of neurons

Our dynamical analysis has now covered the range of behaviors discovered in different nerve cells from a wide range of species. Furthermore, this analysis enables us to specify which aspects of membrane biophysics make each cell type unique. This section provides a brief summary of the dynamical taxonomy that emerges from analysis of mammalian neocortical neurons and from bursting.

The mammalian neocortex contains four classes of neurons differentiated with respect to their spiking responses to sustained intracellular current injection: regular-spiking (RS) cells, fast-spiking (FS) cells, bursting or chattering cells, and intrinsic-bursting (IB) cells (Connors and Gutnick, 1990; Gutnick and Crill, 1995; Gray and McCormick, 1996). FS neurons are the simplest of these dynamically, as they can be described by the two equations in (9.10) with $\tau_R = 2.1$ ms. These inhibitory interneurons in the cortex fire at high rates due to their narrow action potentials and their lack of adaptation. RS neurons are the next more complex cells, as they incorporate an I_{AHP} potential that produces spike frequency adaptation and a longer time-constant τ_R that results in broader action potentials (see Fig. 9.11). The dynamical properties of RS neurons therefore require at least the three equations in (10.1). Neocortical bursting cells (see Fig. 10.4C,D) are the most complex dynamically, as they can only be described by four-dimensional dynamics incorporating depolarizing I_T and hyperpolarizing I_{AHP} currents as described by eqn (10.5). The final cortical neuron type, the IB cell, has dynamical properties intermediate between RS and chattering neurons and can be generated from (10.5) by modest parameter changes. Our models for these neocortical neuron classes are closely related dynamically, as all share the same V and R equations (except for the time constant τ_R). They are differentiated by the presence of zero (FS), one (RS), or two (bursting) additional currents that generate the patterning of the spike trains. While neocortical neurons do possess about 12 different ionic currents (Gutnick and Crill, 1995; McCormick, 1998), the essential nonlinear dynamics can be captured in four dimensions.

A taxonomy of bursting neurons was first proposed by Rinzel (1987). Following his classification and that of Bertram *et al.* (1995), the simplest type of bursting is represented by the recurrent excitatory–inhibitory network oscillator in eqn (8.2) (Wilson and Cowan, 1972). In this case bursting is represented by a spike rate that is periodically terminated by inhibitory feedback. A second example of this simple bursting theme is embodied in (8.20) which describes neural competition with adaptation. The next more complicated burster is the Chay and Keizer (1983) model described by the three equations in (10.2). The most complicated bursting neurons discovered thus far are the parabolic bursters exemplified by the Plant (1981) model in (10.3). As we have seen, parabolic bursting requires two fast dynamical variables, V and R, to describe action potential generation plus two slower variables, X and C, to describe the Ca^{2+}-mediated I_T and I_{AHP} currents that produce endogenous membrane potential oscillations even in the presence of TTX. Thus, we have explored the taxonomy of two-, three-, and four-dimensional bursting behaviors proposed by Rinzel (1987). In subsequent chapters it will be shown that networks incorporating neural bursting are extremely important in many forms of motor control.

10.6 Exercises

1. As indicated in the chapter, the excitability of regular-spiking neurons can be controlled by various modulatory neurotransmitters. To explore the range of their possible effects, consider the following modification of eqn (10.1):

$$\frac{dV}{dt} = -\{17.81 + 47.58V + 33.8V^2\}(V - 0.48) - 26R(V + 0.95)$$
$$- gH(V + 0.95) + I$$
$$\frac{dR}{dt} = \frac{1}{5.6}\left(-R + 1.29V + 0.79 + 3.3(V + 0.38)^2\right)$$
$$\frac{dH}{dt} = \frac{1}{99.0}(-H + 11(V + 0.754)(V + 0.69))$$

where the strength of the H-mediated current in the first equation is determined by the parameter g, which is assumed to be under neuromodulatory control. (a) By suitably modifying **Regularspiking.m**, compare the asymptotic spike rates (simulate for 300 ms) for $g = 0, 6$, and 18. For each value of g, obtain the rate for $I = 0.5, 1.0$, and 1.5, and plot all your spike rates on a single graph as a function of I. (b) Prove that there exists a value of g for which these equations will only produce a finite number of spikes and then cease firing for $I \leq 2$. Be sure to indicate your reasoning. (Hint: solve for the value of g for which the steady state can never reach the value of V defined by the threshold bifurcation. Then, prove that this steady state is asymptotically stable.)

2. The simulation of (10.5) in Fig. 10.7 shows bursts of three spikes separated by about 225 ms. Using the simulation in **Chattering.m**, obtain and plot both the spike frequency within each burst and the burst frequency for input currents I from threshold up to $I = 1.5$. Suggest an explanation for the pattern of results you find.

3. Hindmarsh and Rose (1984) have proposed a model of neural bursting using three differential equations. A slightly modified version of their model is:

$$\frac{dV}{dt} = -V^3 + 3V^2 + 0.125 - R - H + I$$

$$\frac{dR}{dt} = -R + 5V^2 - 1$$

$$\frac{dH}{dt} = \frac{1}{250}(-H + 2(V + 1.5))$$

Give an analysis of the neural bursting produced by these equations. In particular: (a) set $H = 0$ and analyze the steady states and solutions of the V and R equations for $I = 0$. (b) Determine the response when $H = 0$ and the stimulus is a 25 ms pulse with amplitude 0.6. Such a pulse is easily produced in MatLab by setting $I = 0.6 \times$ (Tme ≤ 25). (c) Using what you have now learned about the solutions, make a graph of the two-dimensional state space for $H = 0$ with isoclines and show plots of any limit cycle trajectories. (d) Simulate all three equations for constant stimulation $I = 0.6$ and plot the spike trains to show that bursting does occur. Provide an explanation of the bursting, being sure to indicate what type of bifurcation to bursting is involved. This is a further type of neural bursting (Bertram *et al.*, 1995).

4. In this problem you will predict the results of applying TTX to the neocortical bursting model in (10.5). On the assumption that TTX poisons the active properties of the Na$^+$ channels, replace the first of the four equations in the model by:

$$\frac{dV}{dt} = -\{17.81 + 47.58V_{rest} + 33.8V_{rest}^2\}(V - 0.48) - 26R(V + 0.95)$$
$$- 1.7X(V - 1.4) - 13C(V + 0.95) + I$$

where $V_{rest} = -0.754$. (a) Calculate the altered Jacobian and analyze all equilibrium states and their stability for $I = 0$. (b) Prove that there is a Hopf bifurcation for V near the resting value and determine the values of V and I at this bifurcation. (Note that there will be more than one bifurcation point; you must find the appropriate one.) (c) Simulate your TTX equations and plot $V(t)$ for I just above the bifurcation value. Compare the oscillation frequency predicted by the Hopf theorem with your results.

5. The endogenous bursting model in (10.3) simulates both inactivation of I_T and I_{AHP} current to stop bursts. However, bursting can be produced by an inactivating I_T current alone. To examine this, modify the first equation in (10.3) to:

$$\frac{dV}{dt} = -\{17.81 + 47.58V + 33.8V^2\}(V - 0.48) - 26R(V + 0.95)$$
$$- 2X(1 - C)(V - 1.4)$$

The $(1 - C)$ factor here represents full inactivation of the I_T current. Simulate this system by modifying **PlantBurster.m** appropriately. Next, solve for the equilibrium state, determine the Jacobian, and prove that the equilibrium is unstable. Finally, simulate the effects of TTX application by replacing the voltage dependence of the Na$^+$ conductance in the first bracket above with its evaluation near equilibrium:

$$\{1.065\}$$

Show by simulation that the resulting equations still generate a limit cycle.

6. Addition of I_T and I_{AHP} currents to the Morris–Lecar equations (9.11) can also produce a model for cortical bursting. Consider the following equations:

$$\frac{dV}{dt} = -g(V)(V - 1.0) - 2R(V + 0.7) - 0.5(V + 0.5)$$
$$- 0.1X(V - 2.0) - C(V + 0.7) + I$$

$$\frac{dR}{dt} = 0.4(-R + G(V))$$

$$\frac{dX}{dt} = \frac{1}{10}\left(-X + \frac{1}{1 + \exp(-(V + 0.03)/0.01)}\right)$$

$$\frac{dC}{dt} = \frac{1}{50}(-C + X)$$

$$g(V) = \frac{1}{1 + \exp(-(V + 0.01)/0.075)}$$

$$G(V) = \frac{1}{1 + \exp(-(V - 0.1)/0.07)}$$

where the time constant in the dR/dt equation has been made independent of V for simplicity. The X and C mediated currents here play the same role as in (10.3) or (10.5). Using the simulation **MLburster.m**, determine the threshold current I for producing bursts. Next, simulate the effects of TTX by setting $g(V) = 0$, and again determine the threshold for producing an oscillation. What type of bifurcation appears to be involved in the initiation of bursting based on your simulations? Why?

11 *Neural chaos*

As we have developed more background in nonlinear dynamics, we have dealt with increasingly more complex neural systems. In all cases, however, system trajectories have either approached an asymptotically stable steady state or an asymptotically stable limit cycle. Nonlinear dynamical systems may, of course, have several steady states and several limit cycles, which leads to a rich and complex mix of dynamical possibilities. Various neurophysiological applications of these nonlinear dynamical principles have provided models of short-term memory, decision making, action potential generation, and bursting. One might conjecture that we have exhausted the range of dynamical possibilities available to the nervous system and that all complex neural phenomena are emergent properties determined by combinations of these elementary nonlinear phenomena. While there is no doubt that a large range of brain functions can indeed be understood in terms of the nonlinear phenomena studied thus far, we have not exhausted the range of surprising behaviors that nonlinear dynamics has to offer. In this chapter, deterministic, nonlinear systems will be shown to exhibit seemingly unpredictable behavior known as chaos. Neural chaos is a source of important limitations on scientific prediction and even on our own self-knowledge!

11.1 Defining chaos

To motivate this chapter, it is helpful to recall the Poincaré–Bendixon theorem (Theorem 10). In essence, this theorem says that if an annulus can be constructed in a two-dimensional system such that all trajectories enter it, yet it contains no steady states, then a limit cycle must exist within the annulus. Why is this theorem restricted to two-dimensional systems? Advanced mathematics tells us that the topological generalization of an annulus to three dimensions is a torus or 'doughnut', and this concept readily generalizes to higher dimensions. Why cannot the Poincaré–Bendixon theorem be extended to higher dimensions by substituting a region bounded by a torus for the two-dimensional annulus? The key conceptual reason is that in an autonomous system, trajectories can never intersect and cross, and this provides a definitive constraint in two dimensions. However, trajectories in a system with more than two dimensions can pass by without intersecting in an infinite number of ways. Some of these involve complex periodicities or almost periodic behavior (**quasiperiodic trajectories**), while others can be aperiodic and chaotic.

In developing a definition of chaos, let us consider the various possibilities for a trajectory confined to a torus in a higher dimensional system. Assume that the torus contains no equilibrium points, so the trajectory can never come to rest. The fact that the trajectory is confined to a torus means that it must remain bounded for all future time, and we shall see that this is one crucial element in the definition of chaos. The simplest possibility is that the trajectory will approach an asymptotically stable limit cycle, a scenario studied extensively in the past three chapters.

As a second possibility, consider a four-dimensional linear system. Without bothering to write down the equations, let us assume that the eigenvalues of the system are all pure imaginary. Suppose, for example, that they are $\lambda = \pm 2i$ and $\lambda = \pm 5i$. All four variables describing any particular trajectory will now have the form:

$$X_i = A\sin(2t) + B\cos(2t) + C\sin(5t) + D\cos(5t) \tag{11.1}$$

Thus, all variables will be sums of two oscillations, with the parameters A–D being determined by initial conditions. In this case, no matter what the values of A–D may be, every solution is a closed trajectory with frequency $2\pi/10$. This is an oscillation (but not a limit cycle), and it simply represents a coiled spring spiraling around a torus in the four-dimensional state space of the system and connecting up with itself.

Suppose now that a parameter in the four-dimensional system that gave rise to solution (11.1) is changed so that the eigenvalues now become $\lambda = \pm 2i$ and $\lambda = \pm\sqrt{23}i$. Now the solutions will be of the form:

$$X_i = A\sin(2t) + B\cos(2t) + C\sin(\sqrt{23}t) + D\cos(\sqrt{23}t) \tag{11.2}$$

where A–D are again determined by the initial conditions. This example has been chosen so that one frequency is irrational while the other is not. Therefore, the solution cannot be periodic, because there is no period T after which both oscillations in (11.2) return to exactly the same values simultaneously. The trajectories of a dynamical system that has at least two frequencies, at least one being irrational, are called **quasiperiodic**. In the four-dimensional space of the system with solutions (11.2), each trajectory will spiral around a torus without ever rejoining or crossing itself. Nevertheless, (11.2) is not very different from a true periodic solution such as (11.1), hence the term quasiperiodic. One further characteristic that (11.1) and (11.2) have in common is that trajectories starting from very similar initial conditions (similar values of A–D) will remain close together for all future time. The same is true of nonlinear systems with limit cycle oscillations, where neighboring trajectories approach the limit cycle together.

In contrast to all examples above, some nonlinear dynamical systems exhibit complex, aperiodic dynamics that are now called **chaos**. Chaos can be defined by distinguishing it from the alternative dynamical behaviors in higher dimensional systems discussed above:

Definition: A deterministic nonlinear dynamical system in three or more dimensions exhibits **chaos** if all of the following conditions are satisfied for some range of parameters. (1) Trajectories are aperiodic (not quasiperiodic). (2) These aperiodic trajectories remain within a bounded volume of the state space but do not approach any steady states. (3) There is **sensitivity to initial conditions** such that arbitrarily small differences in initial conditions between nearby trajectories grow exponentially in time.

The three conditions in this definition concisely differentiate chaos from other dynamical behaviors. Periodic and quasiperiodic behaviors are explicitly excluded. The sensitivity to initial conditions means that any small difference in measurement of initial conditions of a chaotic system will shortly lead to exponential uncertainty and inability to predict its behavior in the future. The atmosphere of the earth is a chaotic system with the consequence that weather prediction more than a few days in advance is extremely problematic. Chaos was first fully studied by the meteorologist Edward Lorenz (1963, 1993), although Poincaré was aware earlier of nonlinear dynamical systems that have subsequently been shown to be chaotic. Finally, the boundedness of all solutions means that exponential growth of small differences in initial conditions is not simply a consequence of trajectories moving off to infinity away from an unstable equilibrium point. As already discussed, the Poincaré–Bendixon theorem demonstrates that chaos cannot exist in a two-dimensional dynamical system.

The definition of chaos above is most easily understood by example. The first dynamical system in which chaos was explicitly studied was designed to describe air flow in the atmosphere by the meteorologist Lorenz. The Lorenz (1963) equations describe a three-dimensional dynamical system:

$$\frac{dx}{dt} = 10(-x + y)$$
$$\frac{dy}{dt} = -y + 28x - xz \qquad (11.3)$$
$$\frac{dz}{dt} = -\frac{8}{3}z + xy$$

Note how simple these equations appear, as they contain only two nonlinear product terms, xz and xy. Analysis of (11.3) begins with solving the isocline equations for the equilibrium points. As the first isocline equation is simply $x = y$, the remaining two are easily simplified to:

$$27x - xz = 0$$
$$z = \tfrac{3}{8}x^2 \qquad (11.4)$$

These equations can be solved for x to give $x = 0, \pm\sqrt{72}$, so there are three equilibrium states located at $(0, 0, 0)$, $(\sqrt{72}, \sqrt{72}, 27)$, and $(-\sqrt{72}, -\sqrt{72}, 27)$. Analysis of the Jacobian indicates that all three equilibrium points are unstable, being respectively a saddle point and two unstable spiral points. Clearly, therefore, no trajectory can ever approach an equilibrium point of the system. It is also possible to construct an ellipsoidal surface enclosing all three equilibrium points such that all trajectories that cross this surface enter the ellipsoid (see Chapter 14 for the general approach based on Lyapunov functions). Therefore, all trajectories beginning near any equilibrium point must remain bounded for all future time. This suggests the possibility of chaos, although limit cycles or quasiperiodic solutions have not yet been excluded. Let us therefore simulate (11.3) for the initial condition $(10, 10, 40)$. Running MatLab script **Lorenz.m** produces the results depicted in

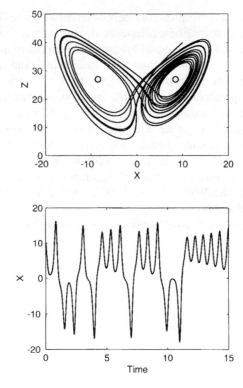

Fig. 11.1 Chaotic trajectory of the Lorenz equations (11.3). The X–Z projection of a trajectory in the three-dimensional state space is plotted above with two of the unstable steady states (circles), and $X(t)$ is plotted below.

Fig. 11.1. This trajectory certainly seems to be far more complex than a limit cycle, but is it chaos?

11.2 Signatures of chaos

To ascertain whether or not equations like (11.3) produce chaos it is necessary to ignore the equations and examine the trajectories themselves. To ascertain whether this time-dependent function is indeed chaotic takes us into the realm of **time series analysis**, the analysis of a sequence of measurements recorded from a system to determine whether the underlying dynamics are chaotic (see Kaplan and Glass (1995) for an excellent introduction to time series analysis). In analyzing a system for chaos, we already know that the time series must fall into one of four categories: a limit cycle (however complex), a quasi-periodic solution, chaos, or random variation caused by noise. The latter possibility does not, of course, occur in (11.3), because this is a deterministic system. A neuroscientist in the laboratory, however, may not know the equations that define the system being studied, and some noise will always be present in the data. Accordingly, let us examine several tests for chaos and see how they differentiate among limit cycles, quasiperiodicity, chaos, and noise.

Three quantitative tests will be applied to a time series to determine whether it is chaotic. These are: the Fourier power spectrum, the Poincaré section, and the Lyapunov exponent. These tests will be applied to each of four different time series: (11.1) with $A = 2, C = 3, B = D = 0$; (11.2) with $A = 2, C = 3, B = D = 0$; the solution $x(t)$ to (11.3), and MatLab's random number generator with numbers normally distributed with mean 0 and standard deviation 1(the **randn** function).

Books on the Fourier transform (e.g. Gaskill, 1978) prove that any finite series of points describing a time series may be exactly represented as a sum of sine and cosine functions with appropriate coefficients. Furthermore, there is a very fast algorithm for calculating these coefficients known as the fast Fourier Transform, which is implemented in MatLab as the function **fft**. The **Fourier power spectrum** is simply the square of the amplitude of each frequency term in the Fourier transform. If a time series is either periodic or quasi-periodic, its Fourier power spectrum will have large peaks at the dominant frequencies. The script **ChaosTester.m** includes time series that are both periodic and quasiperiodic. Running the program and choosing either of these options will produce a Fourier power spectrum with two large peaks reflecting the two frequencies in (11.1) and (11.2). (Choose 0 for the first return map value, which will be discussed below.) Note that the power spectrum is plotted on semi-logarithmic coordinates, so the peaks are about 10^5 times higher than the background noise.

Turning to the simulation of (11.3) using **Lorenz.m**, all computed values for $x(t)$, $y(t)$, and $z(t)$ remain in MatLab's memory after a program is run, so the results can be analyzed by running **ChaosTester.m** after **Lorenz.m**. The variable $z(t)$ will be chosen as the time series for analysis, and 25 is an appropriate value for calculating the first return map. The Fourier power spectrum is plotted at the bottom of Fig. 11.2. There are clearly a vast number of frequencies in the spectrum, and their mean amplitude falls off smoothly as frequency increases. Such an amplitude spectrum is one characteristic signature of chaos.

The second test for chaos requires computation of the **first return map**. Given a time series such as $z(t)$ obtained from solving (11.3), one chooses a value z_0 and measures all the intervals between the times when $z(t)$ returns to this same value as $z(t)$ is increasing. These intervals between successive times T_i when the time series passes z_0 are called **first return times**. A two-dimensional graph can now be generated by plotting T_{i+1} as a function of T_i. This is called a **first return map**. The program (**ChaosTester.m** produces the first return map shown at the top of Fig. 11.2 for $z(t)$ from (11.3) when $z_0 = 25$. Clearly, the return map does not cover all possible values but rather produces a complex shape that is characteristic of chaos. Experimentation with **ChaosTester.m** shows that an oscillation produces a first return map with only one or a small number of points. In the simplest oscillatory case there is only one point, because all first return values are equal to the period. Random noise fills the space eventually, because there is no correlation between successive return times, although there is a bias towards shorter return times when the noise is normally distributed. Finally, a quasiperiodic trajectory will produce a first return map that is a simple ellipse. The reason is that a quasiperiodic trajectory fills the surface of

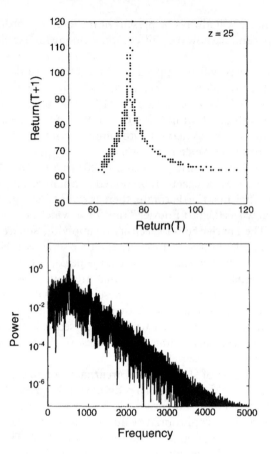

Fig. 11.2 First return map (top) and Fourier power spectrum (bottom) for the Lorenz equations (11.3).

a torus, and the first return map effectively cuts a cross-section of that torus. Thus, chaotic trajectories or time series exhibit a distinguishing structure in their first return maps.

The final chaos test is based on one of the defining elements of chaos: sensitive dependence on initial conditions such that neighboring trajectories diverge exponentially. The average exponent of this rate of divergence is called a **Lyapunov exponent**. To calculate the Lyapunov exponent, it is generally necessary to use the differential equations defining a dynamical model. Suppose we calculate one trajectory of (11.3) that we wish to examine for chaos. We must next measure the rate of divergence of neighboring trajectories all along this one trajectory. Conceptually, this is done by choosing an initial condition that differs by a very small amount Δ, perhaps 10^{-5} from the value used to generate the original trajectory. We now solve the equations for one time iteration δt of the Runge–Kutta routine and measure the Euclidean distance Δ_1 between the new solution and the original one. The ratio Δ_1/Δ is an accurate measure of the divergence of this neighboring trajectory. Thus, if $\Delta_1/\Delta > 1$ the trajectory is diverging from the

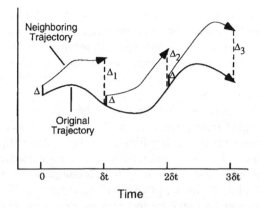

Fig. 11.3 Illustration of Lyapunov exponent computation. A neighboring trajectory, beginning a small distance Δ away from the original trajectory, is computed for one time step, and the resultant distance from the original trajectory, Δ_1 is then determined. The procedure is repeated for the duration of the original trajectory. The results are averaged in (11.5) to determine the Lyapunov exponent.

original trajectory, while if $\Delta_1/\Delta < 1$ the trajectory is converging. These possibilities are associated with instability and asymptotic stability of the original trajectory. To complete calculation of the **Lyapunov exponent**, this process must be repeated at point after point along the trajectory and the results averaged. This is done by choosing successive conditions that are all at distance Δ from successive points on our original trajectory but in the direction of Δ_i. The nature of this computation is illustrated schematically in Fig. 11.3. The assumption in calculating an exponent is that neighboring trajectories converge or diverge as $e^{\lambda \delta t}$, where the Lyapunov exponent is λ. Therefore, the natural logarithms of successive Δ_i/Δ ratios must be averaged and divided by the Runge–Kutta time increment δt. The resulting formula for the Lyapunov exponent λ of a trajectory computed at N time points is:

$$\lambda = \frac{1}{N \delta t} \sum_{i=1}^{N} \ln \left(\frac{\Delta_i}{\Delta} \right) \tag{11.5}$$

There are actually as many Lyapunov exponents as there are dimensions of the dynamical system, three in the case of (11.3). However, chaotic dependence on initial conditions is determined by whether or not the largest of these is positive, and (11.5) provides a means for computing the largest exponent. Computing all of the exponents requires linearizing around every point on the original trajectory and solving for the local eigenvalues, a much more tedious task.

The MatLab program **LyapunovExpt.m** implements the computation of (11.5) for the Lorenz equations (11.3). First, it is necessary to run a Runge–Kutta simulation using **Lorenz.m** to obtain the trajectory to be analyzed, which MatLab will keep in memory. The calculated trajectory can now be analyzed by the program **LyapunovExpt.m**. The program will ask for the value of Δ, which should be small (e.g. 10^{-5}), but not so small as to cause problems with round-off errors in the computation. For the Lorenz equations (11.3) you will find that λ is around 0.9, the exact value depending on the length of trajectory analyzed. As $\lambda > 0$, neighboring trajectories diverge exponentially from one another, and the system is therefore very sensitive to slight variations in initial conditions. This is a defining characteristic of chaos.

To sum up, three separate tests for chaos have been presented. Any chaotic system will have a very complex power spectrum with virtually all frequencies present and with power decreasing with increasing frequency. In contrast, the presence of a few very prominent peaks is indicative of either periodic or quasiperiodic behavior. A chaotic system will also produce a first return map with a complex structure that fills only a small portion of the plot. Periodic systems will produce a small number of isolated points (perhaps only one!), and quasiperiodic systems will generally form a solid closed curve, such as an ellipse produced by cutting the surface of a doughnut. Finally, a chaotic system must have a Lyapunov exponent $\lambda > 0$. Asymptotically stable trajectories, such as limit cycles, will have $\lambda < 0$. If the trajectories of a system satisfy all three criteria for chaos, we can be certain that the dynamics are indeed chaotic. When dealing with experimentally measured time series, however, it is sometimes quite difficult to be certain of chaos, because a deterministic physical system with added noise can sometimes produce non-chaotic trajectories that are seemingly chaotic by our tests. Furthermore, measurement of the Lyapunov exponent is generally not possible for experimental data. As the focus in this book is on deterministic mathematical models in neurobiology, problems of testing noisy experimental time series for chaos will not be discussed, and the interested reader is referred to Kaplan and Glass (1995).

11.3   Hodgkin–Huxley equations and chaos

The Lorenz equations were chosen to provide a focus for characterizing chaos, because they are the first and most famous equations that have been shown to have chaotic dynamics. Let us now turn to an example from neurobiology: the Hodgkin–Huxley equations as simplified in (9.7). The only difference is that a sinusoidal component has been added to the stimulus current I, producing the system:

$$0.8\,\frac{\mathrm{d}V}{\mathrm{d}t} = -(17.81 + 47.71V + 32.63V^2)(V - 0.55) - 26.0R(V + 0.92) + I$$

$$\frac{\mathrm{d}R}{\mathrm{d}t} = \frac{1}{1.9}(-R + 1.35V + 1.03) \tag{11.6}$$

$$I = I_0 + A\sin(2\pi\omega t)$$

We have already seen that this equation only produces periodic spike trains generated by limit cycle dynamics when I is a constant in the suprathreshold range. Is there any combination of I_0, A, and ω that will produce chaos? One's first intuition might be 'of course not', because chaos cannot occur in a system with fewer than three dimensions, and (11.6) contains only two differential equations. This is incorrect, however, for (11.6) is

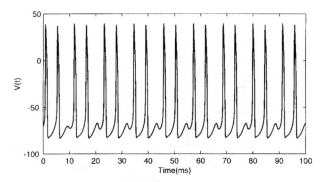

Fig. 11.4 Spikes generated by (11.6) for $A = 0.1$. This represents a more complex limit cycle in which every third spike is missing.

in fact equivalent to a four-dimensional system, as the sine term is the solution of two additional linear, first order differential equations. Thus, (11.6) has sufficiently high dimensionality to permit chaos to occur.

To explore the range of possible solutions, let us set $I_0 = 0.075$ and $\omega = 264.6\,\text{Hz}$ and determine how the solutions vary with amplitude A. If you run the script **HHWchaos.m** with $A = 0$, you will see that a spike train is generated with a frequency of about 170 Hz. Running the program again with $A = 0.4$ will produce a spike train at 264.6 Hz as a result of locking to the sinusoidal stimulus. Between these two extremes, the patterns get much more interesting. An amplitude $A = 0.1$ generates the spike train plotted in Fig. 11.4. The spike train is still periodic, but the period now involves two spikes followed by a sub-threshold response. It can be verified that this is simply a more complex limit cycle oscillation by running the tests for chaos. For $A = 0.1$ you will find that the power spectrum is very complex and would be almost impossible to distinguish from chaos. This is because the limit cycle waveform is so highly nonlinear that there are a large number of higher harmonic frequencies present. However, the first return map with a criterion of $V = 0.0$ has only two points (with a tiny scatter due to simulation errors), thus demonstrating that this is a limit cycle. Finally, the Lyapunov exponent, which can be determined using **LyapunovHHW.m**, is found to be $\lambda = -0.16$. This verifies that the spike train in Fig. 11.4 is an asymptotically stable limit cycle rather than chaos. Chaos in dynamical systems is frequently approached through a series of bifurcations as some parameter varies. In the most common case, the period doubles, then doubles again, etc. For the Hodgkin–Huxley model, however, the bifurcations are rather more complex than period doubling, but the principle of approaching chaos through a series of bifurcations still holds. Reducing the value of A in these simulations by a small amount will reveal several more such bifurcations (e.g. at $A = 0.04$).

Let us now repeat the simulation with $A = 0.007$. This produces the erratic spike train depicted at the bottom of Fig. 11.5. This certainly appears to be chaotic, but the three chaos tests must be applied to be certain. A slightly different strategy can be adopted to generate the first return map in this case, because the stimulus frequency, 264.6 Hz, defines a natural period at which to sample the system. In other words, rather than choosing a value of $V(t)$ and determining the interval between recurrences of that value,

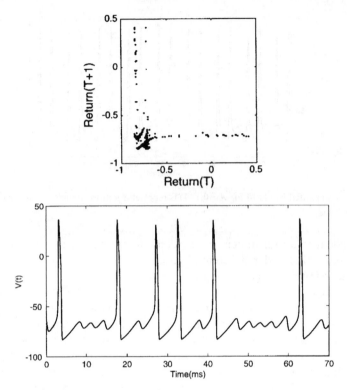

Fig. 11.5 Return map (above) and spike train (below) produced by (11.6) in response to 264.6 Hz sinusoidal modulation with amplitude $A = 0.007$. This spike train is chaotic.

it is natural to choose the known interval that is the sinusoidal stimulus period and record the values of $V(t)$ at those time instants. If the period is p, $V(t)$ is plotted versus $V(t+p)$, and the resulting map provides the same information as a first return map. For example, under conditions where each stimulus cycle triggers a phase-locked spike, the system has a simple limit cycle, and the $V(t)$ versus $V(t + p)$ map will be a single point. For the more complex limit cycle resulting from the more complex limit cycle in Fig. 11.3, the plot will have two points, etc. This variant for producing a first return map is implemented in **ForcedChaosTester.m**. When running this program the length of the sinusoidal period must be specified in terms of the number of samples produced by the simulation, and an offset which sets the phase for sample selection must also be chosen. In this case the period is 189, i.e. $1000\,\text{ms}/(264.6 \times 0.02\,\text{ms})$, where $0.02\,\text{ms}$ is the time increment in the simulation, and a convenient phase is $1/4$ of this (47), so samples are obtained at the peak phase of the forcing function. If this program is used to analyze the spike train generated when $A = 0.007$, you will obtain the first return map shown at the top of Fig. 11.5. (Note that several thousand returns are needed to produce a detailed map.) This return map clearly has the type of structure that is a signature of chaos. Furthermore, the Fourier power spectrum produced by the program is also consistent with the existence of chaos. Finally, running **LyapunovHHW.m**, which incorporates (11.6), shows that the Lyapunov exponent $\lambda = +0.16$, so neighboring trajectories diverge exponentially on average. As the time

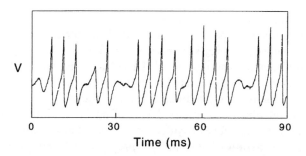

V

| | | |
0   30   60   90

Time (ms)

Fig. 11.6 Chaotic spike train produced by a squid axon in response to sinusoidal modulation at 264 Hz (Everson, 1987). Variability in spike timing and amplitude are similar to those produced by eqn (11.6) in the previous figure.

scale is milliseconds, this value of λ will cause divergence by a factor of about e^9 in 100 ms! Thus, all three criteria for chaos are satisfied, and it can be concluded that (11.6) produces chaotic spike trains over certain ranges of sinusoidal forcing.

Although (11.6) represents a simplification of the Hodgkin–Huxley equations, it has been shown by similar simulations that the original equations do in fact exhibit chaotic dynamics when they are forced sinusoidally. Furthermore, Everson (1987) and others (see Degn *et al.*, 1987) have demonstrated chaotic spike trains in periodically stimulated squid axons. Chaotic data resulting from 264 Hz sinusoidal stimulation of a squid axon are plotted in Fig. 11.6 (Everson, 1987). There is reasonable qualitative agreement between these experimental data and the spike train produced by (11.6) and shown in Fig. 11.5.

11.4 Implications of neural chaos

In addition to the example above, many neural systems have been shown to exhibit chaotic dynamics under appropriate circumstances. What are the potential implications of this fact? The sensitivity to initial conditions in chaos implies lack of long-term pre-dictability when there is any uncertainty about the initial conditions (and there is always some measurement error in the laboratory!). If a neural preparation is in a chaotic regime, the experimenter will never be able to predict details of the spike trains beyond a rather brief interval determined by the Lyapunov exponent. This observation holds true regardless of the level of detail or number of ion channels that the scientist attempts to control in the preparation. There are only two possible alternatives: content oneself with detailed prediction only for a very brief period, or predict only the mean spike rates, etc., over a longer period. The same situation is faced by the meteorologist in attempting to predict the weather. Due to chaos in the atmosphere, detailed prediction of daily tem-peratures and precipitation is only reasonably accurate for about 4–5 days in advance. However, monthly means of temperature and precipitation can usually be predicted with considerable accuracy years in advance. Neuroscientists who believe that it is necessary to understand in detail all of the ionic currents of every neuron in a network to make pre-dictions should recognize that greater descriptive detail in a chaotic neural system will not generally lead to greater predictive power.

At the level of animal behavior, it has been noted that many small prey animals flee in highly erratic, unpredictable paths when chased (Humphries and Driver, 1970). Such behavior, which enhances the prey's chances of escaping, may well represent an adaptive utilization of neural chaos in motor control. This idea has been extended even further by Miller (1997) in an attempt to explain the evolutionary roots of creative intelligence. The basic notion is that dominant individuals in primate societies can enhance their control of subordinates by introducing a certain degree of unpredictability into their behavior. Such 'protean behavior' may have provided a basis for the evolution of creative thinking, because the rapid generation of many unpredictable alternatives is a necessary ingredient of creativity. It is therefore plausible that chaotic neural dynamics in some of the highest cortical areas of the frontal lobes may provide the generative element behind creativity.

On a more philosophic level, what does neural chaos have to say about our own brains and thought processes? Although there is little direct evidence yet, it is likely that at least some of the neurons in our brains are in a chaotic dynamical regime some of the time. Indeed, there is already evidence for chaos among neurons in monkey cortex (Rapp *et al.*, 1985), and simulations have suggested that cortical networks with recurrent excitation and inhibition may exhibit chaos (Van Vreeswijk and Sompolinsky, 1996). This may well be one source of unpredictability in human behavior. The implications of neural chaos go deeper than this, however. First, chaos may provide a limitation on each individual's ability to predict his/her own behavior in detail. Although we know very little about the neural processes involved in thinking or consciousness, we do know that they involve neurons that might sometimes operate in chaotic regimes. Even though the brain may be totally deterministic, therefore, we may sometimes have no idea why we suddenly perform an unexpected act or make a snap decision! Thus, the old free-will versus determinism controversy in philosophy may have its resolution in neural chaos: a totally deterministic brain may nevertheless produce behaviors that are not predictable either by that brain itself or by any other brain on the planet!

11.5 Exercises

1. Consider the FitzHugh–Nagumo equations (8.8) with the following input current:

$$I_{input} = 0.85 + 0.07 \sin(2\pi t/3.69)$$

Modify the script **FitzHugh.m** to simulate the response to this input for as long a duration as computer memory will permit (chaos testing requires long time series). Test the resulting spike train for chaos, determining the first return map and the Lyapunov exponent. (You will have to paste the appropriate dynamical equations into a copy of **LyapunovExpt.m** to do this.)

2. Equations (10.2) for a simple bursting neuron produce erratic spike trains for inputs I near 0.2. Find a value of I that produces chaos in these equations, plot the first return map, and determine the Lyapunov exponent. (This will again require pasting the equations into a copy of **LyapunovExpt.m**.)

3. The data in Fig. 3.4 show that muscle force is only linear over part of its range. Because of this, Barenghi and Lakshminarayanan (1992) developed the following nonlinear muscle model and studied its response to sinusoidal stimulation:

$$\frac{dx}{dt} = y$$

$$\frac{dy}{dt} = -0.1y - \frac{25}{16}x + \frac{5}{8}x^3 - \frac{1}{16}x^5 + A\sin(2\pi0.2t)$$

Simulate these muscle equations for $A = 2$, $A = 8.0$, and $A = 8.905$, for as long a temporal interval as possible. Plot the return maps and determine the Lyapunov exponents in each case. Which of these stimulus amplitudes causes chaos? (The authors suggested that the model might explain irregular muscle tremor.)

4. The following three equations describe the Hindmarsh–Rose neuron model with adaptation, and they can exhibit very different behaviors under different conditions of stimulation. The equations are:

$$\frac{dV}{dt} = -V^3 + 3V^2 + R - z + I$$

$$\frac{dR}{dt} = -R + 1 - 5V^2$$

$$\frac{dz}{dt} = 0.015(-z + 4(V + 1.6))$$

where V, R, and z are respectively the voltage, the recovery variable, and the adaptation variable. Characterize the responses of this neuron model for three different levels of input current: (a) $I = 2.0$; (b) $I = 2.5$; and (c) $I = 3.0$. You will need to combine Runge–Kutta simulations with all tests for chaos, etc. In your simulations, use a time increment $\Delta t = 0.04$, and initial conditions $V = -1$; $R = -1$; $z = 2.0$ (these conditions will minimize some otherwise tedious transient effects). For each simulation show a plot of 300 ms of the spike train, and plots of the power spectrum and first return map. For the first return map you should use as long a simulation as your computer memory will permit in order to obtain between 200 and 500 return values. Determine the Lyapunov exponents, and indicate what type of dynamical behavior each value of I produces.

12   *Synapses and synchrony*

We have investigated the dynamics behind a range of neurons from the squid axon to bursting cells in neocortex. This provides the basis for studying interactions between neurons that are synaptically interconnected. This chapter will explore the simplest such interactions, namely those between pairs of identical neurons connected by either excitatory or inhibitory synapses. It might be thought that few really novel results would emerge from such simple two-neuron systems, but there are surprises in store. For example, swimming behavior in the mollusk Clione is controlled by mutually inhibitory neurons that produce a prolonged swimming cycle in the absence of any excitatory drive! It will also be seen that thalamic bursting during slow wave sleep is synchronized by mutual inhibition rather than by excitation.

In order to gain insight into the dynamical reasons for these phenomena, we shall first study a powerful mathematical simplification of interactions between a pair of neurons. This is the phase oscillator concept introduced by Cohen *et al.* (1982). Once the abstract phase oscillator concept has been developed, it will be shown to predict the behavior of coupled Hodgkin–Huxley neurons. An added benefit is the simulation and analysis of the neural swimming system of the mollusk Clione. The phase oscillator model will then be developed further by incorporating a more accurate description of synaptic interactions. This leads to the surprising prediction that mutual inhibition is the most effective method of synchronizing neurons. Finally, inhibitory coupling responsible for synchronizing bursting thalamic neurons during deep sleep will be analyzed.

12.1   Phase oscillator model

Many types of neurons generate an ongoing spike train or sequence of spike bursts in response to a constant stimulus. From a dynamical perspective, this means that such neurons generate a limit cycle oscillation in response to constant stimulation. If two such neurons are reciprocally coupled by excitatory or inhibitory synapses, we are confronted with the problem of interacting nonlinear oscillators. Although such coupled oscillators are extremely difficult to analyze with full generality, an elegant simplification with significant predictive power was discovered by Cohen *et al.* (1982). The key insight of Cohen *et al.* (1982) (see also Rand *et al.*, 1988) was that only the phase of each oscillator need be considered when the coupling between the oscillators is weak. The reason for this is that the amplitude and waveform of each limit cycle will be largely unaffected under weak coupling, although phase relations and frequencies can be changed. These observations are evident in the discussion of phase shifting in Chapter 9.

To develop the concept of a **phase oscillator**, let us consider two coupled oscillators, an abstraction of two interacting neurons. Each oscillator will be described by a single differential equation governing the rate of change of its phase θ with time. For a single oscillator with no coupling, the equation is:

$$\frac{d\theta}{dt} = \omega \quad \text{so} \quad \theta(t) = \omega t (\text{mod } 2\pi) \tag{12.1}$$

where ω is the frequency of the limit cycle oscillation. The solution of this equation is modulo 2π because the phase is periodic over that range. Suppose now that two such oscillators are coupled together using the synaptic coupling H:

$$\frac{d\theta_1}{dt} = \omega_1 + H_1(\theta_2 - \theta_1)$$
$$\frac{d\theta_2}{dt} = \omega_2 + H_2(\theta_1 - \theta_2) \tag{12.2}$$

where ω_1 and ω_2 are the frequencies of the two oscillators in the absence of coupling. H_1 and H_2 must both be periodic with period 2π for these equations to meaningfully describe phase coupling. The assumption that the coupling functions depend only on phase difference is an approximation that is exact in the limit of weak coupling (Kopell, 1988; Ermentrout, 1994). As phase continuously varies during each cycle of the oscillation, it is only necessary to consider the phase difference between the two oscillators to determine whether they have a constant phase relationship. Let us define the variable $\phi = \theta_2 - \theta_1$ and generate a new equation by subtracting the first from the second equation in (12.2). The result is:

$$\frac{d\phi}{dt} = \omega_2 - \omega_1 + H_2(-\phi) - H_1(\phi) \tag{12.3}$$

Even without considering explicit forms for H_1 and H_2, we can obtain some very general results concerning the effects of oscillator coupling in (12.3). First, however, two definitions are necessary. Two oscillators are **phase locked** when the phase difference between them is constant and independent of time. (This is $1:1$ phase locking; $2:1$, $3:1$, etc. phase locking can also occur as in Fig. 9.12.) Two oscillators are **synchronized** if they are phase locked at zero phase difference. Given the definition of phase locking, a solution to (12.3) will be phase locked if and only if $d\phi/dt = 0$ is an asymptotically stable steady state. Applying the usual stability analysis, therefore, phase-locked solutions must obey the equation:

$$\omega_2 - \omega_1 + H_2(-\phi) - H_1(\phi) = 0 \tag{12.4}$$

Asymptotic stability of the solution is guaranteed if:

$$\frac{d}{d\phi}[H_2(-\phi) - H_1(\phi)] < 0 \tag{12.5}$$

where the derivative is evaluated at the equilibrium value of ϕ. Equation (12.5) is the one-dimensional special case of a Jacobian.

Let us clarify these generalities by solving (12.4) and (12.5) for a particular choice of the H function. Cohen *et al.* (1982) chose the simplest function with period 2π, namely:

$$H_i(\phi) = a_i \sin(\phi + \sigma) \tag{12.6}$$

where the synaptic strength a_i can vary for the two directions of coupling. The additional phase shift $\sigma \geq 0$ is assumed to be produced by synaptic or conduction delays in the coupling (Kopell, 1988; Kopell and Ermentrout, 1988). The special case where $\sigma = 0$ considered by Cohen *et al.* (1982) describes coupling via electrical gap junctions between neurons or extremely rapid short-range coupling that produces a negligible phase shift. Two justifications may be offered for the choice of eqn (12.6). First, (12.6) is the lowest order in the Fourier series expansion of any 2π periodic function (Cohen *et al.*, 1982; Rand *et al.*, 1988). Second, calculations of the H functions for both the Wilson–Cowan (1972) oscillator and the Morris–Lecar (1981) equations show that H assumes a form similar to (12.6) as a first approximation (Ermentrout and Kopell, 1991; Ermentrout, 1994). The actual calculation of H involves averaging the effects of small perturbations over one cycle of the oscillation as a function of phase and is discussed in detail by Ermentrout and Kopell (1994).

Substitution of (12.6) into (12.3) gives:

$$\frac{d\phi}{dt} = w_2 - w_1 - a_1 \sin(\phi + \sigma) - a_2 \sin(\phi - \sigma) \tag{12.7}$$

where we have used the identity $\sin(-x) = -\sin(x)$. Equation (12.4) for the phase-locked states now becomes:

$$w_2 - w_1 - a_1 \sin(\phi + \sigma) - a_2 \sin(\phi - \sigma) = 0 \tag{12.8}$$

The trigonometric identity $\sin(x + y) = \cos(x)\sin(y) + \sin(x)\cos(y)$ transforms (12.8) into:

$$w_2 - w_1 - (a_1 + a_2)\cos(\sigma)\sin(\phi) + (a_2 - a_1)\sin(\sigma)\cos(\phi) = 0 \tag{12.9}$$

The condition for asymptotic stability of a phase-locked state may now be determined from (12.5):

$$\frac{d}{d\phi}[H_2(-\phi) - H_1(\phi)] = -(a_2 - a_1)\sin(\sigma)\sin(\phi) - (a_1 + a_2)\cos(\sigma)\cos(\phi) < 0 \tag{12.10}$$

where ϕ is a solution to (12.9).

The general solution to eqn (12.9) can be written down (see 12.17 below), but much insight may be gained by considering several physiologically important special cases. The first case of interest occurs when the frequencies of the independent oscillators are equal, $w_1 = w_2 = w$. As these frequencies are determined by external signals to the individual oscillators, it is easy to produce this condition physiologically. Equation (12.9) may now be solved with the result:

$$\phi = \arctan\left(\frac{(a_2 - a_1)\sin(\sigma)}{(a_1 + a_2)\cos(\sigma)}\right) \qquad (12.11)$$

Assuming that the synaptic delay $0 < \sigma < \pi/2$ and that $a_1, a_2 \geq 0$ (i.e. excitatory connections), three different solutions to (12.11) may be distinguished:

$$\begin{aligned} \phi &= 0, \pi \quad \text{for } a_1 = a_2; \\ \phi &> 0 \quad \text{for } a_1 < a_2; \\ \phi &< 0 \quad \text{for } a_1 > a_2; \end{aligned} \qquad (12.12)$$

When $a_1 = a_2$ so that $\phi = 0$, the asymptotic stability condition (12.10) becomes:

$$-(a_1 + a_2)\cos(\sigma) < 0 \qquad (12.13)$$

so this is an asymptotically stable steady state in which the two oscillators are synchronized. It is easy to show that the other steady state in this case, $\phi = \pi$, is unstable. In fact, the periodic nature of the arctan function guarantees that there will be two solutions to (12.11) in each of the cases listed in (12.12), but only the smaller, asymptotically stable value of ϕ has been listed in the second and third cases.

The remaining solutions to (12.11) in (12.12) will be particularly relevant to the study of lamprey swimming, as described in the next chapter. In the first case where $a_1 < a_2, \phi > 0$ is an asymptotically stable, phase-locked solution as long as (12.10) is satisfied. For example, suppose $a_2 = 2a_1$ and $\sigma = 2\pi/25$ (4% of a period). Now (12.11) gives $\phi = 0.085$, which is 4.9°, and inspection of (12.10) shows that this solution is asymptotically stable for any values of the coupling coefficients satisfying our conditions. As $\phi = \theta_2 - \theta_1$, this phase-locked solution requires that $\theta_2 > \theta_1$, so the second oscillator leads the first in this case. If coupling strengths were reversed so that $a_1 > a_2 > 0$, it is easy to see that the phase-locked solution $\phi < 0$ becomes asymptotically stable. In either case, the frequency at which the oscillators synchronize is given by the rate of change of phase from (12.2):

$$\frac{d\theta_1}{dt} = w + a_1 \sin(\phi + \sigma) \qquad (12.14)$$

where w is the common frequency of the uncoupled oscillators. For the particular parameters discussed above $\phi = 0.085$, $\sin(\phi + \sigma) = 0.33$ so the coupled oscillators lock at a frequency greater than their uncoupled frequencies. This frequency increase due to coupling is not surprising given that the coupling, although weak, is excitatory. This demonstrates that a short synaptic delay in oscillator coupling plus an asymmetry in coupling strength leads to a phase-locked solution in which one oscillator lags the other but the oscillation frequency is increased.

The solutions to eqn (12.9) above were based on the assumption that $w_1 = w_2 = w$. As a second important example, suppose that the frequencies are not identical but the coupling strengths are, so $a_1 = a_2 = a$. The solution of (12.9) is now:

$$\phi = \arcsin\left(\frac{\omega_2 - \omega_1}{2a\cos(\sigma)}\right) \quad \text{if} \quad \left|\frac{\omega_2 - \omega_1}{2a\cos(\sigma)}\right| < 1 \qquad (12.15)$$

and the phase-locked frequency in (12.2) will be:

$$\frac{d\theta_1}{dt} = \omega_1 + a\sin(\phi + \sigma) \qquad (12.16)$$

The inequality in (12.15) arises from the fact that the sine function is bounded by ± 1. This means that a phase-locked solution to (12.9) is only possible if the frequency difference between the uncoupled oscillators is sufficiently small relative to the sum of the coupling coefficients multiplied by $\cos(\sigma)$. When this is not satisfied, the two oscillators will drift in relative phase, as we shall see shortly. Assuming that the inequality is satisfied, (12.15) will have two solutions differing in phase, but only one will be asymptotically stable. If $\omega_1 = \omega_2$ the oscillators will be synchronized if $a > 0$. On the other hand, if the connections between oscillators are inhibitory, then the coupling coefficient $a < 0$. Now when $\omega_1 = \omega_2$ the asymptotically stable phase solution will be $\phi = \pi$ assuming that $\sigma < \pi/2$ (see eqn 12.10). Thus, inhibitory coupling leads to phase locking in antiphase rather than synchronization in this case.

Before considering several neural examples, let's note the most general solution to (12.10), which is:

$$\phi = \arcsin\left(\frac{\omega_2 - \omega_1}{A}\right) + \beta \quad \text{where}$$

$$A = \sqrt{(a_2 + a_1)^2(\cos\sigma)^2 + (a_2 - a_1)^2(\sin\sigma)^2} \qquad (12.17)$$

$$\beta = \arctan\left(\frac{(a_2 - a_1)\sin\sigma}{(a_2 + a_1)\cos\sigma}\right)$$

This equation will again have two phase-locked solutions, one unstable and one asymptotically stable, if $|\omega_2 - \omega_1|/A < 1$. It can be shown that (12.11) and (12.15) are special cases of (12.17).

12.2   Pairs of synaptically coupled neurons

This discussion of oscillator interactions has employed the very general phase oscillator approach, and the choice of the coupling function $H(\phi)$ in (12.6) may seem too simple to provide physiologically relevant insights. To see just how well the phase oscillator model actually works, let us consider the simplest possible physiological example: two neurons reciprocally coupled by either excitatory or inhibitory synapses.

To examine the effects of synaptic interactions, it will be convenient to begin with a neuron incorporating a simulated I_A current described by eqn (9.10). As this neuron can fire spikes at arbitrarily low rates, this will enable us to explore the effects of spike rate on synaptic interactions most easily. Synaptic inputs can be added to this equation using the formulation from Chapter 2. If the time constant of the synaptic potential is τ_{syn} and its equilibrium potential E_{syn}, synaptic inputs to (9.10) produce the equation:

$$\frac{dV}{dt} = -\{17.81 + 47.58V + 33.8V^2\}(V - 0.48) - 26R(V + 0.95) + I$$
$$\quad - kg(V - E_{syn})$$
$$\frac{dR}{dt} = \frac{1}{5.6}\left(-R + 1.29V + 0.79 + 3.3(V + 0.38)^2\right) \quad\quad (12.18)$$
$$\frac{df}{dt} = \frac{1}{\tau_{syn}}(-f + H_{step}(V_{pre} - \Omega))$$
$$\frac{dg}{dt} = \frac{1}{\tau_{syn}}(-g + f)$$

where

$$H_{step}(x) = \begin{cases} 1 & \text{if } x > 0 \\ 0 & \text{if } x \leq 0 \end{cases}$$

The synaptic potential is generated by variation of the synaptic conductance variable g with time constant τ_{syn}. Changes in the conductance g are driven by the variable f, which is in turn driven by the voltage of the presynaptic neuron, V_{pre}, using the step function $H_{step}(x)$. $H_{step}(x) = 1$ whenever $V_{pre} > \Omega$, the threshold for postsynaptic conductance changes. The magnitude of the postsynaptic potential is controlled by the parameter k and the equilibrium or reversal potential E_{syn}. For these simulations, let us choose the values $\tau_{syn} = 2\,ms$ and $\Omega = -0.20$ (i.e. $-20\,mV$). For excitatory synapses, $E_{syn} = 0$, while at inhibitory synapses $E_{syn} = -0.92$ (i.e. $-92\,mV$).

The use of two differential equations to describe the conductance change at a synapse was introduced by Rall (1967, 1989) and is now in common use. From (2.10), it can be seen that two such equations will produce a response proportional to $(t/\tau_{syn}^2)\exp(-t/\tau_{syn})$ when the stimulus is a brief pulse, and this is Rall's **alpha function**. In Runge–Kutta simulations, direct computation of the two differential equations for the alpha function is about twice as fast as direct use of the exponential. Normalized plots of the synaptic conductance change $g(t)$ for three different values of τ_{syn} are shown in Fig. 12.1. As can be seen and easily derived, the peak of the alpha function occurs at $t = \tau_{syn}$, so fast and slow synaptic potentials can be created by varying this parameter. Although employing two differential equations to describe synaptic conductance changes might seem unnecessarily complex, it has been shown to be crucial to the understanding of certain neural interactions (Van Vreeswijk *et al.*, 1994). The reason is that conductance changes described by the alpha function will peak after the end of the initiating spike if τ_{syn} is sufficiently long (see Fig. 12.1), and this is typical physiologically. As spikes have a brief duration of about $1.0\,ms$, $H_{step}(x)$ will produce about a $1.0\,ms$ pulse (gray rectangle in Fig. 12.1) to the df/dt equation if the threshold for synaptic activation, Ω, is set to a value midway along the rising phase of the spikes. This generates a very smooth alpha function and avoids the need to store a record of the time at which each incoming spike arrived at the synapse. Note also that the area under the alpha function is identical for all values of τ_{syn}.

Spikes, decisions, and actions

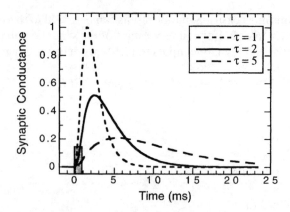

Fig. 12.1 Synaptic conductance changes produced by the alpha function for $\tau = 1$, 2, and 5 ms. The rectangle indicates the duration of the inducing spike.

Putting action potential generation and synaptic inputs together in (12.18) creates what is known as an **isopotential neuron**. This is a cell in which the membrane potential V remains identical throughout both the soma and the dendritic tree. More realistic models with multiple dendritic compartments will be developed in Chapter 15. Nevertheless, isopotential neurons are in common use in neural modeling (e.g. Wang and Rinzel, 1992; Skinner *et al.*, 1994; Rolls and Treves, 1998), and a lot can be learned by studying them.

Let us now explore phase locking between two neurons, each described by eqn (12.18). The MatLab program **EPSPinteractions.m** simulates the case where the two are reciprocally coupled by excitatory synapses, with each receiving an independent input I. Running the program with stimulus levels of 0.4 and 0.3 for the two neurons, and excitatory coupling strength $k = 0$, shows that the two neurons respond at independent frequencies $\omega_1 = 37$ Hz and $\omega_2 = 23$ Hz. Rerunning the program with the same two stimuli but with $k = 4$ produces phase locking at 45 Hz with the neuron receiving the weaker input exhibiting a modest phase lag (about $2\pi/7$) relative to the other. If stronger EPSPs are produced by setting $k = 6$, phase locking at 52 Hz is obtained with a shorter phase lag of about $2\pi/10$. These trends are predicted by the phase oscillator equations (12.15) and (12.16). Furthermore, (12.15) predicts that phase locking will be impossible if the excitatory coupling is too weak. You can verify this by running **EPSPinteractions.m** with stimuli of 0.4 and 0.3 but coupling $k = 2$. Thus, the qualitative predictions of the phase oscillator model are supported by simulations of two interacting neurons, although quantitative details differ due to the generality of the model.

The phase oscillator model also predicts that two neurons with inhibitory coupling will phase lock about 180° out of phase. This can be tested using an inhibitory synapse in (12.18) by setting $E_{syn} = -0.92$, which has been done in MatLab script **IPSPinteractions.m**. The time constant has been shortened to $\tau_{syn} = 1$ ms in this simulation for reasons that will become clear shortly. If the script is run with stimulus levels of 1.1 and 1.0, the uncoupled neurons $(k = 0)$ will oscillate at $\omega_1 = 118$ Hz and $\omega_2 = 107$ Hz

respectively. Setting the inhibitory strength $k = 5$ in the program now produces phase
locking at 96 Hz at a phase difference near π, but weak coupling with $k = 2$ again fails to
produce phase locking. This phase difference, reduction in frequency, and lack of phase
locking for weak coupling are all predicted by eqns (12.15) and (12.16) if the coupling
coefficient $a < 0$ to represent inhibition.

12.3   An inhibitory swimming network

Although these examples of two coupled neurons certainly elucidate the nature of phase
locking, it might seem hard to imagine that two interacting neurons could actually produce functionally significant behavior. It is therefore striking that the mollusk Clione
swims by controlling each of two wing-like flaps using just two types of neurons: one for
dorsal flexion and the second for ventral flexion (Satterlie, 1985). Even more striking is the
fact that these neurons are mutually inhibitory, yet they can generate an ongoing oscillation in response to a brief stimulus. Figure 12.2A shows the alternation of one spike per
cycle in the antiphase firing of one dorsal and one ventral neuron (Satterlie, 1985). If you
run the MatLab simulation **Clione.m** with a stimulus of 0.5 and inhibitory coupling
strength $k = 9$, you will produce the spike trains depicted in Fig. 12.2B. The two neurons
spike in antiphase, each generating one spike per cycle just as in the Clione data. Even
more surprising is the fact that stimulation was only delivered to one neuron for 1.0 ms

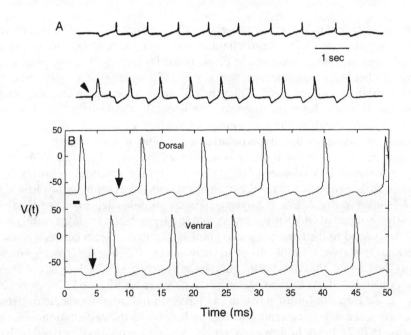

Fig. 12.2 Action potentials in dorsal and ventral neurons controling swimming in Clione. Data in A were
triggered by a brief stimulus at the arrow (reproduced with permission, Satterlie, 1985). Simulation in B was
triggered by a brief stimulus to the dorsal neuron (horizontal bar). Note hyperpolarization produced by
mutual inhibition (arrows).

(short black bar), so reciprocal inhibition here produces a self-perpetuating swim cycle! If the inhibitory coupling is eliminated by setting $k = 0$, just one spike will be triggered in the stimulated neuron, but there will be no further neural activity. The Clione cycle rate (near 1 Hz) is much slower than the simulation, but individual Clione spikes are 50–100 ms in duration, so all of the model time constants could simply be scaled to fit the data.

How can reciprocal inhibition with no apparent source of excitation produce ongoing bursting during Clione swimming? The key to the answer lies in the fact that Hodgkin–Huxley type neurons described by eqn (9.7) have been used in this simulation. These equations have been coupled by adding the two synaptic conductance equations from (12.18) to each:

$$\frac{df}{dt} = \frac{1}{\tau_{syn}}(-f + H_{step}(V_{pre} - \Omega))$$

$$\frac{dg}{dt} = \frac{1}{\tau_{syn}}(-g + f)$$

(12.19)

where

$$H_{step}(x) = \begin{cases} 1 & \text{if } x > 0 \\ 0 & \text{if } x \leq 0 \end{cases}$$

In addition, the synaptic current term $-g(V - E_{syn})$ with $E_{syn} = -0.92$ has been added to the dV/dt equation in (9.7). The mechanism underlying Clione swimming can now be obtained from an examination of the phase plane for eqn (9.7). First, however, let us observe another surprising phenomenon: if you use a hyperpolarizing stimulus of -0.2 with an inhibitory gain of $k = 9$, **Clione.m** will again generate an ongoing swimming rhythm, and this has also been observed experimentally (Satterlie, 1985)! This provides the key to understanding the mechanism of rhythm generation. If the neurons are uncoupled by setting $k = 0$, a hyperpolarizing stimulus of -0.2 will produce one spike following termination of the hyperpolarization, as shown in Fig. 12.3A. As indicated, there is a **postinhibitory rebound** (PIR) of the membrane potential following hyperpo- larization. Spike generation via PIR can be understood by examining the phase plane for (9.7). As shown in Fig. 12.3B, the hyperpolarizing stimulus depresses the $dV/dt = 0$ iso- cline (only a portion of which is shown) so that the asymptotically stable resting state R is shifted downward to the hyperpolarized point H. As all equilibria between R and H are asymptotically stable, V simply drops to the value at H. When the hyperpolarization is abruptly terminated, the equilibrium point immediately jumps back to R, but the system trajectory is still at H. As this is a region of the phase space where dV/dt is very large but $dR/dt = 0$, the trajectory must move in the direction indicated by the arrow, thus gen- erating one spike before returning to rest at R. This is the explanation of PIR spike generation in (9.7). Postinhibitory rebound in the squid axon was described by Hodgkin and Huxley (1952) under the rubric **anode break excitation**.

The explanation of the Clione swimming rhythm should now be clear. Each time one neuron spikes, the brief hyperpolarizing IPSP in the other cell (arrows in Fig. 12.2B)

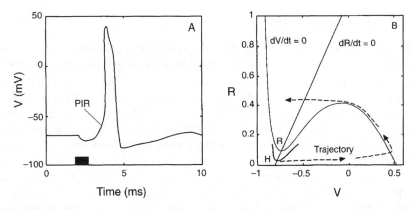

Fig. 12.3 Initiation of a spike in (9.7) via postinhibitory rebound (PIR). The phase plane in B shows that hyperpolarization shifts the equilibrium state from R to H, so a sudden shift of the equilibrium back to R produces the system trajectory shown.

results in postinhibitory rebound, which generates a spike, so the process repeats itself. In order to generate PIR, it is crucial that the IPSP be brief in duration and sufficiently strong, as otherwise the hyperpolarized cell will simply track the slowly moving equilibrium point (which is always asymptotically stable in this case) back to rest. In Clione, each IPSP is of about the same duration as each spike (100 ms), so the simulation of Clione has an inhibitory time constant $\tau_{syn} = 1.0$ ms, the duration of spikes produced by (9.7). If the value is changed to $\tau_{syn} = 1.5$ ms and **Clione.m** is rerun with $k = 9$, no rhythm can be generated at any stimulus intensity, because trajectories simply track the asymptotically stable equilibrium following stimulus termination.

Reciprocal inhibition without a source of excitation can thus produce an antiphase neural oscillation provided that the neurons involved exhibit PIR and the IPSPs are sufficiently brief, being of about the same duration as the spikes themselves. These physiological conditions are fulfilled by Clione (Satterlie, 1985) and by our model. Note, however, that inhibitory coupling between two neurons described by (12.18) will not produce ongoing swimming, because these neurons do not exhibit a PIR. This is a consequence of the shape of the isoclines in the phase plane shown in Fig. 9.9.

This unusual mode of PIR rhythm generation was predicted by Perkel and Mulloney in 1974 based on mathematical simulations. So here is another example in which nonlinear dynamics produced a counter-intuitive prediction that was subsequently verified by neurophysiologists. Spike generation via PIR has also been observed in mammalian and human neocortical and thalamic neurons (Foehring and Wyler, 1990; Crunelli and Leresche, 1991; Silva *et al.*, 1991), and its role in sleep synchronization will be examined shortly. A final note on Clione: as the resting state is asymptotically stable throughout the swimming cycle, hysteresis switching can occur between the swimming limit cycle and rest. Such switching requires a relatively slow external inhibitory input, however, as rapid inhibition would simply reset the phase of the oscillation. Thus the Clione swimming mechanism has a form of short-term memory built in.

12.4   Inhibitory synchrony

Thus far, neural simulations have verified the predictions of the phase oscillator model. However, the phase oscillator model does fail in some surprising ways. If **EPSPinteractions.m** is run with $I_1 = 0.7$, $I_2 = 0.65$, and synaptic strength $k = 8$, the two neurons will phase lock at 182 Hz (well above the free running frequencies as expected), but phase locking will be in antiphase with a lag near π. So much for excitatory interconnections causing neural synchrony! There is one final surprise in the analysis of phase locking. Up to this point, inhibitory coupling has caused phase locking about 180° out of phase, and this is crucial to the functioning of the Clione swimming network. However, several neural modeling studies have demonstrated that near perfect synchronization is actually more typical of reciprocal inhibition (Lytton and Sejnowski, 1991; Wang and Rinzel, 1992, 1993; Traub *et al.*, 1996), and reciprocal inhibition appears to be the physiological basis of synchronization in a number of systems (Aram *et al.*, 1991; Steriade *et al.*, 1993; Kim *et al.*, 1997). To see that reciprocal inhibition can lead to synchrony, change the inhibitory time constant to $\tau_{\text{syn}} = 2$ ms in **IPSPinteractions.m** and run the program with $I_1 = 1.1$, $I_2 = 1.0$, and synaptic strength $k = 5$. Whereas out-of-phase oscillations were obtained above when $\tau_{\text{syn}} = 1$ ms, the longer synaptic time-constant produces the almost perfect synchrony evident in Fig. 12.4. (The very slight phase lead of the top neuron is due to its stronger input current; it vanishes if $I_1 = 1.05$.)

What can explain the counter-intuitive observations that reciprocal excitation may produce antiphase locking, while reciprocal inhibition may result in synchrony? Wang and Rinzel (1992, 1993) found that slow synaptic conductances plus PIR were sufficient to produce inhibitory synchrony. However, the inhibitory synchrony evident in Fig. 12.4 occurs between two neurons described by eqn (9.10), so neither exhibits PIR. Thus, the time course of the synaptic conductances seems to be the prime determinant of inhibitory

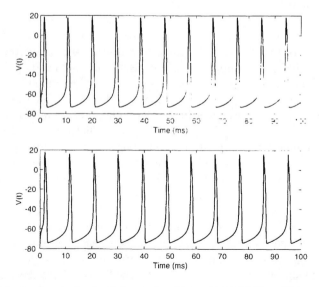

Fig. 12.4   Synchronized spike trains produced by mutual inhibition with $\tau_{\text{syn}} = 2$ ms.

synchrony. Following this lead, Van Vreeswijk *et al.* (1994) developed an elegant explanation of excitatory and inhibitory phase locking that shows how τ_{syn} governs the stability of phase-locked solutions. As developed above, the phase oscillator model makes the assumption that synaptic interactions can be described by an instantaneous pulse with a synaptic delay σ. This assumption is implicit in the description of $H(\phi)$ in (12.6). However, it is evident from the various alpha functions plotted in Fig. 12.1 that this approximation is reasonable only when τ_{syn} is very small. For large τ_{syn} the resultant synaptic conductance change will extend not only throughout one cycle of the oscillation, but it may produce a long enough tail to extend across many cycles.

Van Vreeswijk *et al.* (1994) sought to retain the simplicity and elegance of the original phase oscillator model while improving it by incorporating the effects of synaptic conductance changes that persist over time. They chose the following alpha function $P(t)$ to describe the time course of postsynaptic effects:

$$P(t) = \left(\frac{k}{\tau_{syn}^2} \right) t \exp\left(\frac{-t}{\tau_{syn}} \right) \tag{12.20}$$

where k is the synaptic strength, being positive or negative for excitatory or inhibitory synapses respectively. The factor $(1/\tau_{syn}^2)$ guarantees that the area under the alpha function will be unity for all values of τ_{syn}. Equation (12.20) can be derived from (12.19) by removing the step function from the df/dt equation and replacing it with appropriate initial conditions (see Chapter 2). A new phase interaction function $H_{syn}(\phi)$ can now be computed from the $\sin(\phi)$ interaction function of the phase oscillator model in (12.6) by convolution with $P(t)$ (see Chapter 2):

$$H_{syn}(\phi) = \frac{k}{\tau_{syn}^2} \int_0^\infty \sin(2\pi\omega(\phi - t)) t \exp\left(\frac{-t}{\tau_{syn}} \right) dt \tag{12.21}$$

This formula states that if the effect of an instantaneous pulse is given by the simple periodic function $\sin(2\pi\omega\phi)$ with frequency ω, the effect of a synaptic potential produced by (12.20) is computed by integrating it across an infinite number of cycles of the oscillation. The integral in (12.21) can be solved exactly using integral tables (Gradshteyn and Ryzhik, 1980) or a symbolic mathematics program with the result:

$$H_{syn}(\phi) = \frac{k[1 - (2\pi\omega\tau_{syn})^2] \sin(2\pi\omega\phi) - 4\pi k\omega\tau_{syn} \cos(2\pi\omega\phi)}{[1 + (2\pi\omega\tau_{syn})^2]^2} \tag{12.22}$$

The basic phase oscillator equation (12.3) shows that we need to evaluate $[H_{syn}(-\phi) - H_{syn}(\phi)]$. Substituting (12.22) instead of (12.6) into (12.3) yields:

$$\frac{d\phi}{dt} = \frac{-2k\left[1 - (2\pi\omega\tau_{syn})^2\right] \sin(2\pi\omega\phi)}{\left[1 + (2\pi\omega\tau_{syn})^2\right]^2} \tag{12.23}$$

where it has been assumed that $\omega_1 = \omega_2 = \omega$.

So long as $2\pi\omega\tau_{syn} \neq 1$, it is apparent from inspection that the equilibrium states of (12.23) occur for $\phi = 0, \pi$. To determine the stability of these states, we must compute the one-dimensional Jacobian by differentiating the right-hand side of (12.23) with respect to ϕ and evaluating the result at $\phi = 0, \pi$. This calculation shows the synchronized state $\phi = 0$ to be asymptotically stable if:

$$-k\left[1 - (2\pi\omega\tau_{syn})^2\right] < 0 \qquad (12.24)$$

For inhibitory coupling $k < 0$, so synchronization will occur if:

$$\tau_{syn} > \frac{1}{2\pi\omega} \qquad (12.25)$$

Thus, inhibitory synchronization results when the synaptic time constant is large, thus causing a slow conductance change. Inhibitory coupling only leads to asymptotic stability at phase $\phi = \pi$ if the inequality in (12.25) is reversed, indicating that the inhibitory conductance change must be very fast. This is why a short synaptic time constant $\tau_{syn} = 1$ ms was used in the program **IPSPinteractions.m** to produce antiphase locking, while $\tau_{syn} = 2$ ms produced the inhibitory synchrony in Fig. 12.4.

For excitatory interconnections $k > 0$ in the phase oscillator model, and this results in a reversal of the inequality in (12.25) for synchronization. Thus, only brief EPSPs will produce synchrony, while longer ones will cause phase locking at $\phi = \pi$. From their analysis plus modeling studies, Van Vreeswijk *et al.* (1994) conclude: 'If the rise time of the synapse is longer than the duration of an action potential, inhibition not excitation leads to synchronized firing.' (p. 313). As this is generally the case in the nervous system, inhibition appears to be the most common mechanism of synchronization. Recently, Ringel *et al.* (1998) have shown that inhibitory networks can also generate propagating waves of neural activity under certain conditions.

The qualitative reason that slow IPSPs lead to synchronization of mutually inhibitory neurons is roughly the following. After each neuron fires, it generates a slow rising but prolonged IPSP that prevents the other neuron from firing. The only stable solution is then for the two neurons to fire synchronously, thereby avoiding one another's slowly rising IPSPs. This theme is illustrated at the top of Fig. 12.5, where spikes (black vertical bars) generate the IPSPs in their inhibitory partners as shown by the arrows. When the IPSPs are very fast, however, each neuron prevents the other from firing at the same time, so anti-phase locking is the only possible solution. An analogous explanation, depicted at the bottom of Fig. 12.5, shows why slow EPSPs produce out-of-phase locking: each neuron of the pair will be stimulated to fire at the peak of the EPSP received from its neighbor.

12.5   Thalamic synchronization

The thalamus, which contains cell groups that relay information from sense organs to the cortex, becomes capable of generating synchronized bursts of activity during deep sleep

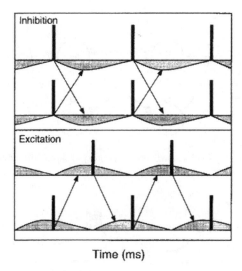

Time (ms)

Fig. 12.5 Schematic explanation of inhibitory synchronization (above) and excitatory antiphase locking (below) between pairs of neurons. Vertical bars represent spikes that produce the gray IPSPs (above) or EPSPs (below) indicated by arrows.

and also during certain epileptic seizures known as absence seizures. In the LGN (lateral geniculate nucleus), which receives input from the retina and relays it to visual cortex, these synchronized oscillations are the result of interactions between the LGN relay cells and cells in the PGN (peri-geniculate nucleus) (Crunelli and Leresche, 1991; Steriade *et al.*, 1993; Kim *et al.*, 1995, 1997). As shown by the circuit diagram in Fig. 12.6, axon collaterals of the relay cells (E) excite neurons in the PGN, and these PGN neurons (I) in turn provide feedback inhibition onto the relay cells. This is a network with feedback inhibition between adjacent relay cells, so one might intuitively feel that synchronized oscillations should be impossible. However, several different factors that we have already analyzed combine to produce stable oscillations and thalamic synchrony during deep sleep (when there is no retinal input).

A dynamical characteristic of the individual neurons in Fig. 12.6 provides the first key to network behavior. During deep sleep the membrane properties are altered (presumably by modulatory neurotransmitters) so that the individual neurons produce a burst of spikes as a result of postinhibitory rebound (PIR). As these bursts of spikes are triggered by I_T currents, let us adapt the bursting neuron in (10.3) to simulate LGN and PGN cells during sleep. The only change necessary is to reduce the magnitude of the change in Ca^{2+} conductance in the dX/dt equation. Thus, all other equations in (10.3) remain the same, except that the dX/dt equation is changed to:

$$\frac{dX}{dt} = \frac{1}{30}(-X + 6.65(V + 0.86)(V + 0.84))$$ (12.26)

the only change being a reduction of one parameter to 6.65. This makes the equilibrium point of the system asymptotically stable so that the model neuron is no longer an

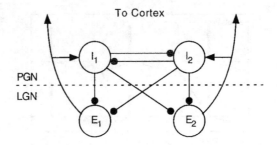

Fig. 12.6 LGN–PGN inhibitory feedback network. Excitatory synapses are shown as arrows and inhibitory as solid circles. Note the presence of inhibitory connections between I_1 and I_2 (gray).

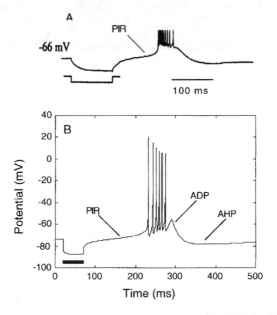

Fig. 12.7 Postinhibitory rebound (PIR) bursting. The data in A (spike amplitudes clipped) are from a PGN neuron (reproduced with permission, Kim *et al.*, 1997), which may be compared with the neural simulation in B. Brief hyperpolarizing stimuli are plotted below each potential record.

endogenous burster. However, the model incorporating (12.26) will generate a burst of spikes caused by PIR. A model PIR burst is compared with data for an LGN neuron in Fig. 12.7. The script **LGN2cell.m** with both synaptic conductance factors equal to zero reproduces this result.

The interaction between one LGN cell (E_1) and one PGN cell (I_1) constitutes the next ingredient in the network dynamics. The script **LGN2cell.m** contains equations for both E_1 and I_1 with synaptic connections incorporated using (12.19). If you rerun **LGN2cell.m** with the excitatory synaptic strength $k_E = 2$ but the inhibitory strength $k_I = 0$, I_1 will generate a burst as a result of EPSPs from E_1. Running the program again with $k_E = 2$ and $k_I = 20$ now completes the inhibitory feedback loop, and the two neurons generate a continuing oscillation in response to the initial brief hyperpolarization of E_1. The

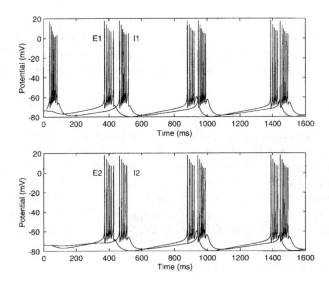

Fig. 12.8 Synchronized bursting in the LGN–PGN network of Fig. 12.6. Brief stimulation of neuron I_1 triggers a synchronized network oscillation at about 2 Hz.

oscillation is similar to the top two spike trains in Fig. 12.8. The following sequence of events cause this oscillation: E_1 excites I_1 to fire a spike burst, but this leads to strong hyperpolarization of E_1 due to IPSPs from I_1. As a result, E_1 generates another burst of spikes triggered by PIR, and thus the sequence repeats itself.

The discussion of negative feedback in Chapter 4 showed that oscillations generally require delays in the feedback loop, and the same is true for the interactions between E_1 and I_1. In the simulation, $\tau_{syn} = 40$ ms, which is within the range typical of GABAb inhibition in the LGN and PGN (Crunelli and Leresche, 1991 Kim *et al.*, 1997). Reducing the synaptic time constant to $\tau_{syn} = 10$ ms and running **LGN2cell.m** with $k_E = 2$ and $k_I = 20$ will abolish the oscillation. In this negative feedback loop, therefore, it is the slow IPSPs that produces the feedback delay necessary to permit oscillations to occur.

The synchronized bursting of the entire LGN–PGN network in Fig. 12.6 can now be understood. The spike trains in Fig. 12.8 show the result of activating the entire synaptically coupled network with a brief 50 ms depolarizing pulse delivered to I_1. As long as τ_{syn} is in the range 40–70 ms (see Exercise 6), the pulse will trigger an E_1–I_1 oscillation as described above. The inhibitory interconnections from I_1 to E_2 and I_2 cause them to begin bursting about 300 ms later as a result of PIR. Synchronization among all four neurons is now achieved as a result of the inhibitory cross-coupling with slow inhibitory time constants in agreement with (12.25). This can be seen by running **LGNsynchrony.m** with $k_E = 2, k_I = 20$, and I–I synaptic weighting $k_{II} = 4$ (this script takes several minutes to run). Note that no attempt has been made to adjust parameters to the detailed characteristics of LGN or PGN neurons, so the oscillations occur at a lower frequency (about 2 Hz) than is produced by models with appropriately tailored parameters (e.g. Destexhe *et al.*, 1993). However, synchronized sleep LGN oscillations can occur at rates comparable to the model (Steriade *et al.*, 1993). Also, the bursting observed experimentally does eventually die out, while the model bursting continues indefinitely. However, burst cessation can be explained by the operation of an additional Ca^{2+} current, I_{CAN} (Partridge and Swandulla, 1988; Bal and McCormick, 1993; Destexhe *et al.*, 1993). This current has also been reported to increase oscillation frequency in model simulations (Destexhe *et al.*, 1993).

The LGN–PGN network in Fig. 12.6 is, of course, only a small segment of the entire neural ensemble within these thalamic nuclei. However, it would be easy to extend the network by incorporating a large spatial array of adjacent E and I neurons. Without even doing so, however, it is possible to draw one more interesting conclusion about the dynamics of such a network. Once E_2 and I_2 begin firing, one would expect inhibition from I_2 to trigger PIR bursting in neighboring E_3 and I_3 of an extended network. This sequence of events would repeat itself, leading to a traveling wavefront of neural activation sweeping across the network. This is exactly what is observed experimentally following localized stimulation of the LGN–PGN network (Kim *et al.*, 1995)!

12.6   Synopsis of phase locking

As many aspects of phase locking and neural synchrony have been explored in this chapter, it may be useful to summarize key points. The phase oscillator introduced by Cohen *et al.* (1982) has proven extremely valuable in providing explanations of many aspects of neural phase locking. The simplest phase oscillator model (12.3) uses $H(\phi)$ from (12.6) and assumes that synaptic effects may be described by pulses following a delay σ. This model predicts many aspects of phase locking and synchrony in more detailed neural networks as long as the synaptic time constant is sufficiently fast. If postinhibitory rebound (PIR) is included, this model predicts the anti-phase locking and ongoing oscillation of the Clione swimming network as shown in Fig. 12.2.

The surprising fact that neural synchronization is frequently due to reciprocal inhibition rather than to excitation can be explained by an extension of the phase oscillator model. As shown by Van Vreeswijk *et al.* (1994), replacement of instantaneous synaptic events with ones described by the alpha function in (12.20) leads to counter-intuitive predictions that depend upon the synaptic time constant: when τ_{syn} is sufficiently large, reciprocal inhibition causes synchrony, and mutual excitation causes anti-phase locking. All these aspects of the mathematical analysis were finally applied to the LGN–PGN network and shown to predict the synchronized, PIR-driven bursting that is actually observed in slow wave sleep. The model developed here is similar in spirit to several in the literature (Wang and Rinzel, 1993; Destexhe *et al.*, 1993, 1994). It is interesting that similar synchronized bursting occurs in epileptic episodes known as absence seizures. The surprising conclusion is that these seizures might best be controlled by *reducing* the strength of reciprocal inhibition in thalamic networks (Gloor and Fariello, 1988; Crunelli and Leresche, 1991; Steriade *et al.*, 1993; Kim *et al.*, 1997).

12.7   Exercises

1. Use the program **EPEPinteractions.m** to determine the characteristics of phase locking when the two neurons are stimulated by currents $I_1 = 0.5$ and $I_2 = 0.4$. (a) Determine the

spike rates of both neurons for coupling strengths $k = 0, 4, 6$. Also estimate the phase lag in each of these cases. (b) Determine the smallest value of k (accurate to 0.1) that is just sufficient to produce phase locking. Set $k = 4$ and $I_1 = 0.5$. Now determine the smallest value of I_2 (to the nearest 0.01) that will produce phase locking and find the frequency. Compare your results with the predictions of the phase oscillator model in (12.15) and (12.16).

2. Explore the conditions governing swimming in Clione using the script **Clione.m**. First determine the minimum synaptic coupling strength k (to the nearest 0.1) that will produce ongoing swimming. Next determine the largest time constant τ_{syn} (set at 1.0 ms in the program) that will still result in swimming. Do this for both the minimum synaptic coupling strength that you just found and for $k = 10$ and 20. Finally, determine the spike frequency for each of these three values of k. Does spike frequency vary significantly as a function of k? Suggest a dynamical explanation based on postinhibitory rebound.

3. Consider the following equations for three coupled phase oscillators with identical frequency inputs:

$$\frac{d\theta_1}{dt} = \omega + H_1(\theta_2 - \theta_1)$$

$$\frac{d\theta_2}{dt} = \omega + H_2(\theta_1 - \theta_2) + H_1(\theta_3 - \theta_2)$$

$$\frac{d\theta_3}{dt} = \omega + H_2(\theta_2 - \theta_3)$$

Assume that the coupling functions H have the forms:

$$H_1(\phi) = a \sin(\phi) \quad \text{and} \quad H_2(\phi) = b \sin(\phi)$$

First, reduce the system to two equations in the phase differences $\phi_1 = \theta_2 - \theta_1$ and $\phi_2 = \theta_3 - \theta_2$. Determine the possible phase-locked states and their stability as functions of a and b.

4. Instead of the alpha function $P(t)$ in (12.20) suppose that the postsynaptic time course is given by an exponential function:

$$P(t) = \frac{k}{\tau_{syn}} \exp\left(\frac{-t}{\tau_{syn}}\right)$$

What is the phase coupling equation that is produced by this function (i.e. the analog of 12.23)? For both excitatory ($k > 0$) and inhibitory ($k < 0$) coupling, determine what phase-locked states are asymptotically stable.

5. Determine the value of τ_{syn} (nearest 0.1 ms) at which **IPSPinteractions.m** switches from antiphase to synchronous oscillations. For all simulations let the two stimulating currents be 1.07 and 1.0. (a) For an inhibitory synaptic strength $k = 5$, what is the transition value of τ_{syn}? (b) If $k = 10$, how does the value of τ_{syn} for synchronization change? Interpret your results based on synchronization models developed in the text.

6. Use the script **LGN2cell.m** to explore the conditions on synaptic connection strengths and time constants necessary for this two-neuron, negative feedback loop to generate sustained oscillations. (a) For $k_E = 2$, $k_I = 20$, determine the maximum and minimum values of τ_{syn} (nearest ms) that will support a bursting limit cycle. (b) Keeping $\tau_{syn} = 40$ ms and $k_E = 2$, what is the smallest value of k_I (to the nearest integer) that will sustain a limit cycle? Explain your results in terms of underlying dynamical principles.

7. Use **LGNsynchrony.m** to determine how network parameters affect synchronous bursting. (a) Let $k_E = 2$, $k_I = 20$, and $k_{II} = 4$ (the I–I cell synaptic connections). Determine the smallest value of τ_{syn} (nearest ms) that permits synchrony. (b) For the same synaptic weights, determine the largest value of τ_{syn} (nearest ms) that permits synchrony. (c) What type of solution is obtained for the following parameters: $k_E = 2$, $k_I = 4$, and $k_{II} = 20$, $\tau_{syn} = 40$ ms? (Note that this program takes about 2 min to run each simulation.)

13   *Swimming and traveling waves*

The analysis of phase locking and synchrony in the previous chapter provides a basis for understanding many motor control systems. One such example has already been discussed: the swimming network of Clione. Indeed, phase-locked limit cycles generated by neural networks form the basis of virtually all rhythmic motor behavior: breathing, swimming, running, chewing, etc. (Stein *et al.*, 1997). Our focus will be on swimming, as some of the most elegant and thorough analyses of motor control lie in this area. The reason for this is straightforward: a swimming animal typically has a density roughly equal to that of the surrounding water, so maintenance of balance against the force of gravity is not a factor in swimming. Balance is a major problem for walking quadrupeds and even more so for bipeds, as the tumbles of any young child demonstrate.

We shall first consider the control of swimming by neural circuits in the spinal cord of the lamprey, a primitive aquatic vertebrate rather like an eel (lampreys are technically cyclostomes, not fish). Lamprey swimming involves not just the generation of nonlinear oscillations in individual segments of the spinal cord but also the coordination of these oscillators along the cord so as to generate smooth traveling waves of neural activity propagating along the body. The conditions for generation of traveling waves lead to a formalism that can also be applied to phase locking in quadruped locomotion.

The chapter closes with a more detailed simulation of swimming in the marine mollusk tritonia. The purpose of this second example is to show that neural network behavior may profitably be studied at different levels of physiological and mathematical detail. Thus, the neurons in the lamprey network will be described at the spike rate level, while those in tritonia will involve simulation of the individual action potentials. Comparison of tritonia and lamprey motor control networks also serves to highlight common principles subserving motor control in widely different species (Pearson, 1993).

13.1   Lamprey central pattern generators

A major concept underlying rhythmic motor behavior is that of a motor central pattern generator. A **central pattern generator** may be defined as a neural network that will produce stereotypical limit cycle oscillations in response to a constant spike rate input. The constant input, sometimes termed a **command signal**, typically serves to trigger the oscillation and to determine its frequency and hence the speed of locomotion. Command signals generally originate from higher motor control centers in vertebrates (although they may also result from sensory input that triggers reflex movements), and they reflect the hierarchical nature of motor control. The classic example is the demonstration by Shik *et al.* (1966) that a cat whose motor cortex has been isolated from the spinal cord can be made to walk or run by electrical stimulation at an appropriate point in the midbrain. Thus, the brain does not 'micro-manage' locomotion but only needs to decide the speed and direction of motion, leaving the neural details of oscillatory pattern generation and coordination to central pattern generators in the spinal cord.

Let us begin our study of central pattern generators with an examination of the lamprey spinal cord. The lamprey is a primitive, eel-like vertebrate which swims by generating traveling waves of neural activity that pass down its spinal cord, thus propelling the animal forward (Gray, 1968; Grillner *et al.*, 1995). Several stages of swimming are depicted in Fig. 13.1B, and a simulation of lamprey swimming can be viewed by running the MatLab script **Swimming_Lamprey.m**. This swimming activity has been intensively studied, and several excellent accounts of the neurophysiology are available (Cohen *et al.*, 1992; Grillner *et al.*, 1995; Grillner, 1996). It has been shown that an intact spinal cord isolated from the rest of the body will generate rhythmic bursts of neural activity appropriate for swimming in response to constant stimulation (which originates in the brainstem in the intact animal). Further experiments have shown that even small sections of the spinal cord are capable of generating rhythmic bursts of spikes in which activity on

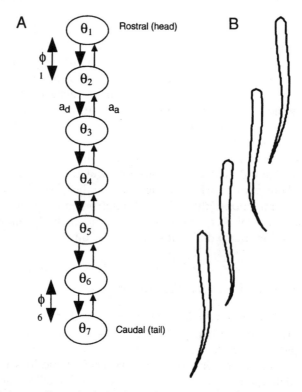

Fig. 13.1 Coupled phase oscillators in the lamprey spinal cord (A), and body shape at four instants during lamprey swimming (B).

one side alternates with that on the other. Such oscillatory networks in the spinal segments cause alternate contraction and relaxation of the body muscles on opposite sides of the body during swimming.

Four types of neurons are present in each segment of the lamprey spinal cord: excitatory interneurons (E), lateral inhibitory interneurons (L), crossed inhibitory interneurons (C), and motoneurons (M). The interconnections among these neurons are depicted in Fig. 13.2A. Several considerations allow us to simplify this neural network. First, the motoneurons are driven by the E neurons and only function to provide output from the segmental oscillator to the muscles; they need not be simulated. Each half of the network contains neurons of all three types, and each receives constant input from command cells in the brainstem during swimming. Within the segment, the E neurons excite both the C and L neurons. The E neurons on each side are also interconnected locally by excitatory

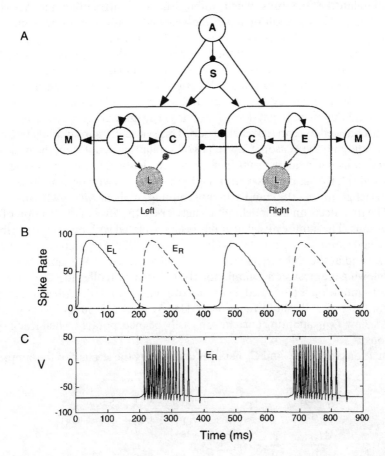

Fig. 13.2 Lamprey spinal central pattern generator. A depicts network connections within and between left and right body sides, both of which are controlled by a command neuron (A) and modulated by a serotonin-releasing neuron (S). E neurons have recurrent excitatory connections, while C cells inhibit all neurons on the other side of the body. Bursting produced by the network equations (13.1) is plotted in B as spike rates, and individual spikes in each burst are approximated in C.

synapses, thus providing a source of local positive feedback. The role of the C neurons is to inhibit all of the neurons on the opposite side of the network, so it is these neurons that produce the alternation of bursting activity via antiphase locking. The L neurons in turn provide delayed inhibition to the C neurons, and it was originally thought that they played a crucial role in terminating the inhibition generated by the C neurons. However, recent neural modeling studies have showed that the L neurons play a small role in the network, and neural bursting is actually terminated by powerful after-hyperpolarizing (I_{AHP}) currents in both the E and C cells (Hellgren *et al.*, 1992; Wallén *et al.*, 1992). Indeed, the most recent simulation of the lamprey oscillator by Lansner *et al.* (1997) eliminates the L neurons entirely, so L cells will be ignored here as well. In the resulting network excitation will cause the E neurons on one side of the spinal segment to excite one another, thereby activating motoneurons and contracting muscles on that side of the body. These E cells simultaneously stimulate the C neurons, which inhibit all neurons on the other side. Activity on this one side terminates as a result of spike frequency adaptation resulting from a slow but powerful I_{AHP} current. This releases neurons on the other side from C cell inhibition; they become active in response to their brainstem inputs; and the cycle repeats itself. As will be seen, modulation of the I_{AHP} current by serotonin (5-HT) is responsible for the extended frequency range of the network (Harris-Warrick and Cohen, 1985; Lansner *et al.*, 1997).

Simulation of the lamprey spinal oscillator will require just four groups of neurons: E and C cells for the left and right sides of the body. The simulation can be further simplified by using just one cell to represent each group. This symmetry subsampling technique was used when studying the Wilson–Cowan (1972) oscillator, and it is mathematically exact under conditions where all neurons in a group are identical in their properties and connectivity. Indeed, the lamprey oscillator has been simulated with both multiple neurons of each type (Hellgren *et al.*, 1992) and with subsampling (Wallén *et al.*, 1992) with very similar results. The final decision in modeling this network is the level of description of the individual neurons. The lamprey oscillator has been simulated with neurons described at the individual spike and ion current level (Grillner *et al.*, 1998; Hellgren *et al.*, 1992; Wallén *et al.*, 1992) and at the spike frequency level (Buchanan, 1992; Williams, 1992; Ekeberg, 1993). Following the most recent simulations by Lansner and colleagues (1997), after which the present simulation is patterned, neuronal responses will be described at the spike rate level using the Naka–Rushton function (2.11). As discussed in Chapters 6 and 8, the effects of a slow I_{AHP} adapting current can easily be incorporated when using spike rate descriptions.

The equations for the E_L and C_L neurons comprising the left half of the lamprey model are:

$$9\frac{dE_L}{dt} = -E_L + \frac{100[A_L + 6E_L - C_R]_+^2}{(64 + gH_{EL})^2 + [A_L + 6E_L - C_R]_+^2}$$

$$\frac{dH_{EL}}{dt} = \frac{1}{\tau_H}(-H_{EL} + E_L)$$

$$9\frac{dC_L}{dt} = -C_L + \frac{100[A_L + 2E_L - C_R]_+^2}{(64 + gH_{CL})^2 + [A_L + 2E_L - C_R]_+^2}$$

$$\frac{dH_{CL}}{dt} = \frac{1}{\tau_H}(-H_{CL} + E_L)$$

$$(13.1)$$

Equations for the right half are of the same form with the subscripts L and R interchanged. The equations in (13.1) represent two neurons described by the Naka–Rushton function (2.11) (recall that the brackets $[\ ]_+$ equal zero when the enclosed argument is negative). The brainstem input is A_L, and the parameters g and τ_H (defined below in 13.3) represent the gain and time constant of the I_{AHP} currents for both neurons. As will be seen, both g and τ_H are dynamically altered through serotonin modulation.

Although this network is eight-dimensional when both sides are considered, the principles underlying its operation are easy to understand. If you run MatLab script **Lamprey.m** with a stimulus $A = 7$, you will obtain the results plotted in Fig. 13.2B. Bursts of spikes are alternately generated by E_L and E_R, and each burst begins at a high rate and then decays as the slow I_{AHP} current produces spike rate adaptation. A function **Make_Spikes.m** has been used to approximate the individual spikes corresponding to the continuous spike rate variable E_R, and these are illustrated in Fig. 13.2C.

From the shape of the bursting pattern, one might conjecture that it is the I_{AHP} currents which cause the termination of each burst, as there can be no inhibition from the other side of the animal when those cells are silent due to suppression by the contralateral C neuron. This can be verified by running **Lamprey.m** again after first setting the variable Ginhib $= 0$ (it lies between rows of asterisks) in the program. This simply cuts off the contralateral inhibition entirely, so the left and right halves of the network become independent. With the input $A = 7$, the bursting pattern in Fig. 13.2B is again observed, except that E_L and E_R now burst almost in phase (the slight phase shift results from a small bias in the initial conditions). This will be even more obvious with a somewhat stronger stimulus, such as $A = 10$. This decoupled half of the lamprey segmental oscillator is sometimes called a **half-center** to indicate that it controls one of two opposed sets of muscles and inhibits the neurons controlling the other set.

As one side of the oscillator will burst even when decoupled from the other, (13.1) can be reduced from eight variables to just two: E_L and H_{EL}. The C_L neuron can be eliminated from consideration along with its I_{AHP} current because its output now has no effect on any other cell in the decoupled network. Under these conditions, (13.1) reduces to:

$$9\frac{dE_L}{dt} = -E_L + \frac{100[A_L + 6E_L]_+^2}{(64 + gH_{EL})^2 + [A_L + 6E_L]_+^2}$$

$$\frac{dH_{EL}}{dt} = \frac{1}{\tau_H}(-H_{EL} + E_L)$$

(13.2)

These equations can now be analyzed using phase plane techniques (the phase plane is plotted in Fig. 2 of the **Lamprey.m** MatLab output). Summarizing the results of such an analysis for $2 \leq A < 17$ (approximately) there is a single steady state that is an unstable spiral point. Due to the nature of the Naka–Rushton function in (13.2), $0 \leq E_L \leq 100$, so the same limits must be true for H_{EL} which is driven by E_L. The Poincaré–Bendixon theorem (Theorem 10) can now be used prove that (13.2) must possess an asymptotically stable limit cycle for this range of A values.

It is notable that the limit cycle for an isolated half of the lamprey is based on principles analogous to the Wilson–Cowan (1972) oscillator discussed in Chapter 8. In both oscillators recurrent excitation drives the upswing at the beginning of each burst. Following this, recurrent inhibition driven by the excitatory activity terminates the bursting in the

Wilson–Cowan (1972) oscillator, and an analogous role is played by the slow but powerful I_{AHP} current in the lamprey oscillator. The importance of recurrent excitation for the lamprey oscillator can be seen by again running **Lamprey.m** with Ginhib = 0 to uncouple the left and right sides. If the strength of recurrent excitation is reduced by setting the program variable $EE = 2$, the limit cycle vanishes, and the system decays to an asymptotically stable steady state.

As each side of the lamprey oscillator is capable of independent oscillation, what then is the role of the crossed inhibition mediated by the C neurons in (13.1)? Basically there are two answers. First, when the two sides are coupled via C neuron crossed inhibition, the sides burst out of phase, which is necessary for the motor control of swimming. Such antiphase locking between interacting oscillators was analyzed in the previous chapter. Second, the crossed C cell inhibition does extend the upper end of the dynamic range from about $A = 17$ to $A = 24$. This is explained by the observation in Chapter 8 that mutually inhibitory neurons with I_{AHP} currents can generate alternating bursts of activity even in the absence of explicit recurrent excitation (see Fig. 8.11).

The effect of serotonin (5-HT) in controlling burst frequency is the final element of lamprey central pattern generation. Harris-Warrick and Cohen (1985) demonstrated that serotonin modulates the swimming frequency in the isolated lamprey spinal cord over about a tenfold range. If the **Lamprey.m** simulation is repeated for $2 \le A \le 24$ the burst frequency curve plotted as a solid line in Fig. 13.3 will be obtained. This range from about 0.5 to 10.0 Hz is the frequency range of bursting observed in the isolated lamprey spinal cord (Wallén *et al.*, 1992; Hellgren *et al.*, 1992). Above this range both the experimental preparation and the model in (13.1) generate steady firing behavior rather than bursting. In the model this is due to a supercritical Hopf bifurcation. The onset of bursting at low stimulus levels, however, occurs at a saddle-node bifurcation, which permits the bursts to be initiated with large amplitude and low frequency.

What is the physiological basis for the almost linear relationship between stimulus intensity A and burst frequency? Lansner *et al.* (1997) have argued cogently that it is due to modulation of the I_{AHP} currents by serotonin. Lansner *et al.* (1997) suggest that

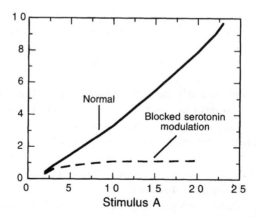

Fig. 13.3 Burst frequencies produced by (13.1) for a range of stimulus values A. Blocking serotonin prevents the network from generating higher frequency bursts.

serotonin-secreting neurons bathe the lamprey network at low stimulus intensities, thereby producing a relatively slow and weak I_{AHP} current. At higher stimulus levels, however, serotonin release is inhibited by the stimulus, which results in stronger and faster I_{AHP} currents. This is shown diagrammatically in Fig. 13.2A, where the brainstem command neuron A inhibits the serotonin releasing neuron S. The effects of serotonin on (13.1) are simulated by varying the I_{AHP} current parameters τ_H and g as functions of the stimulus level A according to the formulas:

$$g = 6 + (0.09 + A)^2$$
$$\tau_H = \frac{400}{1 + (0.2 + A)^2} \tag{13.3}$$

These relationships result in an increase in the gain g and a decrease in the time constant τ_H of the I_{AHP} current as A increases, which produces the burst frequency modulation results of the Lansner *et al.* (1997) network model. To see the effects of blocking serotonin modulation of the network, set the variable HT5 = 0 in **Lamprey.m**, and run the program for a range of stimuli A. The burst frequency is now almost constant as shown by the dashed line in Fig. 13.3.

The lamprey segmental oscillator embodies a number of basic neural principles. As exemplified by the Lansner *et al.* (1997) model, each half of the segmental oscillator will burst as a result of two factors: recurrent excitation initiates the burst, while spike frequency adaptation produced by the I_{AHP} current terminates it. Antiphase oscillations of the left and right sides are produced by the crossed C neuron inhibition. Finally, burst frequency increases with command neuron activation because serotonin modulation of the I_{AHP} current is reduced by command neuron inhibition. Thus, an interplay among ionic currents, excitatory and inhibitory network connectivity, and neuromodulation produces the swimming rhythm of lamprey central pattern generators.

13.2   Traveling waves and swimming

Command signals to central pattern generators in individual lamprey segments are sufficient to evoke oscillatory responses with alternate left and right activity, but this alone cannot produce swimming. The oscillations in successive segments must also be coordinated so that waves of contraction ripple along the body producing smooth forward movement as shown by the MatLab animation **swimming_Lamprey.m**. Experiments using the isolated lamprey spinal cord have shown that waves of neural activity travel along the cord even in the absence of muscles or sensory feedback (Cohen *et al.*, 1982). Thus, the neural architecture within the spinal cord must be capable of generating traveling waves by itself.

As the lamprey spinal cord contains approximately 100 segments, modeling the entire array with 800 differential equations based on (13.1) would be formidable. Fortunately, the phase oscillator model of Cohen *et al.* (1982) can be used to predict conditions under which neural connections between segments will lead to the traveling waves of neural activity observed experimentally. Let us therefore consider chains of coupled phase oscillators as depicted in Fig. 13.1A. Each of the N oscillators (only seven are depicted) is coupled to its neighbors in the adjacent segments. This represents a simplification, because intersegmental coupling extends beyond nearest neighbors in the lamprey spinal cord (Grillner *et al.*, 1995; Grillner, 1996; Cohen *et al.*, 1992). Nevertheless, nearest neighbor coupling can elucidate the principles underlying traveling wave generation during lamprey swimming. Swimming is produced by a wave of segmental oscillator activity traveling from the rostral (head) to the caudal (tail) of the animal in forward swimming or in the reverse direction for backward swimming. Furthermore, experiments have shown that there is always approximately one wavelength of neural activity per body length of the animal independent of swimming speed (Grillner *et al.*, 1995; Cohen *et al.*, 1992). These observations will be explained by studying a chain of coupled phase oscillators.

A **traveling wave** in a chain of coupled neural oscillators is defined as a phase-locked solution in which there is a constant phase difference $|\phi| < \pi/2$ between adjacent segments. In the lamprey with 100 segments per body length or wavelength, therefore, $|\phi| \approx \pi/50$. The intersegmental phase differences are defined as $\phi_i = \theta_{i+1} - \theta_i$. This means that a traveling wave solution with $\phi_i < 0$ corresponds to a wave traveling from head to tail of the animal, while $\phi_i > 0$ corresponds to a wave traveling from the tail back to the head.

The N equations for a lamprey spinal network with nearest neighbor coupling represent a natural generalization of (12.2), and equations analogous to (12.3) are derived by setting $\phi_i = \theta_{i+1} - \theta_i$ for $i = 1$ to $N - 1$ (see Fig. 13.1A). The sinusoidal form for coupling functions in (12.6) will again be used. Assuming that all descending coupling (rostral to caudal) is identical, and all ascending coupling (caudal to rostral) is also identical, only two coupling coefficients a_d and a_a a are required along with a synaptic delay σ. Thus, the equations for the N coupled phase oscillators are:

$$\frac{d\theta_1}{dt} = \omega_1 + a_a \sin(\theta_2 - \theta_1 + \sigma)$$

$$\frac{d\theta_i}{dt} = \omega_i + a_a \sin(\theta_{i+1} - \theta_i + \sigma) + a_d \sin(\theta_{i-1} - \theta_i + \sigma) \quad i = 2, \ldots, N-1 \qquad (13.4)$$

$$\frac{d\theta_N}{dt} = \omega_N + a_d \sin(\theta_{N-1} - \theta_N + \sigma)$$

Pairwise subtraction and substitution of $\phi_i = \theta_{i+1} - \theta_i$ now leads to an $(N-1)$-dimensional system for the phase differences:

$$\frac{d\phi_1}{dt} = \Delta\omega_1 + a_a[\sin(\phi_2 + \sigma) - \sin(\phi_1 + \sigma)] + a_d \sin(-\phi_1 + \sigma)$$

$$\frac{d\phi_i}{dt} = \Delta\omega_i + a_a[\sin(\phi_{i+1} + \sigma) - \sin(\phi_i + \sigma)]$$

$$+ a_d[\sin(-\phi_i + \sigma) - \sin(-\phi_{i-1} + \sigma)] \quad i = 2, \ldots, N-2 \qquad (13.5)$$

$$\frac{d\phi_{N-1}}{dt} = \Delta\omega_{N-1} - a_a \sin(\phi_{N-1} + \sigma) - a_d[\sin(-\phi_{N-1} + \sigma) - \sin(-\phi_{N-2} + \sigma)]$$

where $\Delta\omega_i = \omega_{i+1} - \omega_i$.

As in the previous chapter, asymptotically stable steady states of (13.5) correspond to phase-locked solutions of (13.4). For different values of the frequency differences, $\Delta\omega_i$, which are determined by command signals from the lamprey's brain, many different phase-locked solutions are possible. However, our goal is to find traveling wave solutions. To determine conditions under which (13.5) has a traveling wave solution, set $d\phi_i/dt = 0$ and substitute $\phi = \phi_i$ for all i. This constant phase difference defines a traveling wave. The resulting equations for the steady state traveling wave then become:

$$\Delta\omega_1 + a_d \sin(-\phi + \sigma) = 0$$

$$\Delta\omega_i = 0 \qquad i = 2, \ldots, N - 2 \qquad (13.6)$$

$$\Delta\omega_{N-1} - a_a \sin(\phi + \sigma) = 0$$

Thus, a traveling wave solution requires that the frequency differences $\Delta\omega_i$ must vanish for $i = 2, N - 2$, so all frequencies must be the same except for the first and last. The first and last equations in (13.6) yield distinctive conditions for traveling waves, because these two equations describe the boundaries (head and tail) of the chain of oscillators. Note that $\Delta\omega_1$ and $\Delta\omega_{N-1}$ cannot both be zero unless $\sigma = 0$ or π. As a first traveling wave solution, therefore, let us choose $\Delta\omega_{N-1} = 0$ and solve the first and last equations in (13.6) with the result:

$$\Delta\omega_1 = -a_d \sin(-\phi + \sigma)$$

$$\phi = -\sigma \qquad (13.7)$$

As $\Delta\omega_1 = \omega_2 - \omega_1 = -a_d \sin(2\sigma)$, ω_1 must be slightly larger than all remaining ω_i (which are identical) for a traveling wave solution to exist. This assumes that $2\sigma < \pi$, which is appropriate for synaptic delays between segments. Thus, we have obtained a traveling wave solution in which the wave moves from the head to the tail, generating forward swimming in the lamprey. Backward swimming can be generated by setting $\omega_1 = 0$ and again solving (13.6).

The stability of solution (13.7) to the coupled oscillator system (13.5) is determined by the following Jacobian matrix in which the equilibrium solution $\phi = -\sigma$ has been substituted:

$$\overset{\leftrightarrow}{A} = \begin{pmatrix} -a_a - b & a_a & 0 & \cdots & 0 \\ b & -a_a - b & a_a & & \\ 0 & b & -a_a - b & a_a & \vdots \\ \vdots & & & \ddots & \\ 0 & & \cdots & b & -a_a - b \end{pmatrix} \qquad (13.8)$$

where $b = a_d \cos(2\sigma)$.

The eigenvalues of this matrix will have negative real parts as long as $a_a > 0$, $a_d > 0$, and $\sigma \leq \pi/4$. Thus, the traveling wave solution to (13.5) is asymptotically stable.

We have now proved that nearest neighbor coupling between the segmental oscillators in the lamprey spinal cord can generate traveling waves producing either forward or backward swimming behavior. All aspects of this swimming behavior are produced by the segmental oscillators and their coupling within the spinal cord. The lamprey's brain only needs to send a constant command signal (i.e. a constant spike rate) to all segments to produce the desired swimming rate plus a slightly larger signal to either the first or last segment to determine the direction of swimming. The remaining aspect of lamprey swimming to be explained is the fact that the traveling wave of neural activity in the spinal cord has a wavelength equal to the body length of the animal. As explored in Exercise 2, this can be done by choosing appropriate nonzero values for both $\Delta\omega_1$ and $\Delta\omega_{N-1}$. The lamprey spinal cord actually includes both nearest neighbor connections as modeled here and longer range connections. The role of these longer range connections in producing traveling waves with a wavelength equal to the body length of the organism is discussed by Cohen and Kiemel (1993) and Ermentrout and Kopell (1994).

13.3 The swimming lamprey

The mathematical analyses above incorporate all of the principles thought to underlie neural control of lamprey swimming. Ekeberg (1993) demonstrated this in truly elegant fashion by simulating the entire lamprey spinal chord plus the biomechanics of muscle contraction in alternately curving the animal's body first left and then right. His simulations incorporated the mass of the animal, sensory feedback to the spinal networks from stretch receptors, and drag forces of water on the body. Some of the biomechanical principles of water drag forces had been developed by Gray (1968).

Although the biomechanical details go beyond the scope of this book, it is possible to get the flavor of lamprey swimming from a few simple considerations. Let us start with the neural activity generated by the segmental oscillator as shown in Fig. 13.2B. As activity in E_L and E_R causes alternate activation of the left and right motor neurons M, the appropriate waveform for control of the muscles is $E_L - E_R$. This is plotted by the solid line in Fig. 13.4. As one cycle of this oscillation forms a traveling wave with one wavelength per body length, the abscissa in Fig. 13.4 is body length in cm, the lamprey averaging 30 cm in length (Ekeberg, 1993). To simplify computation of traveling waves, the neural activity has been approximated by the sum of two sine waves shown by the dashed line in the figure. From the discussion of muscle models in Chapter 3, let us assume that the neural activity sets the threshold length and therefore the force generated by the muscles. Given Newton's second law of motion, $F = MA$, the change in muscle length will be inversely proportional to mass per body segment. As discussed by Ekeberg (1993), lamprey body mass is approximately proportional to the width of the body, which is constant for about 10 cm from the head and then tapers linearly to the tail. Accordingly, the same level of neural activity will produce a greater bend in the angle between body segments near the tail than near the head. Finally, the drag forces of water result in the lamprey's forward progress being less than the distance of wave travel along its body per

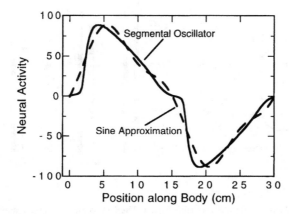

Fig. 13.4 Neural activity of the lamprey segmental oscillator ($E_L - E_R$) in (13.1) compared with an approximation based on the sum of two sines.

unit time. The MatLab script **Swimming_Lamprey.m** approximates these factors to produce a computer animation of a swimming lamprey as seen from above, and several frames of the animation are depicted in Fig. 13.1B. Although the biomechanics of turning to the left or right have not been simulated, the MatLab script **LampreyTurn.m** with a small difference in stimulus levels to the left and right sides (e.g. 7 and 5) will produce more prolonged neural activity on one side of the lamprey than on the other, which is the neural basis of turning. This completes the analysis of swimming behavior in lampreys, from serotonin modulation of I_{AHP} currents to the complete animal! For further details of the neural modeling and biomechanics of swimming, see Gray (1968) and Ekeberg (1993).

13.4 Quadruped locomotion

Terrestrial quadruped locomotion is a much more complex problem than lamprey swimming and is accordingly much less well understood. One reason for this is that quadrupeds have several different gaits that are used for different speeds of locomotion, walking and galloping being the two extremes. A second reason is that quadruped locomotion must be stable with respect to the force of gravity, or the animal will fall. Nevertheless, it is possible to gain some insight into phase locking for stable quadruped locomotion by developing a phase oscillator network for control of four limbs.

During locomotion, each limb of a quadruped is controlled by an oscillator in the spinal cord that generates alternate flexion and extension of the limb. During flexion the limb is pulled up and swung forward, while during extension the limb makes contact with the ground and supports the body weight. Furthermore, sensory feedback is not necessary to generate the flexion–extension cycle (Pearson, 1993; Shik *et al.*, 1966). Walking and trotting obviously require coordination among the four oscillators controlling the four

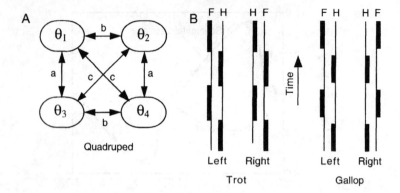

Fig. 13.5   Phase oscillator model for quadruped gaits. The four oscillators and synaptic connection parameters a–c are depicted in A. The phase relationships between left and right front (F) and hind (H) limbs in trotting and galloping are illustrated in B.

limbs, so let us examine the network of four phase oscillators in Fig. 13.5A. The equations are:

$$\frac{d\theta_1}{dt} = \omega + a\sin(\theta_3 - \theta_1 + \sigma) + b\sin(\theta_2 - \theta_1 + \sigma) + c\sin(\theta_4 - \theta_1 + \sigma)$$

$$\frac{d\theta_2}{dt} = \omega + a\sin(\theta_4 - \theta_2 + \sigma) + b\sin(\theta_1 - \theta_2 + \sigma) + c\sin(\theta_3 - \theta_2 + \sigma)$$

(13.9)

The synaptic coefficients a, b and c determine the strengths of the front–hind, left–right, and diagonal coupling respectively (see Fig. 13.5A). Equations for only two of the phase oscillators, those controlling the left (θ_1) and right (θ_2) front legs, are necessary here. The equations for the hind legs become redundant, because quadruped trotting and galloping both require that the hind leg on the same side of the animal be 180° out of phase with the front leg. A simplified view of these two gaits is depicted in Fig. 13.5B. In trotting, each diagonal pair of legs (e.g. left front and right hind) is in phase but is 180° out of phase with the other diagonal pair. In galloping the phase locking switches so that the front legs are in phase but 180° out of phase with the hind legs. The oscillator equations for the hind legs are thus redundant because of the equalities:

$$\theta_4 = \theta_2 + \pi, \qquad \theta_3 = \theta_1 + \pi \qquad (13.10)$$

These equations hold true (as idealizations) for both trotting and galloping, and substitution into (13.9) yields:

$$\frac{d\theta_1}{dt} = \omega - a\sin(\sigma) + b\sin(\theta_2 - \theta_1 + \sigma) - c\sin(\theta_2 - \theta_1 + \sigma)$$

$$\frac{d\theta_2}{dt} = \omega - a\sin(\sigma) + b\sin(\theta_1 - \theta_2 + \sigma) - c\sin(\theta_1 - \theta_2 + \sigma)$$

(13.11)

where use has been made of the identity $\sin(x + \pi) = -\sin(x)$ for any x. This reduces the network to a pair of coupled oscillators like those already considered. Proceeding as before and setting $\phi = \theta_2 - \theta_1$ gives:

$$\frac{d\phi}{dt} = (c - b)\sin(\phi + \sigma) + (b - c)\sin(-\phi + \sigma) \qquad (13.12)$$

Setting $d\phi/dt = 0$, and using trigonometric identities for $\sin(\pm\phi + \sigma)$, shows that the Phase-locked equilibrium states must satisfy:

$$(b - c)\cos(\sigma)\sin(\phi) = 0 \qquad (13.13)$$

As the synaptic delay $\sigma > 0$, the solutions to (13.13) will be $\phi = 0, \pi$ so long as $b \neq c$.

This shows that the four legs of the phase oscillator quadruped will be phase locked into one of two gaits as long as the left–right coupling coefficient b is different from the diagonal coupling coefficient c. The solution $\phi = 0$ produces a gallop in which the front two legs are in phase, while $\phi = \pi$ corresponds to a trot in which the front two legs are out of phase (and therefore given 13.10, the diagonal legs are in phase). To determine which gait is asymptotically stable, let us carry out a linearized stability analysis on (13.12) for $\phi = 0$. The resulting linearized equation shows that $\phi = 0$ will be asymptotically stable if:

$$(b - c)\cos(\sigma) > 0 \qquad (13.14)$$

If the inequality in this expression is reversed, then $\phi = 0$ is unstable, but $\phi = \pi$ is asymptotically stable. Assuming $\sigma < \pi/2$, which is appropriate given the brief duration of synaptic transmission relative to a single cycle of running, the balance between the left–right coupling strength b and the diagonal coupling strength c determines which gait will be stable. These coefficients are presumably regulated by excitatory and inhibitory command signals from higher brain centers.

Before leaving the quadruped example, it is worth mentioning that inhibitory coupling can lead to phase locking as effectively as excitatory coupling can. If $b = 0$ in (13.14), inhibitory coupling between diagonal limbs ($c < 0$) leads to asymptotic stability of the solution $\phi = 0$. The extent to which different quadruped gaits are determined by inhibitory as opposed to excitatory coupling is at present unknown.

13.5 Tritonia swimming

Recent research indicates that some rather general principles underlie the control of swimming in both vertebrates and invertebrates (Getting, 1989a,b; Pearson, 1993; Stein *et al.*, 1997). As a basis for comparison with the vertebrate lamprey, let us briefly examine the neural network underlying swimming in the marine mollusk Tritonia. The discussion that follows is based on both the experimental and neural modeling work of Getting and colleagues (Getting and Dekin, 1985; Getting, 1988, 1989a,b, 1997).

Spikes, decisions, and actions

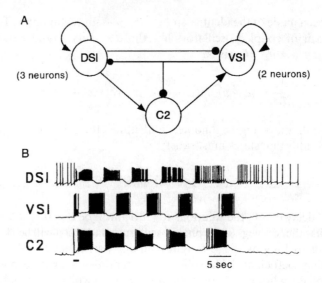

Fig. 13.6 Tritonia swimming network. The three cell types with excitatory (arrows) and inhibitory (solid circles) interconnections are illustrated in A. B shows five cycles of rhythmic firing behavior triggered by a brief stimulus (bar at bottom left) (reproduced with permission, Getting and Dekin, 1985).

The Tritonia swimming network is illustrated in Fig. 13.6A, while the neural activity of the three neuron types is depicted in Fig. 13.6B (data from Getting and Dekin, 1985). Following brief activation of the dorsal swim interneurons (DSIs), the network produces 4–7 cycles of rhythmic activity. As is evident from Fig. 13.6B, activation of the DSIs leads to activation of the C2 neurons and finally to activation of the ventral swim interneurons (VSIs); then the cycle repeats itself. Swimming activity is produced as a result of DSI cells exciting motor neurons that produce dorsal flexions of the mollusk's tail, while VSI cells excite other motor neurons causing ventral flexion. Thus, the mollusk swims away.

The anatomy of the network, depicted in Fig. 13.6A, illustrates the interconnections among all three classes of neurons. The DSI cells (three in the network) are mutually excitatory and also excite the lone C2 neuron while inhibiting the two VSIs. The VSI are also mutually excitatory. Simply looking at the anatomical diagram, it might seem difficult to suppose that this network could generate a sustained oscillation. As with the lamprey, the key to understanding lies in the interplay between the anatomical connections and the intrinsic ionic currents within the different cell types. Both the DSI and C2 exhibit prominent I_{AHP} currents that produce prominent spike frequency adaptation, while the VSI cells have a prominent I_A current. Thus, the DSIs driving the network adapt sufficiently so that activity from the VSI cells (driven by the DSI via the C2 neuron) will quench their activity. The VSIs, despite their mutual excitation, inhibit their C2 activation source, and their I_A current causes their firing to reduce to such a low level that the DSIs can shut them off.

A simulation of the Tritonia swimming network requires a number of decisions to be made, and the simulation developed here is similar in spirit to those of Getting (1989a,b) and Frost *et al.* (1997). First, let us choose to simulate the individual neurons at the level of ion currents and individual spikes. Second, the network can be simplified by treating the individual neurons of the DSI and VSI types as identical and using symmetry subsampling as in the lamprey model. In addition, the simulation must incorporate I_{AHP} currents in the DSI and C2 neurons and an I_A current in the VSI cells. These modeling decisions are embodied in the MatLab script **Tritonia.m** which uses neural equations from Chapters 9 and 10. With the I_{AHP} current magnitude set to 1.62, the program produces the spike trains plotted in Fig. 13.7 (it takes about 4 min). As with the experimental data in Fig. 13.6B, brief stimulation of the DSIs leads to a repetitive series of bursts in which the DSI and VSI cells fire in antiphase, while the C2 firing overlaps the DSI–VSI transition. The experimental bursting data are over 10 times slower than the model (0.15 Hz versus 2.4 Hz), but this can be rectified by rescaling the time constants appropriately.

The network behavior raises two prima facie puzzling issues: (a) how do the DSIs become re-excited after the cessation of VSI inhibition, and (b) why does activity cease after a finite number of cycles (six in this case)? The answer to the first question is that the model DSI generate very long duration EPSPs via their recurrent excitatory connections, and this permits excitation to survive through the period of active inhibition by the VSI cells, whose IPSPs are considerably shorter (see Fig. 13.7). Getting and Dekin (1985) have reported that synaptic potentials in Tritonia vary in duration from 0.65 s up to 20.0 s, so a model synaptic time constant $\tau_{syn} = 320$ ms for the DSI–DSI connections is certainly well within range, even when the model is scaled by a factor of about 20. Bursting in the model finally ceases as a result of the very slow increase in hyperpolarization mediated by an I_{AHP} current with a time constant $\tau_{AHP} = 1250$ ms. This means that when the magnitude of this current is sufficiently large (1.62 in this case), the resting state of the network is always asymptotically stable, although it takes about 3000 ms for transient activity to die out once the network is excited. This interplay between the duration of the DSI–DSI

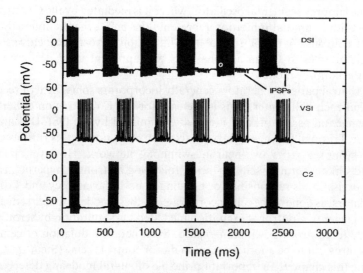

Fig. 13.7 Simulation of tritonia swimming network in Fig. 13.6A.

recurrent excitation and the strength of the slow I_{AHP} current is explored further in Exercise 5.

A command neuron believed to drive the Tritonia swimming network has recently been discovered (Frost and Katz, 1996; Frost et al., 1997). As this neuron appears to receive excitatory connections from C2 and inhibitory connections from the VSI cells, it bursts along with the other neurons under physiological conditions. Thus, this cell could be incorporated into the network quite easily, and the same principles would explain how it becomes re-excited following inhibition.

13.6 Principles of central pattern generation

This chapter has attempted to fulfill the promise inherent in the title of this book by integrating ion currents and spikes into an explanation of the behavior of an organism. In doing so, several common principles of central pattern generator organization have emerged. First, motor pattern generation emerges from a complex interplay between intrinsic membrane properties of the constituent neurons and network properties defined by their synaptic interconnections. As emphasized by Getting and Dekin (1985, p. 7) with respect to Tritonia: 'Pattern generation ... emerges as a property of the network as a whole. No single cell nor synapse is capable, in isolation, of producing the oscillatory burst pattern. The ability of this network to generate patterned activity depends on the interaction of both synaptic connectivity and the intrinsic cellular properties of each neuron.' Our analysis has shown that the same is true of the lamprey segmental oscillator and even of the two-neuron Clione swimming system.

Second, virtually all central pattern generators incorporate mutual inhibition between the two half-centers controlling opposed sets of muscles. This inhibition is present between the DSI and VSI cells in Tritonia (see Fig. 13.6A) and between the left and right halves of the lamprey segmental oscillator, where it is mediated by the C neurons (see Fig. 13.2A). In Clione, mutual inhibition represents the only synaptic interaction between component neurons. A major role of this inhibition is to force neurons controlling antagonistic muscles to fire in antiphase, as was apparent from decoupling the half-centers of the lamprey model.

Finally, central pattern generators generally incorporate some form of recurrent excitation within each half-center. This is the case for the E neurons on either side of the lamprey segmental oscillator, and it is also exemplified by the DSI–DSI and VSI–VSI excitation in Tritonia. Clione might seem to be an exception to this rule, as there are no excitatory synapses in its two-neuron swimming network. However, postinhibitory rebound acts like a form of seif-excitation following a strongly hyperpolarizing event. Recurrent excitation is responsible for bursting in both the lamprey and Tritonia swimming oscillators, as none of the cells incorporated in these models can burst due to intrinsic membrane currents. Control of excitation mediated burst duration by serotonin enables the lamprey to control its swimming speed. Such neuromodulation of central pattern generators turns out to be another common motor control theme (Stein et al., 1997).

In closing this chapter, an important principle of neural modeling deserves emphasis. Three different central pattern generators have been modeled and analyzed here: Clione, lamprey, and Tritonia. The modeling itself has been carried out at three different levels of generality. The Clione and Tritonia models incorporated detailed descriptions of multiple ion currents and resulted in simulations at the level of individual spikes. The lamprey segmental oscillator, on the other hand, was simulated at the spike rate level, with only the I_{AHP} current being explicitly included. Finally, the very general phase oscillator model used to analyze traveling waves in the lamprey ignores all details of the segmental oscillators except that they do in fact oscillate. The deep insight provided by these examples is that neural network behavior can be elucidated through studies conducted at several different levels of generalization. For example, the lamprey oscillator has been simulated both at the individual ion current and spike level (Grillner *et al.*, 1988; Hellgren *et al.*, 1992; Wallén *et al.*, 1992) and at the spike frequency level (Buchanan, 1992; Williams, 1992; Ekeberg, 1993), and the results have been similar, although not identical. In Ekeberg's (1993) simulation of the entire spinal cord and biomechanics of lamprey swimming, description at the spike frequency level was necessary in order for the problem to remain conceptually and computationally tractable. Bigger simulations utilizing more complex descriptions of individual neurons are not necessarily better. Sometimes, the sheer size of a simulation obscures an understanding of the dynamical and physiological principles upon which network operation is based. As Kopell (1988, p. 400) phrased it: 'The level of detail needed to account for a particular phenomenon depends on the phenomenon. In working with a "realistic", i.e., detailed model, one can lose sight of why it is working, and whether it would continue to work if some ad hoc part of the description is modified.'

Excellent sources of material on other central pattern generators include the edited volumes by Selverston (1985), Cohen *et al.* (1988), and Stein *et al.* (1997). All are excellent sources for both deepened understanding of motor control and modeling projects.

13.7 Exercises

1. Determine the constant phase lag ϕ and frequency ω for backward swimming in (13.6) by setting $\Delta\omega_1 = 0$ and solving the resulting equations.

2. As discussed in the chapter, lamprey swimming involves a traveling wave with a wavelength equal to the body length of the animal. As there are about $N = 100$ segments, that means that the constant phase lag between segments must be $\phi = \pm\pi/50$. Using this value, solve (13.6) for the values of $\Delta\omega_1$ and $\Delta\omega_{99}$ that will produce this wavelength for (a) forward and (b) backward swimming.

3. Insects have six legs arranged roughly hexagonally, and one of their main gaits during locomotion involves locking three alternate legs in phase and moving them while the remaining three legs provide a stable triangular platform to support the body. Based upon the example of quadruped locomotion in the chapter, develop a phase oscillator model for insect walking by reducing the six equations to two for the activity of adjacent legs. Assume that there is only coupling between adjacent oscillators and that this reciprocal coupling has strength k. For what values of k will the two sets of three legs oscillate in anti-phase?

4. Consider the equation for a lamprey half-center in (13.2). Assume a stimulus $A = 10$, and use (13.3) to determine the parameter values g and τ_H. Now determine the bifurcation

value for the excitatory connectivity factor (set at 6 in 13.3) for a half-center to generate a limit cycle. What type of bifurcation does the system go through as the excitatory connectivity is increased?

5. Use the script **Tritonia.m** to explore the dependence of network bursting on the following two parameters: I_{AHP} current strength and τ_{syn} for the DSI–DSI recurrent excitatory connections (termed TauED in the script). (a) Keep TauED $= 320$ ms and determine the number of bursts obtained for the following values of I_{AHP} current magnitude: 1.5, 1.6, 1.7, and 1.8. (b) For a current magnitude of 1.6, determine the number of bursts for each of the following DSI–DSI excitatory synaptic time constants: 400, 320, 300, and 250 ms. Under what conditions does the network appear to generate an asymptotically stable bursting pattern as opposed to a short transient set of bursts? (This program will take about 5 min per simulation.)

14   *Lyapunov functions and memory*

All of the nonlinear dynamical problems encountered so far have been analyzed by linearizing around the equilibrium points of the system to determine stability. While this is by far the most useful method in practice and one that has enabled us to understand phenomena as complex as bursting and neural synchrony, it is not the most general method. Linearized stability analysis of nonlinear systems fails in two important ways. First, it tells us nothing in cases where the associated linear system has a pair of pure imaginary eigenvalues (see Theorem 8). In this case higher order nonlinear terms determine the stability. Second, the linearized analysis tells us nothing about the range of initial conditions for which trajectories will decay to an asymptotically stable steady state. All that linearized analysis can guarantee is that trajectories starting in a sufficiently small neighborhood will decay to an asymptotically stable steady state, but no estimate of neighborhood size is provided.

In a series of deep and elegant theorems published in 1892, the Russian mathematician Lyapunov solved the problems inherent in linearized stability analysis by developing a geometric interpretation of the state space trajectories defined by the dynamics. His results permit us in principle to analyze the stability of any linear or nonlinear dynamical system whatsoever, and they provide a means for estimation of the neighborhood within which all trajectories decay to the origin (the 'domain of attraction' see below). Finally, Lyapunov function theory leads to a generalization of the conservation of energy concept so important in theoretical physics. As a result, it is sometimes possible to solve for trajectories in state space analytically even when the temporal dependence of solutions can only be determined by simulation. As we shall see, Lyapunov functions make it possible to analyze neural networks for long-term memory, such as those found in the hippocampus. An excellent, in-depth treatment of Lyapunov's work may be found in the book by La Salle and Lefschetz (1961).

14.1   Geometry and evolution of state functions

In order to understand Lyapunov functions, it will first be necessary to develop a number of geometric concepts associated with the states of a dynamical system. Although these concepts generalize to any size system, it will help in visualizing them if we restrict examples to two-dimensional dynamical systems defined by the phase plane variables x and y. A **state function** of the system is any scalar function of the system variables that has continuous partial derivatives throughout the state space. Using $U(x, y)$ as a two-dimensional example, suppose that:

$$U = 3x^2 + xy + y^2 \qquad (14.1)$$

The partial derivatives of U, denoted by $\partial U / \partial x$ and $\partial U / \partial y$, are obtained by treating all variables other than the differential variable as constant while performing the differentiation. Thus, the two partial derivatives of U are:

$$\frac{\partial U}{\partial x} = 6x + y, \quad \text{and} \quad \frac{\partial U}{\partial y} = x + 2y \qquad (14.2)$$

As both partial derivatives are continuous for all x and y, U is a state function of the system. However, the function $U = \sqrt{(x+y)}$ is not a state function, as the partial derivatives are discontinuous wherever $(x + y) = 0$.

State functions are valuable in analyzing differential equations because they create a landscape of hills and valleys in the state space. We will mainly be concerned with valleys and whether trajectories flow downward to steady states located at the bottom. First, therefore, let us define a mathematical valley with a steady state at the bottom. The key concept is that of a **positive definite** state function in a region R of state space:

Definition: A state function $U(x)$ in an N-dimensional state space x is **positive definite** in a region R surrounding an internal singularity x_0 if:

(a) $U(x_0) = 0$ and
(b) $U(x) > 0$ in R if $x \neq x_0$.

Consider a simple dynamical system with a unique singular point at the origin:

$$\frac{dx}{dt} = -x - y - 3xy^2$$

$$\frac{dy}{dt} = -y + x \qquad (14.3)$$

This might be considered to be a nonlinear version of the retinal negative feedback circuit in Chapter 3, as (14.3) differs only by the addition of the nonlinear term in the dx/dt equation. Both of the following functions satisfy the definition of positive definite state functions around the origin:

$$U_1 = x^2 + y^2$$

$$U_2 = x^2 + y^2 - y^4 \qquad (14.4)$$

For U_1 the region R includes the entire plane, while the region within which U_2 is a positive definite state function is limited by $|y| < 1$. Evidently, positive definite state functions are not unique, so many different ones can be defined in different regions enclosing each singular point of a dynamical system.

Given a positive definite state function U defining a mathematical valley around a singularity, let us determine whether trajectories of the system flow downhill to the singularity at the bottom. This requires calculation of the temporal change of a positive definite state function along trajectories of the system. As $U(x, y)$ does not explicitly include time, it can vary with time only as a result of the time variation of x and y as the system evolves. This can be calculated using the partial derivatives of U:

$$\frac{dU}{dt} = \frac{\partial U}{\partial x}\frac{dx}{dt} + \frac{\partial U}{\partial y}\frac{dy}{dt} \qquad (14.5)$$

As dx/dt and dy/dt are always determined by the dynamical equations of the system as explicit functions of x and y, (14.5) can be evaluated exactly at any point in the phase plane. Taking the dynamics from (14.3) and U_1 from (14.4) as an example, evaluation of (14.5) produces the result:

$$\frac{dU_1}{dt} = 2x\frac{dx}{dt} + 2y\frac{dy}{dt}$$

$$= -2x^2 - 6x^2y^2 - 2y^2 \qquad (14.6)$$

Thus, at any point in the state space dU_1/dt can easily be calculated. Equation (14.5) readily generalizes to an N-dimensional system with variables x_i as follows:

$$\frac{dx_i}{dt} = F_i(x_1 \cdots x_N)$$

$$\frac{dU}{dt} = \sum_{i=1}^{N} \frac{\partial U}{\partial x_i} F_i \qquad (14.7)$$

14.2 Lyapunov functions and asymptotic stability

These ideas may now be related to the asymptotic stability of an equilibrium point in the following intuitive way. A positive definite state function in a region surrounding an equilibrium point at x_0 defines a mathematical valley with x_0 as its lowest point. If all trajectories of the system flow like water down to the bottom of the valley, all will end up at x_0. This, however, satisfies the definition of asymptotic stability: all trajectories within a neighborhood of a singular point approach it as $t \to \infty$. These ideas are made precise in the **Lyapunov Function Theorem**:

Theorem 12 (Lyapunov Functions): Consider an N-dimensional dynamical system defined by the equation:

$$\frac{d\vec{x}}{dt} = \vec{F}(\vec{x})$$

Let $U(x)$ be a positive definite state function of the system in a region R surrounding an equilibrium point at x_0. If either of the following conditions is satisfied by dU/dt evaluated along trajectories, then x_0 is asymptotically stable:

(a) dU/dt is negative definite within R, or

(b) $dU/dt \leq 0$ in R, but except at $x = x_0$, all trajectories passing through points where $dU/dt = 0$ return to regions where $dU/dt < 0$.

A state function U satisfying the conditions of Theorem 12 is called a **Lyapunov Function** in honor of its discoverer. A state function is **negative definite** in a region surrounding a singular point if $U < 0$ within R except at the singularity where $U = 0$.

The Lyapunov function U defines a mathematical valley with x_0 as its absolute minimum point. Condition (a) of the theorem states that all trajectories will flow down to x_0 like water seeking its lowest point if $dU/dt < 0$ along these trajectories. Our evaluation of dU_1/dt in (14.6) produced a negative definite function in the (x, y) plane, so trajectories of the system in (14.3) satisfy condition (a) of the theorem for any region enclosing the origin. It follows from the Lyapunov Function Theorem that the equilibrium point at the origin of (14.3) is asymptotically stable.

Condition (b) generalizes Theorem 12 to the case where trajectories pass through regions where $dU/dt = 0$ so long as they subsequently re-enter regions where $dU/dt < 0$. To complete our metaphor of trajectories as water flowing down the sides of a state space valley, condition (b) means that trajectories may briefly enter a lake where flow is level so long as they eventually enter an outlet stream again flowing downhill from the lake. Recall discussion of the Hopf bifurcation theorem in Chapter 8. The van der Pol equation (8.13) was analyzed exactly at the bifurcation point where the linearized eigenvalues were purely imaginary. The equations at this point are:

$$\frac{dx}{dt} = y$$

$$\frac{dy}{dt} = -\omega^2 x - x^2 y \tag{14.8}$$

It is easy to verify that $(0, 0)$ is the only steady state and that linearized analysis cannot be applied because of the pure imaginary eigenvalues. However, the positive definite state function $U = (\omega^2 x^2 + y^2)$ has the time derivative along trajectories:

$$\frac{dU}{dt} = -x^2 y^2 \tag{14.9}$$

Thus, $dU/dt < 0$ except on the x and y axes, where $dU/dt = 0$. When $y = 0$, $dy/dt \neq 0$ except at the origin, so trajectories at points where $y = 0$ must move into regions where $dU/dt < 0$. Similarly, when $x = 0$, $dx/dt \neq 0$ (again excluding the origin), so trajectories must again move into regions where $dU/dt < 0$. This satisfies condition (b) of Theorem 12, so U is a Lyapunov function for eqn (14.8) in the (x, y) plane, and the steady state at the origin is asymptotically stable. A Lyapunov function has thus enabled us to prove the asymptotic stability of an equilibrium point under conditions where a linearized analysis fails.

Lyapunov functions define a topographic landscape associated with the trajectories of a dynamical system, and this enables us to develop an even more powerful theorem concerning asymptotic stability. First recall that the linearized stability analysis embodied in Theorem 8 only proves that trajectories originating sufficiently close to an asymptotically stable steady state will approach it as $t \to \infty$. However, Theorem 8 says nothing about the meaning of 'sufficiently close.' Lyapunov functions remove this restriction because they define all points of the valley within which the equilibrium point lies. Intuitively, therefore, you might guess that any trajectories originating within the walls of the valley would flow downhill to the equilibrium point. This intuition is basically correct, but to make it precise we must define the **domain of attraction** or **domain of asymptotic stability** (synonymous terms) of a singularity at x_0.

Definition: The **domain of attraction** of an asymptotically stable singular point at x_0 is the region of the state space defined by the set of all states a, such that if $x(t) = a$ for any t, then $x(t) \to x_0$ as $t \to \infty$.

As trajectories cannot intersect in an autonomous system, the domain of attraction is simply the set of all points on all trajectories that approach the equilibrium point at x_0 asymptotically. Earlier examples, such as short-term memory networks, have shown that many nonlinear systems have more than one asymptotically stable equilibrium point, so the domain of attraction of any given equilibrium is generally only a portion of the state space. Otherwise stated, a nonlinear dynamical system can have many valleys surrounding different equilibrium states with ridges of mountains separating them.

Let us assume that we have found a Lyapunov function U for an equilibrium point of a dynamical system at x_0. By definition a Lyapunov function must satisfy Theorem 12,

---

Therorem 13: Let $U(x)$ be a Lyapunov function in a region R surrounding an equilibrium point at x_0. Then there is a range of constants $\alpha > 0$ defining the regions $U(x) < \alpha$ with the properties:

(a) $U(x) < \alpha$ is a region enclosing x_0 but no other singularities; and
(b) $U(x) < \alpha$ lies entirely within R.
Then $K = \max(\alpha)$ defines a region D_A bounded by $U(x) = K$ within which all trajectories approach x_0 asymptotically.

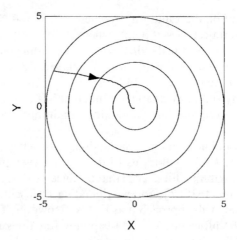

Fig. 14.1 Circular contours of U_1 in (14.4), a Lyapunov function for (14.3). One trajectory of (14.3) is shown by the heavier line and arrow.

so U will delimit a region R within which the conditions of the theorem hold. From this information one can prove that a constraint $U < K$ defines a subregion D_A contained within the domain of asymptotic stability of x_0.

The key concept implied by this theorem is that of a closed contour in two dimensions, or a spheroidal surface in higher dimensions, that encloses the equilibrium point. So long as the region within this contour lies entirely within R, Theorem 12 guarantees that all trajectories within the region will decay asymptotically to the equilibrium point. Regarding U as defining a topographic valley with equilibrium at the bottom, the constant K in Theorem 13 is the highest point on the valley walls from which nothing can flow out of the valley. Although you might guess that D_A would encompass the domain of attraction, this is generally not true. This is because D_A depends not only on the dynamical equations of the system but also on the choice of Lyapunov function, which is not unique, as the following example will show. Thus D_A will usually define only a subregion of the total domain of attraction of a given equilibrium.

The feedback dynamics in eqn (14.3) have already been analyzed using the Lyapunov function U_1 from (14.4). From the evaluation of dU_1/dt in (14.6), it was shown that U_1 is a Lyapunov function within the entire (x, y) phase plane. Contours defined by $U_1 = \alpha$ are simply circles surrounding the origin, as shown in Fig. 14.1. Therefore, the domain of attraction of the singularity at the origin is the entire phase plane: all trajectories arising from all possible initial conditions approach the origin asymptotically. One such trajectory is shown moving downhill across the contours of constant U_1 by the heavy line in Fig. 14.1.

Estimation of the domain of attraction for (14.3) using U_2 from (14.4) gives a much different result. U_2 itself is only positive definite in the strip $|y| < 1$. The derivative of U_2 is:

$$\frac{dU_2}{dt} = -2x^2 - 6x^2y^2 - 2y^2 + 4y^4 - 4xy^3 \tag{14.10}$$

This function is clearly negative definite very near the origin, where the two quadratic terms dominate, and so it is a second Lyapunov function for (14.3). However $dU_2/dt > 0$ for all points $x = 0$, $|y| > 1/\sqrt{2}$. Without further ado, it is apparent that the domain within which U_2 is a Lyapunov function must be restricted to some closed contour of U for which $|y| < 1/\sqrt{2}$.

This example leads to two conclusions. First, Lyapunov functions for any particular dynamical system are not unique. Indeed, there will frequently be an infinite number of Lyapunov functions for any given singularity. For example, the Lyapunov function U_2 in (14.4) can be generalized to:

$$U_2 = x^2 + y^2 - ay^{4n} \tag{14.11}$$

for any $a \geq 0$ and any integer $n \geq 1$. All of this infinite class of functions are positive definite state functions in regions of varying size surrounding the origin, and all have negative definite derivatives near the origin. For each of these functions, however, the estimate of the domain of attraction obtained from Theorem 13 will be different. We proved above, using U_1 as a Lyapunov function, that the domain of attraction for the unique equilibrium point of (14.3) comprises the entire phase plane. This demonstrates that the vast majority of Lyapunov functions that might have been chosen will only prove that a small subspace is part of the domain of attraction of the equilibrium. Nevertheless, all Lyapunov functions are guaranteed to provide a rigorous boundary on a finite region within the generally larger domain of attraction.

Two further points should be made here concerning Lyapunov functions before moving on to consider their application to neural problems. First, Lyapunov also proved several instability theorems analogous to Theorem 13. These will not be considered here, as they are almost never used in practice, but those interested are referred to the excellent treatment by La Salle and Lefschetz (1961). Second, Theorem 12 states that if a Lyapunov function exists within a region surrounding a singular point, then that point must be asymptotically stable. Lyapunov (1892) also proved that if a singular point is asymptotically stable, then a Lyapunov function must exist within some surrounding region. Thus, all asymptotically stable singular points have associated Lyapunov functions, and the existence of a Lyapunov function is both necessary and sufficient for asymptotic stability.

14.3 Divisive feedback revisited

The development of Lyapunov function theory above leaves one crucial question unanswered: how does one construct a Lyapunov function for a given dynamical system? There is good news and bad. First, the bad: there is no truly general algorithm for constructing Lyapunov functions. Now the good news: several fairly simple techniques work in most situations. One of these techniques will be illustrated here using divisive or shunting neural feedback as an example. Divisive feedback has been implicated as part of the retinal light adaptation circuitry (Wilson, 1997), and very similar mechanisms appear to regulate neural responses in cortical gain control circuits (Wilson and Humanski, 1993; Heeger, 1992; Carandini and Heeger, 1994). Based on eqn (6.3), a very simple example of such a circuit is described by the equations:

$$\frac{dx}{dt} = -x + \frac{10}{1+y}$$

$$\frac{dy}{dt} = -y + 2x$$

(14.12)

where x and y may be thought of as retinal bipolar and amacrine cell responses respectively and the value 10 in the first equation is the stimulus intensity (see Chapter 6). The equilibrium states of this system are $x=2$, $y=4$; and $x=-2.5$, $y=-5$. Only the first equilibrium state has physiological significance, and we proved in Chapter 6 that all trajectories originating from initial conditions $x \geq 0$, $y \geq 0$ must forever remain in the first quadrant. Thus we shall restrict consideration to the equilibrium $x = 2$, $y = 4$.

Linearization of (14.12) was used in Chapter 6 to prove that the equilibrium state in the first quadrant is asymptotically stable, but that analysis revealed nothing about the domain of asymptotic stability. Let us therefore see what we can learn by applying Lyapunov function theory. To do so, let us first state a simple theorem about functions of a particular form:

Theorem 14: Suppose that $F(x,y)$ and $G(x,y)$ are continuous functions with continuous derivatives defined over a region of the x, y plane. Suppose further that the set of points where $F = 0$ and $G = 0$ is finite. Then the function:

$$U(x,y) = \frac{1}{2}F^2 + \varepsilon FG + \frac{1}{2}G^2$$

is positive definite in some region surrounding each point $F=0$, $G=0$ as long as $|\varepsilon| < 1$.

A proof of this theorem is easy. If $\varepsilon = \pm 1$, $U = (1/2)(F \pm G)^2$, and $U = 0$ only when $F = -G$ (or $F = G$). As the only term in U that can ever be negative is εFG, reducing the absolute value of ε below that required to make U a perfect square guarantees that U will be positive definite.

To apply Theorem 14 to the divisive feedback network in (14.12), let us equate F and G with the right-hand sides of the two equations:

$$F = -x + \frac{10}{1+y}$$

$$G = -y + 2x$$

(14.13)

Note that $F = 0$ and $G = 0$ are the isocline equations for (14.12), which is the insight behind this Lyapunov function method (Wilson and Humanski, 1993). Letting $\varepsilon = 0$, the following function is obtained from Theorem 14:

$$U = \frac{1}{2}\left(-x + \frac{10}{1+y}\right)^2 + \frac{1}{2}(-y + 2x)^2 \tag{14.14}$$

Because $F = 0$ and $G = 0$ are the isocline equations for (14.12), U is a positive definite state function around the equilibrium point at $x = 2$, $y = 4$. Because of the definitions of F and G in (14.13), $F = dx/dt$ and $G = dy/dt$, so the derivative of U along trajectories takes the form:

$$\frac{dU}{dt} = F^2 \frac{\partial F}{\partial x} + FG\left(\frac{\partial F}{\partial y} + \frac{\partial G}{\partial x}\right) + G^2 \frac{\partial G}{\partial y} \tag{14.15}$$

A simple generalization of Theorem 14 shows that any function of the form $aF^2 + bFG + cG^2$ will be *negative definite* so long as:

$$a < 0, \quad c < 0, \quad |b| < 2\sqrt{ac} \tag{14.16}$$

From (14.13):

$$\frac{\partial F}{\partial x} = -1, \quad \frac{\partial G}{\partial y} = -1, \quad \left(\frac{\partial F}{\partial y} + \frac{\partial G}{\partial x}\right) = -\frac{10}{(1+y)^2} + 2 \tag{14.17}$$

Thus, by (14.16) dU/dt in (14.15) is negative definite as long as $y > \sqrt{2.5} - 1$, a value derived from the third equation in (14.17). This completes the demonstration that U is a Lyapunov function for the neural system in (14.12) within the region R defined by $y > \sqrt{2.5} - 1$.

The Lyapunov function in (14.14) may now be used to obtain an estimate of the domain of asymptotic stability of the singularity at $x = 2$, $y = 4$ for the divisive feedback system in (14.12). Theorem 13 indicates that we need only plot contours of the Lyapunov function U and find the largest one that forms a closed curve surrounding the singularity and lying entirely within the region R where U is a Lyapunov function. The program **LyapunovFB.m** plots the contours of U from (14.14) as shown in Fig. 14.2. As R is defined by $y > \sqrt{(2.5)} - 1$, this value ($y \approx 0.582$) can be substituted into eqn (14.14), and the minimum value of $U(x)$ determined. This occurs at $x \approx 1.497$. Therefore, the largest contour of $U(x, y)$ that encloses a region lying entirely in R is $U(1.497, 0.582) = 14.54$. This contour defines the region D_A in Theorem 13 and is indicated in Fig. 14.2 along with three representative trajectories of (14.12). Note that D_A is actually a very conservative estimate of the domain of asymptotic stability, which in fact includes the entire first quadrant. However, D_A is the best estimate obtainable from the Lyapunov function in (14.14).

The approach to constructing Lyapunov functions using Theorem 14 is generally useful given the flexibility in choosing a value of ε such that $|\varepsilon| < 1$. In two-dimensional systems for which linearized stability analysis using Theorem 8 demonstrates the asymptotic

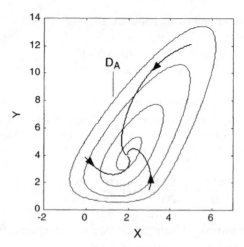

Fig. 14.2 Contours of the Lyapunov function (14.14) for (14.12). The outer contour is an estimate of the region D_A from Theorem 13 that is within the domain of asymptotic stability. Three trajectories of (14.12) are plotted by heavier lines with arrows.

stability of an equilibrium point, there will always be a value of ε for which U in Theorem 14 is a Lyapunov function in a region surrounding the singularity, an observation explored in the problems. Theorem 14 can also be generalized to higher dimensional systems, but the number of coefficients analogous to ε grows rapidly.

14.4   Conservative systems and predator–prey interactions

Nonlinear dynamical systems in neuroscience possess equilibrium points that are either asymptotically stable or else unstable in virtually all cases. In theoretical physics, however, enormously important insights have been obtained through application of the law of conservation of energy, which was discovered by the same Helmholtz who is famous for his pioneering research into visual and auditory function (Helmholtz, 1909). The law of conservation of energy simply states that the energy (a state function) of an isolated physical system will remain constant as the system evolves in time according to Newton's laws of motion. Conservation of energy is first of all an idealization, as it requires that the energy conserving system be totally isolated from all external forces. In the limit, therefore, only the universe as a whole conserves energy. Second of all, conservation of energy represents only a small subset of nonlinear dynamical systems that conserve more general quantities. For mathematical completeness, and because of its intellectual importance and elegance, therefore, this section will provide a brief examination of conservative dynamics, which may be viewed as a special or limiting case of Lyapunov function theory.

   Although conservative systems have not played a major role in biological or neural theory, there are nevertheless several elegant and interesting examples that deserve mention. Cowan (1970) developed a conservative formulation of interactions between excitatory and inhibitory neurons, while Goodwin (1963) produced an analogous conservative formulation for the oscillations inherent in cellular biophysics. Rather than developing these examples here, however, let us explore the original example of a conservative biological system: the Lotka–Volterra equations (Lotka, 1924). The equations will be developed here using Volterra's description of predator–prey interactions. Let us assume that a population of wolves, W, survives by catching and eating rabbits, R, on an isolated island (such as Isle Royale National Park). The rabbits, being herbivores, are assumed to multiply exponentially in the absence of predation by the wolves. The wolves, on the other hand, will die out exponentially if there are no rabbits to eat. Assume that the frequency of wolves encountering and eating rabbits is proportional to the product of their populations RW. Then the following equations will describe the population dynamics:

$$0.5\frac{\mathrm{d}R}{\mathrm{d}t} = R - \frac{1}{20}RW$$

$$0.1\frac{\mathrm{d}W}{\mathrm{d}t} = -W + \frac{1}{500}RW$$

(14.18)

The time constants here are in years and have been chosen to produce equilibrium values in a reasonable range. They indicate that it takes roughly 20 wolf–rabbit encounters to produce a rabbit kill but about 500 rabbit encounters to provide enough protein (coniglio alla cacciatori!) to enable the wolves to reproduce.

Analysis of the Lotka–Volterra equations (14.18) indicates that the steady states are $R=0$, $W=0$, and $R=500$, $W=20$. The former represents the case where both species have died out, while the latter represents the only equilibrium balance between predators and prey. The Jacobian of (14.18) shows that $(0, 0)$ is an unstable saddle point, but $(500, 20)$ produces a pair of pure imaginary eigenvalues. Thus, Theorem 8 tells us nothing about the stability of $R=500$, $W=20$, and we must turn to Lyapunov function theory. In this case Theorem 14 does not work (try it if you are skeptical), but eqn (14.18) has a particular property that will enable us to derive a positive definite state function directly. To develop this method, note first that the right side of each equation in (14.18) can be written as a product of a function of x and a different function of y. A general form of which (14.18) is a particular example is therefore:

$$\frac{\mathrm{d}x}{\mathrm{d}t} = F_x(x)F_y(y)$$

$$\frac{\mathrm{d}y}{\mathrm{d}t} = G_x(x)G_y(y)$$

(14.19)

If the second equation here is divided by the first, the following relationship is determined between $\mathrm{d}y$ and $\mathrm{d}x$ along trajectories:

$$\frac{\mathrm{d}y}{\mathrm{d}x} = \frac{G_x(x)G_y(y)}{F_x(x)F_y(y)}$$

(14.20)

Rearrangement and integration now yields the result:

$$\int \frac{F_y(y)}{G_y(y)}\,\mathrm{d}y = \int \frac{G_x(x)}{F_x(x)}\,\mathrm{d}x + K \qquad (14.21)$$

where the constant of integration K has been added because indefinite integrals have been taken. This proves the following theorem concerning **constants of motion:**

Theorem 15: For any dynamical system of the form (14.19) the function:

$$U(x,y) = \int \frac{F_y(y)}{G_y(y)}\,\mathrm{d}y - \int \frac{G_x(x)}{F_x(x)}\,\mathrm{d}x$$

is a **constant of motion** assuming that the integrals can be evaluated. Any trajectory in the phase plane with initial conditions $(x_0,\ y_0)$ will evolve forever along the locus defined by $U(x,\ y) = U(x_0,\ y_0)$, a constant.

Thus, when a two-dimensional system such as (14.19) permits solution for a constant of the motion, we can determine the locus of trajectories in the phase plane exactly even though the temporal evolution of the system can only be approximated using Runge–Kutta methods.

Let us now apply Theorem 15 to the Lotka–Volterra equations (14.18). The first step is to write the functions on the right-hand side as products:

$$\frac{\mathrm{d}R}{\mathrm{d}t} = 2R\left(1 - \frac{1}{20}W\right)$$

$$\frac{\mathrm{d}W}{\mathrm{d}t} = 10W\left(-1 + \frac{1}{500}R\right) \qquad (14.22)$$

where the time constants have also been moved over to the right. The two relevant integrals defined by Theorem 15 are now:

$$\int \frac{1 - 1/20W}{10W}\,\mathrm{d}W = \frac{1}{10}\ln(W) - \frac{1}{200}W$$

$$\int \frac{-1 + 1/500R}{2R}\,\mathrm{d}R = -\frac{1}{2}\ln(R) + \frac{1}{1000}R \qquad (14.23)$$

so the function $U(R,\ W)$ becomes:

$$U(R,\ W) = \frac{1}{10}\ln(W) + \frac{1}{2}\ln(R) - \frac{1}{200}W - \frac{1}{1000}R \qquad (14.24)$$

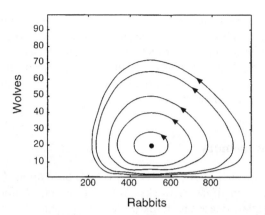

Fig. 14.3 Contours of constant $U(R, W)$ in (14.24), which are trajectories of the Lotka–Volterra equations (14.18). This is a conservative nonlinear system.

This function is plotted in Fig. 14.3 for several constant values using the MatLab script **LotkaVolterra.m**. It is apparent that all trajectories are stable oscillations about the equilibrium point (black dot), which is therefore a center. When the wolf population is low, the rabbits multiply rapidly. This, however, provides more food for the wolves, whose population now multiplies as they gorge on rabbits. In consequence, the rabbit population drops, which leads to a decline in the wolf population. The rabbits now begin to multiply rapidly again, and the cycle repeats itself.

Any oscillation to which Theorem 15 applies is a **conservative oscillation**, because the quantity $U(x, y)$ remains constant along the trajectory. Conservative oscillations are not limit cycles, because there are infinitely many conservative oscillations within any given neighborhood of each such oscillation. Limit cycles, by definition, are isolated oscillations with no other oscillatory solutions within some finite neighborhood. Conservative oscillations in nonlinear systems represent the nonlinear generalization of cosine oscillations surrounding a center in linear systems.

There is a simple relationship between conservative oscillations and Lyapunov functions. If, as in the Lotka–Volterra example above, all contours $U(x, y) = K$ are closed and encircle a center, then $U(x, y)$ can always be transformed into a positive definite state function. First, one subtracts the value of U at the equilibrium point or center. The resulting function will then be either positive or negative definite in a region surrounding the center (this region comprises the first quadrant in the Lotka–Volterra example). If the function is negative definite, multiplication by -1 will convert it into a positive definite function. This produces a mathematical valley with the steady state as its minimum point. Unlike a Lyapunov function, however, trajectories do not flow downhill in this case; rather they circulate at a constant elevation on the valley walls. Systems with a constant of motion thus are generalizations of the concept of conservation of energy, while systems described by Lyapunov functions are ones in which energy is lost through generalized 'frictional' forces.

Before leaving the topic of conservative systems, it should be emphasized that Theorem 14 is not applicable to all conservative systems. In physics, conservative systems are generally Hamiltonian systems for which the constant of motion cannot be obtained using Theorem 14. Such systems have never found application in biology and will therefore be passed over here. Mathematical aspects of Hamiltonian systems are covered in any good text on classical mechanics.

14.5 Long-term memory

The major context in which Lyapunov functions have been applied in neuroscience is in the study of long-term memory networks. Among those developing memory networks that are biologically plausible are Kohonen (1989), Hopfield (1982, 1984), and Grossberg (1987) (see also Cohen and Grossberg, 1983). Back-propagation networks are considerably less plausible biologically and will not be examined here, although Elman *et al.* (1996) have provided some striking insights into development using them. Readers interested in the mathematical foundations of back propagation are referred to the excellent book by Hertz *et al.* (1991).

Long-term memory is generally divided into two categories frequently referred to as declarative and procedural. Declarative memory in turn may be subdivided into episodic and semantic memory. Episodic memory stores information about important events or episodes in our lives along with aspects of the spatial contexts associated with them. Thus, many of us still have vivid recollections of our first date with the person we eventually married including what they wore, where we ate, etc. Such memories must necessarily be encoded in one trial because of the uniqueness of the events themselves. This may be contrasted with those procedural memories that can only be acquired with repeated practice, such as learning how to ski or play the violin.

Different brain areas are involved in the encoding and storage of these different types of memories. Among the brain areas known to be involved in long-term memory, the hippocampus plays a central role in declarative or episodic memory (O'Keefe and Nadel, 1978; Squire and Zola-Morgan, 1991). Classic studies have shown that humans with severe hippocampal damage form no long-term memories after the time of the damage, although episodic memories from earlier periods of their lives remain intact. The hippocampus is thus believed to store and consolidate episodic memories for a period from a few days to a few weeks before the information is transferred back to higher cortical areas (eg. inferior temporal cortex in vision) for permanent storage. Thus, hippocampal damage disrupts any further episodic memory storage but does not destroy memory information already permanently laid down in the cortex.

In 1949 Donald Hebb proposed that long-term memory storage required the strengthening of synapses contingent upon concurrent presynaptic and postsynaptic activity. If neurons i and j fire at rates R_i and R_j during episodic learning, therefore, the **Hebb Rule** implies that the weight of the synaptic connection between them would become:

$$w_{ij} = kR_iR_j \tag{14.25}$$

Years of research into the physiological basis of Hebb synapses has led to a general consensus that NMDA receptors are involved in both the hippocampus and cortex. As summarized by Bliss and Collingridge (1993), NMDA receptors can only be activated if two events occur within about a 100–200 ms time window. First, the presynaptic neuron must release the transmitter glutamate as a result of presynaptic spike activity so that it can bind to NMDA receptor sites. Second, the postsynaptic cell must be depolarized sufficiently to remove Mg^{2+} ions that normally block NMDA receptor channels. Elegant recent experiments have demonstrated that action potentials triggered at the axon hillock of hippocampal neurons actually propagate back up the major dendrites as well as along the axon (Magee and Johnston, 1995, 1997; Johnston *et al.*, 1996; see Chapter 15). This postsynaptic dendritic spike activity is currently thought to provide the depolarization necessary for NMDA receptor activation. Following NMDA receptor activation, further chemical events within the postsynaptic cell lead to synaptic potentiation that lasts for hours or even days in the hippocampus (Bliss and Collingridge, 1993). Although these final postsynaptic chemical events are not yet well understood, it is nevertheless clear that NMDA receptors provide a physiological basis for the Hebb (1949) synapse.

The model of long-term episodic memory to be developed here is based on the circuitry of the CA3 region of the hippocampus. The circuitry of this area is shown in the beautiful anatomical drawings of Cajal (1911) reproduced in Fig. 14.4A and schematically in Fig. 14.4B. In addition to receiving synapses from external axons originating in other parts of the brain, CA3 hippocampal neurons have extensive recurrent axon collaterals that synapse onto many of the neighboring CA3 neurons. These recurrent synapses shown by open arrowheads in Fig. 14.4B, contact the apical dendrites of CA3 pyramidal cells. These recurrent synapses in CA3 will be assumed to be modifiable according to a Hebb rule. In addition to this recurrent excitation, all of the hippocampal output axons in the model also contact an inhibitory interneuron that provides recurrent subtractive inhibition to all of the hippocampal pyramidal cells (neuron G in the diagram). Marr (1971) was one of the first to propose a hippocampal model involving both recurrent excitation via Hebb synapses and inhibition.

During pattern learning, external axon inputs cause some model CA3 pyramidal cells to become active, and the Hebb synapses between each pair of active neurons are then modified. For simplicity in the current implementation, the Hebb Rule was used in the form:

$$w_{ij} = kH(R_i - 0.5M)H(R_j - 0.5M)$$

where

$$H(x) = \begin{cases} 1 & x > 0 \\ 0 & x \le 0 \end{cases} \tag{14.26}$$

In this expression M is the maximum firing rate of the neurons (100 spikes/s in the simulation). Thus the Hebb synapses in the model are strengthened only when both presynaptic and postsynaptic neurons are firing at greater than half their maximum rates. Furthermore, the synapse changes to a fixed value k when modified. The version of the

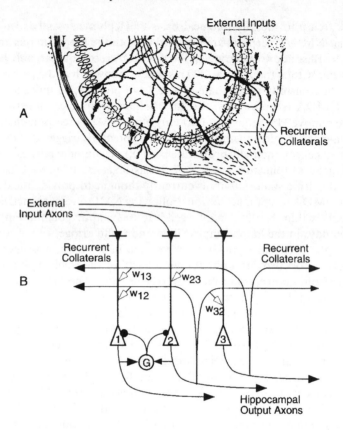

Fig. 14.4 Recurrent CA3 hippocampal network. The anatomical diagram in A is from Cajal (1911), while the schematic in B depicts the connectivity of the neural model in (14.27). Recurrent collaterals contact other pyramidal cells with Hebbian synapses (open arrowheads), and all pyramids generate inhibitory feedback via the G interneuron. For clarity, only a subset of connections is shown.

Hebb Rule in (14.26) thus imposes both a threshold for modification and a saturation on the strength of each synapse, and it requires concurrent presynaptic and postsynaptic activity for modification to occur. This form of Hebb Rule was first employed in the associative memory model of Willshaw *et al.* (1969). Once modified by learning, a Hebb synapse in the model is assumed to remain in its strengthened state for the duration of the simulation. As pairs of neurons are reciprocally interconnected, (14.26) implies that the synaptic connections will be modified identically: $w_{ij} = w_{ji}$. This symmetry of Hebbian modified connections has been assumed by both Hopfield (1982, 1984) and Cohen and Grossberg (1983), and it will be adopted here.

The CA3 network simulated here incorporates 256 pyramidal cells in a 16×16 array plus one interneuron providing feedback inhibition. Each of the 256 pyramidal cells provides recurrent Hebb synapses onto apical dendrites of the other 255 pyramids but not onto itself. Thus, each CA3 pyramid has one synapse from an external axon, 255 modifiable Hebb synapses from the other CA3 pyramids, and one inhibitory synapse from the interneuron. Using spike rate descriptions of the neurons, the dynamical equations for the network are:

$$10\frac{dR_i}{dt} = -R_i + \frac{100(\sum_{j=1}^{256} w_{ij}R_j - 0.1G)_+^2}{\sigma^2 + (\sum_{j=1}^{256} w_{ij}R_j - 0.1G)_+^2}$$

$$10\frac{dG}{dt} = -G + g\sum_{i=1}^{256} R_i$$

(14.27)

where $i = 1, \ldots, 256$. The time constants have been set to the reasonable value of 10 ms, and the semi-saturation constant $\sigma = 10$. The modifiable synaptic weights $w_{ij} = 0.016$ if the synapse has been modified during training and are zero otherwise. Finally, the inhibitory interneuron G has input synaptic weights $g = 0.076$.

As observed by Rolls and Treves (1998), the recurrent collaterals of CA3 cells suggest that CA3 hippocampus is an autoassociative network. This means that stimulation by a portion of any previously learned pattern will cause the network to recall the entire pattern. The CA3 autoassociative network described by eqn (14.27) is implemented in the MatLab script **CA3memory.m**. The network has learned to recognize four different patterns by modification of the Hebb synapses according to (14.26), each pattern being represented by activity in 32 CA3 neurons. If you run **CA3memory.m**, you can choose which of the four patterns to recall. As a stimulus to the network, the program randomly selects a block equal to about one-third of the pattern (10–12 active cells) and degrades this partial image with noise consisting of 20 randomly activated cells. Four external inputs generated in this manner are shown in the left column of Fig.14.5. The program then simulates (14.27) for a total of 100 ms, plotting the active neurons as white on gray throughout the computation. As shown by the results in the right column of Fig. 14.5, the network invariably recalls the entire learned pattern while eliminating all of the noise from the input image. Note that the external input remains on for only the first 20 ms of the simulation, so the final network output not only recalls and completes the pattern, it also retains the recalled information in short-term memory as a pattern of self-sustained neural activation!

If you run **CA3memory.m** several times with each pattern, you will see that a range of different inputs varying in the nature of the added noise will all generate each correct output. This means that the network *generalizes* from novel stimuli to the most similar stimulus in its previous experience provided that the similarity is great enough (about 33% in this model)! In addition to describing the generalization inherent in memory, this may also account for the *deja vu* (literally, 'already seen') experience wherein a novel situation is experienced as being remembered. This is particularly salient in recall of the pattern 'Assoc' illustrated at the bottom of Fig. 14.5, where a novel stimulus containing information about one of the three elements in the memory record evokes the entire memory of the additional associated elements.

Before turning to a mathematical analysis of (14.27), a few final comments concerning CA3 hippocampus and the model are in order. Regarding storage capacity, calculations

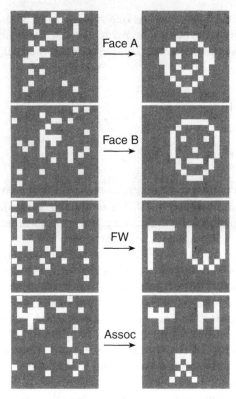

Fig. 14.5 Patterns stored in CA3 model via Hebb synapses (right column) and examples of noisy inputs (left column) that triggered their recall by **CA3memory.m**.

developed by Hertz *et al.* (1991) and by Rolls and Treves (1998) indicate that the 256 neuron CA3 network can store about 12 patterns with 32 neurons active in each. You can save and recall an additional pattern in the CA3 model by running the script **CA3learning.m** (see Exercise 5). Estimates for human hippocampus, which contains about 3×10^6 neurons, range from about 50 000 to 70 000 on the assumption that only 1–2% of CA3 neurons are active in any given memory (Rolls and Treves, 1998). Given a round figure of 1000 minutes awake per day (16.67 hours), this corresponds to storage of a memory every 12 s for 10–14 days, the typical range of storage in the hippocampus. Thus, a scaled-up version of our model would have a capacity appropriate to the human hippocampus but would take impossibly long to simulate on most computers! Finally, note that each pair of patterns in Fig. 14.5 shares several active neurons with each of the other stored patterns. Inhibitory feedback from the G neuron in (14.27) suppresses units weakly activated by this cross-talk. This works effectively so long as the active units common to any pair of patterns represent a relatively small percentage of the entire patterns. Brain areas providing input to CA3 are believed to contain networks that minimize overlap between patterns to be stored in CA3 (Rolls and Treves, 1998).

Let us now see how Lyapunov functions can be used to prove that the stored patterns are asymptotically stable steady states of the dynamical equations (14.27). For simplicity, let us first focus on the two-neuron short-term memory system developed in Chapter 6. In the present context, these neurons are assumed to have formed mutually excitatory connections through previous modification of Hebb synapses. The resulting equations are:

$$10 \frac{dR_1}{dt} = -R_1 + \frac{100(0.25R_2)_+^2}{\sigma^2 + (0.25R_2)_+^2} \equiv F_1(R_1, R_2)$$

$$10 \frac{dR_2}{dt} = -R_2 + \frac{100(0.25R_1)_+^2}{\sigma^2 + (0.25R_1)_+^2} \equiv F_2(R_1, R_2)$$

(14.28)

The modified strength of the Hebb synapses in this case is 0.25, and $\sigma = 10$ as in (14.27). The function designations F_1 and F_2 simply provide a convenient means of referring to the right-hand sides of these equations. It is easy to show that these equations have three equilibrium points at $(0, 0)$, $(20, 20)$, and $(80, 80)$. To find a Lyapunov function for (14.28), let us choose a state function in agreement with Theorem 14 that is positive definite about each equilibrium point:

$$U = \frac{1}{2}(F_1^2 + F_2^2)$$

(14.29)

where F_1 and F_2 are defined in (14.28). The derivative of U along trajectories of (14.28) may now be calculated to be:

$$\frac{dU}{dt} = -F_1^2 - F_2^2 + F_1 F_2 \left(\frac{\partial F_1}{\partial R_2} + \frac{\partial F_2}{\partial R_1} \right)$$

(14.30)

The right-hand side of this expression is again in the form in Theorem 14, except that the right-hand side has been multiplied by -2. Therefore dU/dt will be negative definite throughout any region where the following is satisfied:

$$\frac{\partial F_1}{\partial R_2} + \frac{\partial F_2}{\partial R_1} < 2$$

(14.31)

Note that the absolute value sign is not necessary here, as both partial derivatives must always be positive or zero from the definitions in (14.28). A sufficient, although slightly conservative, condition for (14.31) to be satisfied is:

$$\frac{\partial F_1}{\partial R_2} < 1 \quad \text{and} \quad \frac{\partial F_2}{\partial R_1} < 1$$

(14.32)

Both partial derivatives can be evaluated directly from (14.28) using (6.10). In the first case in (14.32) the result is:

$$\frac{\partial F_1}{\partial R_2} = \frac{25\sigma^2 R_2}{2\left[(0.25R_2)^2 + \sigma^2\right]^2} < 1 \tag{14.33}$$

A plot of (14.33) is shown by the dashed line in Fig. 14.6B, where it is evident that the inequality is satisfied both for $R_2 > 48.5$ and for $R_2 < 8$ (regions shaded in gray). Identical considerations apply to the second inequality in (14.32).

Let us now put all these considerations together. Figure 14.6A plots contours of the positive definite state function U from (14.29), and a three-dimensional surface plot of U is illustrated in Fig. 14.7. The MatLab script **LyapunovCA3.m** produces an animation of the 3D surface representation rotating to provide a better feel for its shape (U has been

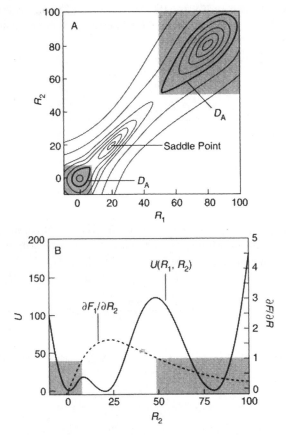

Fig. 14.6 Contour plot (A) and cross-section (B) of the Lyapunov function U in (14.29) for (14.28). Gray regions indicate where $dU/dt < 0$, and the heavier marked contours show the estimates of D_A from Theorem 13.

compressed by plotting $U^{0.3}$ to emphasize its shape near the minima). This positive definite state function has three minima, one at each equilibrium point of (14.28). Within the region defined by $R_1 > 48.5$ and $R_2 > 48.5$ we have shown that dU/dt in (14.30) is negative definite. This region is plotted in gray in Fig. 14.6A. Therefore, U is a Lyapunov function for the equilibrium point at $(80, 80)$. Theorem 13 may now be used to obtain an estimate of the domain of attraction for this equilibrium. As indicated in Fig. 14.6A, this is the largest closed contour $U = \alpha$ that encloses $(80, 80)$ and falls entirely within the gray region where $dU/dt < 0$. All trajectories originating within this droplet-shaped region are guaranteed to approach $(80, 80)$ asymptotically. The same analysis indicates that all trajectories originating within the indicated region surrounding $(0, 0)$ will asymptotically approach $(0, 0)$. Thus, we have obtained finite (rather than infinitesimal as is the case with linearized analysis) estimates of the domains of attraction of these two equilibria. This also proves rigorously that this memory system will generalize, because any input pattern generating an initial condition within the domain of attraction of $(80, 80)$ will lead asymptotically to the same final state, thus recalling the same stored pattern.

Our analysis is not applicable to the third steady state, which is a saddle point at $(20, 20)$, because $dU/dt > 0$ near this point. The Lyapunov function U is moot regarding the fate of trajectories originating within any of the white regions of the phase plane in Fig. 14.6A. This is because contours $U = \alpha$ originating in these regions surround all three steady states, and thus they encompass zones where $dU/dt > 0$. The three-dimensional view of U in Fig. 14.7 suggests that all trajectories will flow down to one of the two asymptotically stable steady states, but the dividing line between the global domains of attraction of the two cannot be ascertained given this particular Lyapunov function. Thus, U defined in

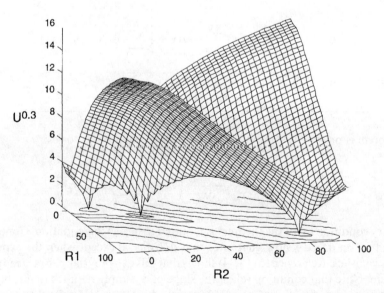

Fig. 14.7 Surface plot of the Lyapunov function (14.29) that was also illustrated in the previous figure. Note the three minima at the steady states.

(14.29) provides rigorous but conservative (i.e. small subregions) estimates of the domains of asymptotic stability.

The two-dimensional system in (14.28) is a special case of a more general N-dimensional neural network formulation due to Cohen and Grossberg (1983):

$$\frac{dx_i}{dt} = \frac{1}{\tau_i(x_i)}\left[b_i(x_i) + \sum_{j=1}^{N} w_{ij}S_j(x_j)\right] \qquad (14.34)$$

where $i = 1,\ldots,N$, and $S_j(x_j)$ is a sigmoidal function such as the Naka–Rushton or logistic function. System (14.28) is a two-dimensional example of (14.34) where τ_i is a constant, and $b_i(x_i) = -x_i$. Equation (14.34) differs somewhat in form from (14.27), as the summation of synaptic inputs in (14.27) occurs within the argument of the sigmoidal function, while (14.34) contains summation over sigmoidal functions. However, the two formulations are very closely related. The Cohen–Grossberg (1983) theorem defines a Lyapunov function for (14.34) under the following conditions:

Theorem 16 (Cohen–Grossberg): Let the following three conditions be true for system (14.34):

(a) $w_{ij} = w_{ji}$
(b) $\tau_i(x_i) > 0$ for all i; and
(c) $dS_i(x_i)/dx_i > 0$ for all i.
Then the following function is a Lyapunov function for (14.34) throughout the state space:

$$U = -\sum_{i=1}^{N} \int^{x_i} b_i(x) \frac{dS_i(x)}{dx}\,dx - \frac{1}{2}\sum_{j,k=1}^{N} w_{jk} S_j(x_j) S_k(x_k)$$

and $dU/dt < 0$ except at steady states of the system.

The theorem is proved by computing dU/dt along system trajectories:

$$\frac{dU}{dt} = -\sum_{i=1}^{N} \frac{1}{\tau_i(x_i)} \frac{dS_i(x_i)}{dx_i}\left(b_i(x_i) + \sum_{j=1}^{N} w_{ij}S_j(x_j)\right)^2 \qquad (14.35)$$

Symmetry condition (a) of the theorem is required to derive this equation. Conditions (b) on τ_i and (c) on dS_i/dx_i now guarantee that $dU/dt < 0$ except when the expression in brackets vanishes. Reference to (14.34) shows that this can only happen at steady states of the system. Note that condition (c) on the slope of S_i simply requires that it be a monotonically increasing function, a condition satisfied by both Naka–Rushton (for $x > 0$) and logistic functions.

As emphasized by Ermentrout (1998), the Cohen–Grossberg (1983) theorem is very general and includes the Hopfield (1984) network and others as special cases. As an example, let us apply Theorem 16 to the network in (14.28) by identifying τ_i as a constant, $b_i(x_i) = -x_i$, and $w_{ij} = 1$. The Lyapunov function in the theorem requires evaluation of integrals of the following form:

$$\int^{x_i} x \frac{\mathrm{d}S_i(ax)}{\mathrm{d}x} \, \mathrm{d}x = x_i S_i(ax_i) - 100x_i + \frac{100\sigma}{a} \arctan\left(\frac{ax_i}{\sigma}\right) \tag{14.36}$$

where S_i is the Naka–Rushton function in (14.28) and the factor of 100 is simply the maximum value of that function. This result was obtained using standard integral tables (Gradshteyn and Ryzhik, 1980). Given this result, the Lyapunov function of Theorem 16 for (14.28) becomes:

$$U(x, y) = \frac{100x(0.25x)^2}{100 + (0.25x)^2} - 100x + 4000 \arctan(0.025x) + \frac{100y(0.25y)^2}{100 + (0.25y)^2}$$

$$- 100y + 4000 \arctan(0.025y) - \frac{100(0.25x)^2}{100 + (0.25x)^2} \frac{100(0.25y)^2}{100 + (0.25y)^2} \tag{14.37}$$

where x and y have been substituted for the variables R_1 and R_2 in (14.28). (Note that (14.37) must have a constant value $-U(80, 80) = 742.8$ added to make it positive definite throughout the state space.) Contours of U may be obtained from the script **CG_Lyapunov.m** and are plotted in Fig. 14.8. Comparison of this figure with Fig. 14.6 indicates that the estimated domain of attraction D_A obtained using Theorem 13 is much larger for the Lyapunov function in (14.37) than for the function in (14.29). Furthermore,

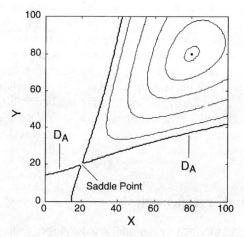

Fig. 14.8 Contours of the Cohen–Grossberg Lyapunov function (14.37). Note that this Lyapunov function produces considerably larger estimates of the domain of attraction, D_A, of (14.28) than those in Fig. 14.6. The contours bounding D_A are closed beyond the region shown.

the Cohen–Grossberg theorem guarantees that the system cannot have a limit cycle, as $dU/dt < 0$ everywhere except at the steady states. These results indicate the power of the Cohen–Grossberg (1983) theorem.

Let us now return to analyze the CA3 hippocampal model defined by (14.27). Because (14.27) is not in the form required by Theorem 16, let us adopt a generalization of the Lyapunov function in (14.29):

$$U = \frac{1}{2}\sum_{i=1}^{256} F_i^2 + \frac{1}{2}H^2 \tag{14.38}$$

where F_i is the function on the right-hand side of the ith dR_i/dt equation in (14.27) and H is the right-hand side of the dG/dt equation. Differentiating U along trajectories of (14.27) yields:

$$\frac{dU}{dt} = -\sum_{i=1}^{256} F_i^2 - H^2 + \sum_{i=1}^{255}\sum_{j=i+1}^{256} F_i F_j \left(\frac{\partial F_i}{\partial R_j} + \frac{\partial F_j}{\partial R_i}\right) + H\sum_{i=1}^{256} F_i\left(g + \frac{\partial F_i}{\partial G}\right) \tag{14.39}$$

Satisfaction of condition (14.32) defines regions within which the double sum terms will be small enough so that $dU/dt < 0$. Similarly, it can be shown that $|g + \partial F_i/\partial G| < 0.7$. Thus U is a Lyapunov function for any equilibrium state in which $R_i > 50$ for M of the neurons (32 in this model) and $R_i = 0$ for the remaining neurons.

The relevance of inhibitory feedback mediated by the G neuron in (14.27) may now be clarified. The model parameters guarantee that all 32 active neurons in a pattern will reach an equilibrium firing rate of 80. You can verify this by substitution back into (14.27) remembering that each of the 32 active neurons receives recurrent excitation from the remaining 31 (i.e. no self-excitation). In addition, all neurons receive an inhibitory signal of $-0.1G = -19.5$ in the steady state. When several patterns are stored in the network, as in **CA3memory.m** different patterns will frequently have a few active neurons in common. As long as two patterns share 15 or fewer units, however, inhibition from the G neuron will completely suppress the response of inappropriate neurons initially excited due to pattern overlap. You may notice this when running **CA3memory.m**, as parts of several patterns are sometimes initially activated by the input but are subsequently suppressed by inhibition. Thus, the network permits up to about a 46% overlap between distinct patterns stored in memory. Obviously the less overlap the better, however, and it has been suggested that cortical circuits in several brain areas are designed to minimize overlap among neural activity patterns before storage in regions such as CA3 hippocampus (Rolls and Treves, 1998).

14.6   Dynamic temporal memories

The autoassociative CA3 hippocampal model just developed evolves to an asymptotically stable steady state following brief stimulation with a noisy portion of a previously learned pattern. While such a memory network may encode a learned face or place, we are also capable of remembering complex temporal sequences, such as skilled motor control (e.g. canoeing or piano playing) or musical themes (e.g. the ABC jingle by which many children learn the alphabet). In these cases, it seems natural to suppose that the memory is encoded in the form of an asymptotically stable limit cycle capable of cycling through the learned time sequence over and over again. Kleinfeld (1986) and Sompolinsky and Kanter (1986) simultaneously suggested the basic way in which this could be accomplished. These authors began with an autoassociative network like the CA3 model but added a second set of synaptic connections that conducted information from presynaptic to postsynaptic neurons with a time delay. (Recall from Chapter 4 that time delays frequently lead to oscillations.) Hertz *et al.* (1989) showed how such time delays could be made compatible with Hebbian synaptic modification, thereby establishing the biological plausibility of the concept. Accordingly, the development here is based on their work.

Suppose that neurons in the CA3 model are interconnected by several sets of modifiable Hebb synapses: for one set axonal conduction and synaptic events occur extremely rapidly (as was assumed above), while for each additional set there is a progressively longer delay before synaptic activation. Such delays are known to be a constituent of the neural apparatus for auditory localization (Carr, 1993), and Hertz *et al.* (1989) summarize evidence for the existence of axonal conduction delays as long as 100–200 ms. At a modifiable synapse at which the presynaptic activity has a delay of Δ, the simple Hebb rule in (14.25) takes the modified form:

$$w_{ij} = kR_i(t)R_j(t - \Delta) \tag{14.40}$$

and an analogous modification can be made to (14.26). Note that synaptic strength w_{ij} depends on contemporaneous pre- and postsynaptic activity; the time lag Δ results from a delay in reaching the synapse. Whereas synapses between neurons for which $\Delta = 0$ will be symmetric: $w_{ij} = w_{ji}$, when $\Delta > 0$ each pair of neurons will be coupled asymmetrically. Thus, if a network is trained to recognize a sequence of patterns A–B–C, synapses modified by learning among the neurons encoding A will be symmetric, while synapses from pattern A to pattern B neurons will be asymmetric with A neurons activating B but not the reverse. This combination of symmetric and asymmetric neural interconnections is the key to sequence learning via Hebb synapses.

These concepts are incorporated in the MatLab script **DynamicMemory.m**. The network is the same as the CA3 network depicted in Fig. 14.4, except that the recurrent axon collaterals activate two sets of synapses: an effectively instantaneous one and a set with a time delay. Based on the development in Chapter 4, the delay is modeled by four exponential delay stages, each with a time constant $\tau = 8$ ms. The network has been trained via Hebbian rules to recall one static pattern (a cartoon face) and one dynamic sequence of three patterns. As you will discover by running the script, the network recalls the face to active working memory in response to an appropriate noisy input, but it settles into a limit cycle replaying the learned temporal sequence when the input is related to any one of the patterns in that sequence. The spatial neural activity states in the three patterns of the dynamic sequence successively spell MO–VE–A2, which may be construed as the neural code for either a skilled muscular movement or a particular movement in a musical piece.

The dynamical basis for the stored limit cycle MO–VE–A2 is quite easy to understand. The symmetric synaptic connections among active neurons in each individual pattern serve to make it temporarily asymptotically stable in the same manner as each of the patterns in the CA3 network above. After a time delay, however, these neurons activate neurons in the next pattern of the sequence via asymmetric connections and cause them to begin firing. Recurrent inhibition in the network mediated by the G neuron now becomes sufficiently strong to set up a competition between the previous and current neural patterns, and the latter wins because of its additional delayed stimulation by the former. This sequence of events is shown in Fig. 14.9, where the response of a neuron unique to each pattern in the sequence is plotted. Thus, the limit cycle encoded in this dynamical memory operates via a sequence of bifurcations in which the asymptotic stability of each stored pattern is ultimately destroyed by competition with the next pattern that it triggers following a delay. (You can increase the competition and make the transitions between patterns in the dynamic memory more abrupt by increasing the value of the inhibitory gain to $g = 0.11$, but this does produce too much inhibition for the static pattern stored in memory to remain stable.)

The autoassociative memory models above incorporate the basic features of excitatory and inhibitory interactions that were developed in Chapters 6 and 7. The recurrent excitatory connections in the CA3 model are the product of learning according to a Hebb rule for synaptic modification. They cooperatively reinforce firing of other neurons in a pattern, which results in pattern completion, recall, and maintenance of the pattern in short-term memory. Inclusion of additional synapses with delays leads to a stored limit cycle in which a sequence of bifurcations result as each stored dynamic pattern loses a winner-take-all competition (mediated by the inhibitory cell) to its successor. Hertz *et al.* (1989) have shown that inclusion of synapses with multiple delays leads to a quite sophisticated dynamic memory. More detailed models of motor memory (Lukashin *et al.*, 1996) and hippocampal dynamics (Samsonovich and McNaughton, 1997) have recently been developed using similar basic principles. Lyapunov functions provide the

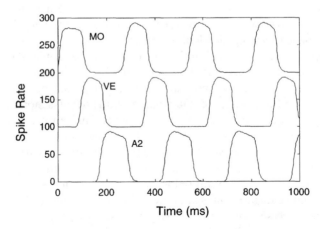

Fig. 14.9 Limit cycle response of three neurons in the dynamic memory program **DynamicMemory.m**. Each neuron is active in only one of the three patterns in the memory cycle (MO, VE, or A2).

mathematical basis for understanding how all these networks achieve their memory storage capabilities.

14.7 Exercises

1. Determine which of the following functions U are positive definite state functions of a system defined on the (x, y) phase plane. Assume that the system has a singularity at the origin. For each function that is positive definite, determine the region R within which this property holds:

$$U = x^2 + xy + y^2$$
$$U = 3x^2 + 4y^2 - y^3$$
$$U = 10x^4 + xy + 10y^4$$

2. Consider the following dynamical system:

$$\frac{dx}{dt} = -x^3 + y^5$$
$$\frac{dy}{dt} = -2y - x^3 y^2$$

Prove that the function U below is a Lyapunov function for the singularity at the origin. Based upon U, estimate the domain of asymptotic stability.

$$U = \frac{x^4}{4} + \frac{y^4}{4}$$

3. Consider the following linear negative feedback system:

$$\frac{dx}{dt} = -3x + 2y$$
$$\frac{dy}{dt} = y - 2x$$

Use Theorem 14 to find a Lyapunov function and prove that the domain of asymptotic stability is the entire (x, y) phase plane. (Note: you will have to find an appropriate range of values of ε to satisfy the theorem; remember that Lyapunov functions are not unique.)

4. Consider the following predator–prey equations for the ecological relationships between wolves (W) and moose (M):

$$\frac{dM}{dt} = M - \frac{1}{16} MW^2$$

$$\frac{dW}{dt} = -W + \frac{1}{81} WM^2$$

The squared terms here might arise from the requirement that it requires two wolves to kill a moose and eating of two moose to permit a wolf to reproduce. After solving for the equilibria, use Theorem 15 to derive a constant of motion for this system. Using MatLab script **LotkaVolterra.m** as a guide, plot several trajectories surrounding the center. (Hint: determine a useful range of values for your constant of motion by first evaluating it at the relevant steady state.)

5. Use the program **CA3learning.m** to determine the effects of storing an additional pattern via Hebbian learning. First, determine the effect of storing another pattern that is very different from any of the four presently in storage. Next, design a pattern that is quite similar to one of those presently stored. For example, design a pattern that is identical to the Ψ shape in the upper left of the 'Assoc' pattern but differs in its remaining elements. How does this interfere with pattern recall? Next, determine how noise level affects pattern recall. What is the lowest noise level in pixels at which the probability of corrupted pattern recall is greater than 50%?

6. Consider the following two-neuron memory system:

$$\frac{dx}{dt} = -5x + \frac{100}{1 + e^{-y+10}}$$

$$\frac{dy}{dt} = -5y + \frac{100}{1 + e^{-x+10}}$$

These equations are similar to (14.28), except that they employ the logistic function rather than the Naka–Rushton function to describe spike rates. First, solve for the steady states, which will require using MatLab procedure **fzero** as discussed in the Appendix. Now use Theorem 16 to derive a Lyapunov function for the system (this will require integral tables). Using **CG_Lyanpunov.m** as a guide, make a contour plot of this function and use it to estimate the domain of attraction of the higher steady state using Theorem 13.

7. The autoassociative memory net for dynamical patterns depends critically on two factors: the strength of excitation via the delayed synapses, and the strength of the recurrent inhibition mediated by the G neuron. In the program **DynamicMemory.m** these parameters are set respectively to Wdelay = 0.008 and g = 0.076 (see 14.27). For each parameter independently determine the smallest value that will still produce a limit cycle following stimulation by the first dynamic pattern with no added noise. For each parameter indicate what is recalled when the parameter becomes too small, and give a dynamical explanation of the result.

15 *Diffusion and dendrites*

All equations considered thus far have been ordinary differential equations, as there is only one independent variable, namely time. However, many problems in neuroscience involve interactions in both space and time. One major example is the spread of a post-synaptic potential along a dendrite to the cell body. A second example is the propagation of an action potential along an unmyelinated axon. These processes are controlled by the diffusion of ions within the dendrite and axon. Even communication between neurons involves synaptic diffusion of neurotransmitters between cells. In this chapter, therefore, the dynamics of diffusion processes will be explored, including the nonlinear diffusion inherent in action potential propagation. As will be seen, solutions to the diffusion equation can be reduced to solutions of ordinary differential equations with which we are already familiar. In a sense, therefore, we have already learned how to solve the diffusion equation.

In its earliest incarnation, the diffusion equation was developed by Fourier (creator of Fourier analysis and an engineer in Napoleon's army) to describe the conduction of heat along a wire. Diffusing ions will, of course, generate electrical currents, and these obey the cable equation, which is mathematically identical to the diffusion equation (and to the heat equation). As Rall (1989) has pointed out, the cable equation was so named because it was derived by Lord Kelvin to predict electrical transmission along the first transatlantic telegraph cable. The first applications of the diffusion or cable equation to neurons were by Hodgkin and Rushton (1946) and by Davis and Lorente de Nó (1947). The most extensive and elegant application of cable models since that time has been the work of Wilfrid Rall (1962, 1967, 1989), who pioneered the study of postsynaptic potential propagation through electrically passive dendritic trees. Most recently, the recognition that many processes in dendrites are inherently active and nonlinear has resulted in the development of compartmental models of dendrites (see Segev *et al.*, 1989). Compartmental models represent approximations to nonlinear diffusion processes in dendrites.

15.1 Derivation of the diffusion equation

To derive the diffusion equation, suppose that there is some chemical species whose concentration C varies with both position and time. Furthermore, let us assume that this chemical is confined within a very long, thin cylinder so that C effectively varies only with the distance x along the cylinder. Thus, the concentration will be a function of x and t. In order to derive an equation for the spatio-temporal variation of $C(x, t)$, let us focus on Fig. 15.1. Here a section of the thin cylinder within which diffusion occurs has been divided into three adjacent compartments, each of length Δx. Focusing on the center

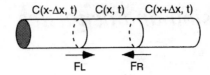

$C(x-\Delta x, t)$ $C(x, t)$ $C(x+\Delta x, t)$

FL   FR

Fig. 15.1   Schematic of three adjacent compartments in a long, thin cylinder along which a substance with concentration $C(x, t)$ is diffusing. F_L and F_R represent fluxes into the center compartment from the adjacent ones.

compartment, it is clear that the chemical can flow across both the left and right boundaries, thereby altering the local concentration, $C(x, t)$. The technical term for the rate of flow across a boundary is **flux**, and the two fluxes will be denoted by F_L and F_R. It has been established empirically that the flux between neighboring compartments in a cylinder is directly proportional to the concentration difference, with the diffusing chemical flowing from the higher concentration to the lower. Moreover, the flux is inversely proportional to the distance Δx over which the concentration difference occurs. Putting these empirical results together gives us the following:

$$F_R = K\frac{C(x + \Delta x, t) - C(x, t)}{\Delta x}$$
$$F_L = -K\frac{C(x, t) - C(x - \Delta x, t)}{\Delta x}$$
(15.1)

where K is the diffusion constant. The second equation has a minus sign, because flow into the center segment will only occur if $C(x - \Delta x, t) > C(x, t)$. Intuitively, the flux will increase as the cross-sectional area of the cylinder increases, so K will be proportional to this area. The net change in concentration $C(x, t)$ due to flux within the cylinder is the sum of the two fluxes in (15.1).

In addition to the changes in $C(x, t)$ that are produced by flux within the cylinder, some of the diffusing ions may leak out of the cylinder through the membrane. Within the ith compartment of the cylinder, the amount that will leak through the membrane will be proportional to C and to the length of each compartment Δx (because membrane surface area depends on Δx). To complete derivation of the diffusion equation, it is only necessary to combine the effects of flux and loss through the membrane. As $C(x, t)$ is a concentration variable, its rate of change will depend on the total amount of chemical in the compartment, which is $\Delta x\, C(x, t)$, so:

$$\Delta x \frac{\partial C}{\partial t} = F_R + F_L - \Delta x\, MC$$
(15.2)

In this equation $\partial C/\partial t$ indicates that this is a partial derivative with respect to t. As Δx is independent of t, it is treated as a constant with respect to the t differentiation. Thus, eqn (15.2) states that the rate of change of concentration in the ith compartment is the sum of the fluxes across the compartment boundaries minus the quantity lost through the membrane. Combination of (15.1) and (15.2) gives:

$$\frac{\partial C}{\partial t} = \frac{K}{\Delta x}\left[\frac{C(x + \Delta x, t) - C(x, t)}{\Delta x} - \frac{C(x, t) - C(x - \Delta x, t)}{\Delta x}\right] - MC \qquad (15.3)$$

In the limit as $\Delta x \to 0$, the expression in square brackets becomes the difference of two derivatives taken at adjacent spatial points. As this difference is divided by Δx, this limit becomes the derivative of a derivative, which is the second derivative. Thus, we have derived the **diffusion equation**, which is:

$$\frac{\partial C}{\partial t} = K\frac{\partial^2 C}{\partial x^2} - MC \qquad (15.4)$$

This equation states that the rate of change of C in time is equal to K times the second derivative of its change with position along the cylinder minus the rate of loss by leakage through the membrane.

In describing dendrites, we generally deal with the membrane potential V rather than the ionic concentrations that produce that potential. Furthermore, it is conventional to divide through by M so that (15.4) assumes the form:

$$\tau\frac{\partial V}{\partial t} = D^2\frac{\partial^2 V}{\partial x^2} - V \qquad (15.5)$$

In this form, known as the **cable equation**, τ is a time constant and D is a constant termed the **length constant** for reasons that will be apparent shortly. The dependence of D on dendritic radius R is easy to derive from the following simple considerations. The proportionality constant K of the flux term in (15.1) and (15.4) will vary directly with the cross-sectional area of the cylinder: $K = g_i\pi R^2$, where R is the dendritic radius and g_i is the conductance inside the cylinder. Loss through the membrane, determined by the constant M, will be proportional to cylinder circumference: $M = g_m 2\pi R$, where g_m is the leakage conductance through the membrane. Because $D = \sqrt{(K/M)}$, it follows that:

$$D = \sqrt{\frac{g_i R}{2g_m}} \qquad (15.6)$$

Thus, the length constant increases with the square root of dendritic radius R.

15.2 Steady state solution

Although the cable equation (15.5) is a partial differential equation involving both x and t as independent variables, it can be solved exactly using the techniques developed in studying ordinary linear differential equations. The first step in solving ordinary differential equations is to determine the steady states by equating all time derivatives to zero.

The same approach can be used to obtain the steady state behavior of the cable equation. Setting $\partial V/\partial t = 0$ in (15.5) produces the equation:

$$D^2 \frac{d^2 V}{dx^2} - V = 0 \tag{15.7}$$

Note that the derivative with respect to x is now an ordinary derivative rather than a partial derivative, because V is independent of t in the steady state. Equation (15.7) is simply a second order linear differential equation which can be solved by finding the eigenvalues (Theorem 2) with the result:

$$V = A\,e^{-x/D} + B\,e^{x/D} \tag{15.8}$$

The constants A and B must be chosen to satisfy the **boundary conditions** for the cable equation. Boundary conditions are the spatial analog of initial conditions in time, as **boundary conditions** specify the values of V (or its spatial derivatives) at the ends (boundaries) of the cylinder along which diffusion is occurring. For the present, let us assume that the potential is maintained constant at $x = 0$ so that $V(0) = V_0$. Assuming that the cylinder is very long with respect to D, which effectively means that it is longer than about $5D$, there is little loss of accuracy if it is treated as effectively infinite. The second boundary condition will then be $V(\infty) = 0$, and (15.8) becomes:

$$V = V_0\,e^{-x/D} \tag{15.9}$$

This solution to the cable equation (15.5) reveals why D is referred to as a length constant. D plays the same role with respect to space in the solution as a time constant does in the solution of an ordinary linear differential equation: V decays to $1/e$ of its maximum at $x = D$.

One of Rall's (1962, 1989) major contributions to the theory of dendritic potentials was to derive conditions under which a complex dendritic tree could be represented as a single equivalent cylinder. The complexity of dendritic trees arises because dendritic segments closer to the cell body (proximal dendrites) split at intervals into several smaller daughter dendritic segments as illustrated schematically in Fig. 15.2. Let us consider a point at which two daughter dendrites with radii of r_1 and r_2 join a parent dendrite of radius R, as depicted in Fig. 15.2B. Rall's insight was to realize that if the flux of a diffusing ion was conserved at such dendritic junctions, then the daughter dendritic cylinders could be collapsed into a single cylinder of the same radius as the parent dendrite. Using the definition of flux in (15.1) in the limit as $\Delta x \to 0$ and the steady state equation for V in (15.9), the flux through the end of the parent dendrite is:

$$K\frac{dV}{dx} = -\frac{K}{D}V \tag{15.10}$$

In terms of Ohm's law, the constant K is the ionic conductance, while dV/dx is the voltage difference between adjacent points along the dendrite, so the product $K\,dV/dx$ is an

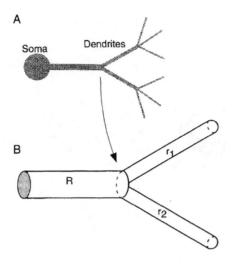

Fig. 15.2 Stereotypical dendritic tree in A with a single branch point magnified in B.

ionic current (if $dV/dx = 0$, there is no current flow). Thus, the flux of a diffusing ion is identical to the resultant electrical current.

The discussion leading to (15.6) indicated that $K = g_i \pi R^2$, namely, flux is proportional to dendritic cross-sectional area. Combining this expression with that for D from (15.6) yields an equation for the dependence of flux on dendritic radius R:

$$K \frac{dV}{dx} = -\sqrt{2g_i g_m} \pi R^{3/2} V \tag{15.11}$$

Thus, the ionic flux or current exiting or entering the end of the parent dendrite is proportional to $R^{3/2}$. Similar considerations may be applied to each of the daughter dendrites to show that the flux where they join the parent will be proportional to their radii raised to the 3/2 power. Rall (1962) observed that if the sum of the fluxes entering the daughter dendrites was equal to the flux exiting the parent, the daughters were mathematically equivalent to an extension of the parent dendrite. Assuming that the constants g_i (internal conductance) and g_m (membrane leakage conductance) are identical for both parent and daughter dendrites, equality of fluxes across the dendritic junction is guaranteed if the following equation is satisfied:

$$R^{3/2} = r_1^{3/2} + r_2^{3/2} \tag{15.12}$$

Note that r_1 and r_2 need not be identical to satisfy the equation. It should also be mentioned that all terminal dendritic branches must be the same electrotonic distance from the soma in order to collapse them into a single cylinder (Rall, 1989). Equation (15.12) can easily be generalized to equality of fluxes across a dendritic junction with N

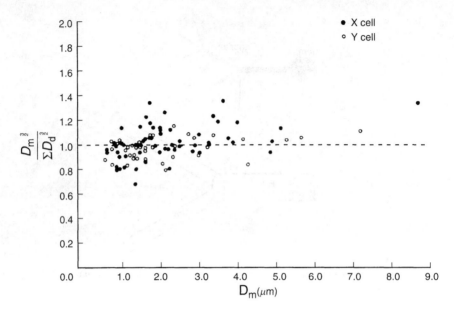

Fig. 15.3 Data (reproduced with permission) from Bloomfield *et al.* (1987) showing that Rall's (1962) requirement in (15.13) is a good approximation. The abscissa is the diameter D of the parent dendrite, while the ordinate is the ratio $D^{3/2}/\sum d^{3/2}$ where d are the daughter dendritic diameters. A ratio of 1.0 (– –) agrees with (15.13). Data on two neuron types are shown.

emergent daughter branches with radii r_n:

$$R^{3/2} = \sum_{n=1}^{N} r_n^{3/2} \tag{15.13}$$

If (15.12) or (15.13) does not hold, there will be a build-up of ionic concentration on one side of the dendritic junction, in which case the dendritic tree cannot be simplified into a single cylinder. However, anatomical measurements such as those by Bloomfield *et al.* (1987) in Fig. 15.3 indicate that (15.13) is a good approximation in many areas of the brain.

15.3   Separation of variables

The steady state solution of the cable equation (15.5) reveals nothing about the dynamical properties of important neuronal events, such as postsynaptic potentials. To obtain dynamical information, (15.5) must be solved directly. In 1807 Joseph Fourier, discoverer of Fourier analysis, had an insight that resulted in an exact solution to (15.5). This was the hypothesis that the dependence on space and time could be represented by a product of functions of x and t, that is:

$$V(x, t) = H(x)G(t) \tag{15.14}$$

To show that this assumption solves (15.5), let us substitute (15.14) for $V(x, t)$. The result is:

$$\tau H \frac{dG}{dt} = D^2 G \frac{d^2 H}{dx^2} - GH \tag{15.15}$$

Note that the assumption in (15.14) reduces the partial derivatives in (15.5) to ordinary derivatives. If the term GH is moved to the left side and (15.15) is divided by GH, the result is:

$$\frac{\tau}{G} \frac{dG}{dt} + 1 = \frac{D^2}{H} \frac{d^2 H}{dx^2} \tag{15.16}$$

The left side of (15.16) is now a function only of t, while the right side is a function only of x. The only way this can possibly be true for all values of t and x is if each side of (15.16) is equal to the same constant, which will be called $-\beta$ (the minus sign is convenient, not essential). Setting each side equal to $-\beta$ now reduces (15.16) to two ordinary differential equations:

$$\frac{\tau}{G} \frac{dG}{dt} + 1 = -\beta$$
$$\frac{D^2}{H} \frac{d^2 H}{dx^2} = -\beta \tag{15.17}$$

These separate equations for t and x may now be further rearranged algebraically to generate the results:

$$\tau \frac{dG}{dt} + (1 + \beta)G = 0$$
$$\frac{d^2 H}{dx^2} + \frac{\beta}{D^2} H = 0 \tag{15.18}$$

Thus, assumption (15.5) that $V = H(x)G(t)$ reduces the cable equation to the two ordinary, linear differential equations in (15.18). This procedure is known as **separation of variables**, as it leads to separate equations in x and t. Although separation of variables might seem like a mathematical trick, mathematicians have proven that the solution to the cable equation is unique. Given this, Fourier's deep insight that $V = H(x)G(t)$ reduces the cable equation to a unique pair of equations that are easily solved. It will be seen in a moment that the seemingly arbitrary constant β is actually determined by the physiology of the desired solution.

In order to solve (15.18) we must specify initial conditions, just as in solving ordinary differential equations. However, the initial condition is now a spatial distribution of concentrations at $t = 0$, $V(x, 0) = H(x)$. In addition, we are again confronted with **boundary conditions**, because it is necessary to specify how the ionic concentration or potential behaves at the ends or boundaries of the cylinder. In arriving at (15.9) above for the steady state, the cylinder was assumed to extend from $x = 0$ to $x = \infty$. On the more reasonable assumption that the cylinder is of total length L and extends from $-L/2$ to $+L/2$, boundary conditions require specification of some functions of $V(-L/2, t)$ and $V(L/2, t)$. The way in which these functions are chosen depends on the physiology of the problem. Suppose that we wish to solve (15.18) for the case where the chemical cannot diffuse through the ends of the cylinder. This is equivalent to specifying that the flux is zero at the end points, termed a **zero flux boundary condition**, so:

$$\frac{dH}{dx} = 0 \quad \text{for } x = \pm L/2 \tag{15.19}$$

Solving the second equation in (15.18) gives:

$$H(x) = A \cos\left(\sqrt{\beta}\,\frac{x}{D}\right) \tag{15.20}$$

where A is an arbitrary constant (sine terms can also be included, see below). In order to satisfy the **zero flux** boundary condition (15.19), it is necessary to choose β to produce an integer number of cycles n of the cosine function along the length L of the cylinder:

$$\beta = \left(\frac{2\pi n D}{L}\right)^2 \tag{15.21}$$

Thus, the boundary conditions constrain β to assume only discrete values determined by the integer n. With this value for b, (15.20) becomes:

$$H(x) = A_n \cos\left(\frac{2\pi n x}{L}\right) \tag{15.22}$$

where the subscript on A_n indicates that a different value of this constant may be chosen for each n.

To complete the solution of (15.5), β from (15.21) is substituted into the dG/dt equation in (15.18), which can now be easily solved for $G(t)$:

$$G(t) = e^{-t/\tau} \exp\left(\frac{-4\pi^2 n^2 D^2 t}{\tau L^2}\right) \tag{15.23}$$

The separation of variables assumption (15.14) indicates that the solution to the cable equation (15.5) is $V(x, t) = G(t)H(x)$, so:

$$V(x, t) = e^{-t/\tau} \sum_{n=1}^{\infty} A_n \exp\left(\frac{-4\pi^2 n^2 D^2 t}{\tau L^2}\right) \cos\left(\frac{2\pi n x}{L}\right) \tag{15.24}$$

Summation over n has been included here, because each GH product for a given value of n independently solves (15.5). Hence, a sum of such products is also a solution. The astute reader may have observed that sine terms are also part of the general solution in

(15.20) and (15.22), and this is correct. Appropriate sine terms would then also be added into the general solution in (15.24) (see Exercise 2).

The initial conditions on a cable equation solution specify a distribution $V(x, 0) = H(x)$. The greatest achievement of Fourier was to prove that the values A_n can be chosen to fit any initial distribution $H(x)$ that is physiologically possible. The representation of a function as a sum of cosines and sines is accordingly termed a **Fourier series representation**. Development of the mathematics of Fourier series is outside the scope of this book, but excellent treatments are available elsewhere (e.g. Gaskill, 1978).

Examination of the solution to the cable equation in (15.24) reveals that the high spatial frequencies defined by large values of n decay much more rapidly than the lower frequencies, because the constant in the time exponential is proportional to n^2. This can be seen in the MatLab animation **Diffusion.m**, which starts with an initial distribution that is the sum of cosine frequencies $n = 1$ and $n = 5$ in (15.24). Consistent with the time dependence of (15.24), the frequency $n = 5$ variation in the initial distribution dies out much faster than the lower frequency $n = 1$ variation. In running the script, you may also notice that the overall level of V declines with time, which is a consequence of the term $e^{-t/\tau}$ that multiplies the entire solution. Reference back to the derivation of (15.5) and the solution by separation of variables shows that this term results from leakage of the diffusing ion through the membrane. In the case of a dendrite, V will ultimately decay back to the resting potential of the neuron.

15.4 Passive dendritic potentials

Dendrites are often sufficiently long relative to the length constant D so that they can be treated as infinite for the purposes of mathematical analysis. This leads to consideration of solutions to the cable equation (15.5) for the case where the boundary conditions are $V(\pm\infty, t) = 0$. This assumes that the potential V represents the deviation of the dendritic potential from the resting potential. Furthermore, we are assuming that locations where postsynaptic events cause local changes in V are sufficiently far from the ends of the dendrite that conditions at the ends make little contribution to the solution. If the dendritic geometry is such that equation (15.13) is satisfied (see Fig. 15.3), we may follow Rall (1962, 1989) and treat the entire dendritic tree as a single long cylinder.

In order to solve the cable equation for an equivalent cylinder model of a dendritic tree, it is necessary to consider how postsynaptic potentials diffuse away from a synapse. Suppose, therefore, that there is a synapse located at $x = 0$ which is briefly active at $t = 0$ resulting in the local postsynaptic potential being elevated to V_0. If the dendrite is infinitely long in both directions, the solution can be obtained from (15.24) in the limit as the dendritic length $L \to \infty$. In doing this, it is appropriate to define a new frequency variable $\omega = n/L$, so ω becomes a continuous variable as $L \to \infty$. In this limit the summation in (15.24) becomes an integral and the solution is:

$$V(x, t) = e^{-t/\tau} \int_0^\infty A(\omega) \exp(-4\pi^2\omega^2\, D^2 t/\tau) \cos(2\pi\omega x)\, d\omega$$

Note that the coefficients A_n in (15.24) become a continuous function $A(\omega)$ in this equation.

For the case of an idealized synaptic input at $x = 0$, Fourier analysis can be used to prove that $A(\omega) = 1$ for all ω. In this case the integral above can be evaluated exactly (Gradshteyn and Ryzhik, 1980), and the result is:

$$V(x,t) = \frac{e^{-t/\tau}}{2D\sqrt{\pi t/\tau}} \exp\left(\frac{-\tau x^2}{4D^2 t}\right) \tag{15.25}$$

It can be verified that this is a solution to the cable equation by substitution back into (15.5) and evaluation of the derivatives. $V(x,t)$ in (15.25) at first appears to behave rather strangely when $t = 0$. If $x = 0$, $V(0,0)$ is infinite, while $V(x \neq 0, 0) = 0$ everywhere. However, (15.25) has a very important property that clarifies the situation. This property becomes obvious if $V(x,t)$ is integrated with respect to x while treating t as constant:

$$\frac{e^{-t/\tau}}{2D\sqrt{\pi t/\tau}} \int_{-\infty}^{\infty} \exp\left(\frac{-\tau x^2}{4D^2 t}\right) dx = e^{-t/\tau} \tag{15.26}$$

Thus, at $t = 0$, the area under $V(x,t)$ is unity, and this area decreases exponentially with time as ions leak outward through the membrane. At $t = 0$, (15.25) is therefore an example of a Dirac (1958) δ function, which was encountered briefly in the discussion of time delays at the end of Chapter 4. In the context of dendritic potentials, $V(x,t)$ in (15.25) represents the potential change at an idealized synapse where a fixed number of ions have entered the postsynaptic cell in an infinitesimally small region. Thus, the local potential is formally infinite, although the total number of ions that have entered is finite as given by the integral in (15.26).

To see how (15.25) can be used to calculate the effects of postsynaptic potentials diffusing passively down a dendrite, suppose that a synapse at point x_1 is active at t_1 and produces a potential change V_1 relative to the resting potential. The appropriate solution to the cable equation for an idealized infinite dendrite is now:

$$V(x,t) = \begin{cases} \dfrac{V_1\, e^{-(t-t_1)/\tau}}{2D\sqrt{\pi(t-t_1)/\tau}} \exp\left(\dfrac{-\tau(x-x_1)^2}{4D^2(t-t_1)}\right) & \text{for } t \geq t_1 \\ 0 & \text{for } t < t_1 \end{cases} \tag{15.27}$$

This is just (15.25) with the origin shifted to $x = x_1$ and the time of the synaptic event shifted to $t = t_1$. The MatLab script **Dendrite.m** produces an animation of $V(x,t)$ in (15.27) as the potential change diffuses outward from a local synaptic site. The simulation incorporates a time constant $\tau = 10\,\text{ms}$ and a length constant $D = 100\,\mu\text{m}$. In evaluating (15.27), it is convenient to begin with $t > t_1$ by a small amount, $10^{-5}\,\text{ms}$ in this case, to avoid problems at $t = 0$. On the assumption that the soma is located at $x = 0$, **Dendrite.m** permits you to explore the effect of synapses at various distances from the site of spike generation. Figure 15.4 plots the relative potential changes at the soma ($x = 0$) produced by equal strength synaptic events occurring 25, 50, and 100 μm away. As is apparent from

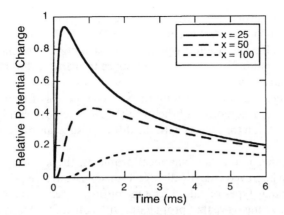

Fig. 15.4 Potential change at soma due to potential propagation along a passive dendrite. Initial synaptic events occurred at 25, 50, and 100 μm from the soma, and curves were computed using (15.27).

the figure, passive diffusion along the dendrite causes the potential change at the soma to be much smaller and to peak later (0.25 ms for $x = 25$ μm; 3.0 ms for $x = 100$ μm) as the distance from soma to synapse increases. As the length constant $D = 100$ μm in this example, it is apparent that synapses located much farther than one length constant from the soma will have a negligible effect in generating spikes (try running **Dendrite.m** with $x = 200$ μm). In fact, this analysis of passive dendrites using cable theory suggests that synapses located more than one length constant from the soma will generally have only a minor, subthreshold biasing effect on the excitability of the neuron. Conversely, synapses close to the soma will have a very powerful and rapid effect on spike activity, which is probably the reason that inhibitory GABA synapses are generally close to the somas of cortical neurons (Colonneir, 1968; Scheibel and Scheibel, 1970). Cumulative effects of multiple synaptic events at various locations are explored in Exercise 3.

15.5 Diffusion–reaction equations

The cable equation (15.5) is a linear partial differential equation describing passive diffusion of ions along dendrites. However, recent research has revealed that apical dendrites of many neocortical and hippocampal pyramidal cells contain active, voltage-sensitive ion channels (Stuart and Sakmann, 1994; Magee and Johnston, 1995; Johnston *et al.*, 1996), similar to those responsible for action potential propagation along axons. To understand active propagation along axons and dendrites, it is necessary to consider nonlinear diffusion processes. Let us therefore consider the simplest nonlinear generalization of (15.5):

$$\tau \frac{\partial V}{\partial t} = D^2 \frac{\partial^2 V}{\partial x^2} + F(V) \qquad (15.28)$$

This is just the linear diffusion equation with a general term $F(V)$ added. The simple case where $F(V) = -V$ is just the cable equation (15.5). If $F(V)$ is nonlinear, however, eqn (15.28) is called a **diffusion–reaction equation**, and striking new phenomena can emerge. This name implies that diffusion of V across space triggers local nonlinear chemical (or ionic) reactions characterized by $F(V)$. In an axon, for example, we shall see that $F(V)$ can describe the voltage-dependent opening of Na^+ channels. Diffusion–reaction systems can also involve several equations of the form (15.28), each describing the diffusion of a different chemical or ion that reacts with the diffusing substances described by the other equations.

As with most other nonlinear dynamical problems, the diffusion–reaction equation (15.28) can seldom be solved analytically. Nevertheless, we are frequently interested in the existence of particular types of solutions. One diffusion–reaction solution of particular importance is a traveling wave that propagates with constant velocity and shape, such as an action potential along an unmyelinated axon. Denoting position by x, time by t, and propagation velocity by v, a solution to (15.28) $V(x, t)$ is a **traveling wave** if:

$$V(x, t) = V(x \pm vt) \tag{15.29}$$

The function on the right side is of constant shape but shifts its position to the left or right with time depending on whether it is a function of $(x + vt)$ or $(x - vt)$.

Under what conditions will a diffusion–reaction equation have traveling wave solutions of the form (15.29)? The answer is obtained by defining a new variable z:

$$z = x + vt \tag{15.30}$$

The transformation of variables defined by (15.30) has the effect of placing us in a coordinate frame that is moving at exactly the speed of the propagating wave that we wish to study. (Note that $z = x - vt$ could also be used.) You might wonder what constant velocity v should be chosen in defining z. The appropriate v will be seen to emerge from an analysis of the diffusion–reaction equation if any solutions of the form (15.29) exist.

To see how transformation to the variable z simplifies the problem of finding traveling wave solutions to the diffusion–reaction equation, notice first that (15.30) transforms the partial derivatives with respect to space and time as follows:

$$\frac{\partial V}{\partial t} = v \frac{dV}{dz}; \quad \frac{\partial^2 V}{\partial x^2} = \frac{d^2 V}{dz^2} \tag{15.31}$$

Thus, transformation to variable z reduces the partial derivatives to ordinary derivatives, and the diffusion–reaction equation (15.28) therefore becomes:

$$\tau v \frac{dV}{dz} = D^2 \frac{d^2 V}{dz^2} + F(V) \tag{15.32}$$

So, the diffusion–reaction equation has been reduced to an ordinary second order differential equation that can be studied using familiar phase plane techniques after transformation into normal form. In particular, (15.32) may have multiple equilibrium states. If so, there may be a unique trajectory that originates at (i.e. infinitesimally near) one steady state for $z \to -\infty$ and approaches a second steady state as $z \to +\infty$. Such a trajectory connecting two equilibrium points is called a **heteroclinic trajectory**. Typically, such a trajectory will exist only for a unique value of the velocity v. Under these conditions the heteroclinic trajectory defines the shape of a traveling wave for the diffusion–reaction equation (15.28). Let us collect these observations into a theorem and then see how it may be applied to action potential propagation.

Theorem 17: Consider the diffusion–reaction equation:

$$\tau \frac{\partial V}{\partial t} = D^2 \frac{\partial^2 V}{\partial x^2} + F(V)$$

where $F(V)$ is nonlinear and transform into the two-dimensional ordinary differential equation in $z = x + vt$:

$$\frac{dV}{dz} = W$$

$$\frac{dW}{dz} = \frac{1}{D^2}(\tau v W - F(V))$$

If this system has more than one equilibrium point, and a heteroclinic orbit joins two equilibria for some critical value $v = v_c$, then the original equation has traveling wave solutions that move at velocity v_c. Furthermore, this heteroclinic trajectory defines the shape of the traveling wave.

15.6 Action potential propagation

Theorem 17 seems abstruse in many ways. After all, it is usually impossible to solve analytically for trajectories in the state space of nonlinear systems. However, it is only necessary to demonstrate the existence of a heteroclinic orbit for some particular v_c; we do not necessarily have to solve for its shape directly. Moreover, in the case of the Hodgkin–Huxley approximation in (9.7), it is possible to solve for the shape of the traveling wave front as well as for its velocity. The analysis developed here has been adapted from FitzHugh's (1969) analysis of spike propagation in the FitzHugh–Nagumo equations.

The diffusion–reaction equations describing the generation and propagation of action potentials are a simple generalization of (9.7):

$$0.8 \frac{\partial V}{\partial t} = D^2 \frac{\partial^2 V}{\partial x^2} - (17.81 + 47.71V + 32.63V^2)(V - 0.55) - 26.0R(V + 0.92)$$

$$\tag{15.33}$$

$$\frac{\partial R}{\partial t} = \frac{1}{1.9}(-R + 1.35V + 1.03)$$

All that has been added to (9.7) is a term describing the diffusive spread of the membrane potential V in the first equation. If (15.33) is transformed to a moving coordinate system with the substitution $z = x + vt$, it is easy to see that the result will be a system of three coupled first order differential equations. However, previous analysis of (9.7) revealed that $dV/dt \gg dR/dt$ during the rising phase of each spike. Accordingly, let us assume that R remains constant at its equilibrium value during the rising phase of the spike and focus on the changes in V. This separation of time scales will enable us to study the propagation of the leading edge of the action potential, although it will not, of course, allow us to examine the recovery phase, which is driven by R. Accordingly, let us set $R = 0.088$, the equilibrium value of R, and transform (15.33) into the moving coordinate frame $z = x + vt$. The results of this transformation converted to normal form are:

$$\frac{dV}{dz} = W$$

$$\frac{dW}{dz} = \frac{0.8vW + (17.81 + 47.71V + 32.63V^2)(V - 0.55) + 2.288(V + 0.92)}{D^2} \tag{15.34}$$

Although the values of v and D have not been specified yet, neither enters into the calculation of steady states, because $W = 0$ is one isocline. Using the MatLab roots function, the steady states are found to be: $(0, -0.704)$, $(0, -0.692)$, and $(0, 0.484)$. Linearized stability analysis reveals that the first and third of these points are unstable saddle points, while the intermediate one is an asymptotically stable spiral point. The isoclines of (15.34) are plotted in Fig. 15.5 using the critical value of v to be derived presently.

As indicated in Theorem 17, there will be a traveling wave solution to (15.33) if there is a critical value of the velocity v that results in a heteroclinic trajectory connecting the first to the third of the equilibria of (15.34). Let us therefore consider the equation for a trajectory in the phase plane, which is obtained by dividing the two equations in (15.34) to obtain a single equation for dW/dV:

$$\frac{dW}{dV} = \frac{(17.81 + 47.71V + 32.63V^2)(V - 0.55) + 2.288(V + 0.92)}{D^2 W} + \frac{0.8v}{D^2} \tag{15.35}$$

The cubic expression on the top can now be written as a product of the three roots determined by solving the isocline equations:

$$\frac{dW}{dV} = \frac{32.63(V + 0.704)(V + 0.692)(V - 0.484)}{D^2 W} + \frac{0.8v}{D^2} \tag{15.36}$$

This cubic differential equation has the following general form:

$$\frac{dW}{dV} = \frac{a(V - r)(V - s)(V - h)}{D^2 W} + \frac{bv}{D^2} \tag{15.37}$$

where r, s, and h represent the resting steady state, stable steady state, and highest steady state respectively.

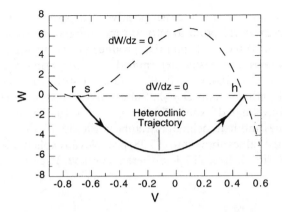

Fig. 15.5 Phase plane for (15.34) with r, s, and h defined in (15.37). The heteroclinic trajectory moves from the saddle point at r to the one at h. This trajectory defines the leading edge of the propagating action potential.

FitzHugh (1969) published the solution to (15.37) describing action potential propagation, and in a different context Grafstein (1963) credited the solution to unpublished work by Huxley (see Exercises). Based on that work, let us prove that the solution to (15.37) is:

$$W = -A(V - r)(V - h) \tag{15.38}$$

Thus, W is a parabola that originates at the resting equilibrium point r and terminates at the highest equilibrium point h (see Fig. 15.5). The constant A is not arbitrary and will be determined as part of the solution. Calculation of dW/dV and substitution of (15.38) into (15.37) yields:

$$-A(V - r) - A(V - h) = \frac{-a(V - s)}{AD^2} + \frac{bv}{D^2}$$

Algebraic rearrangement of this equation leads to the expression:

$$(-2A^2D^2 + a)V = -A^2D^2(r + h) + as + Abv \tag{15.39}$$

The left-hand side of this equation is a function of V, while the right-hand side is just a collection of constants. The only way that (15.39) can be true for all V, therefore, is if A and v are chosen to guarantee that each side vanishes identically. Equating the coefficient of V to zero and the right-hand side to zero produces two equations that are easily solved for A and v:

$$A = \frac{1}{D}\sqrt{\frac{a}{2}} \quad \text{and} \quad v = \frac{D(h + r - 2s)}{b}\sqrt{\frac{a}{2}} \tag{15.40}$$

Note that A and therefore v can be either positive or negative, depending on the sign of $\sqrt{(a/2)}$: positive signs correspond to leftward motion and negative to rightward. This determines the unique values of A and v that produce a traveling wave solution to (15.34). The parameter values are easily determined by comparing (15.37) with (15.36): $a = 32.63; b = 0.8; r = -0.704; s = -0.692$; and $h = 0.484$. Choosing a plausible value for the diffusion length constant $D = 0.25\,\text{mm}, v = 1.47\,\text{mm/ms}$, or $1.47\,\text{m/s}$, a reasonable value for unmyelinated axons. Similarly, $A = 16.16$. These values were used in plotting the isoclines and heteroclinic trajectory in Fig. 15.5.

Substitution of the values above into (15.38) provides an explicit solution for W. Recall, however, that $W = dV/dz$ from (15.34), where $z = x + vt$. Thus, the equation for $V(z)$ is:

$$\frac{dV}{dz} = -A(V - r)(V - h) \qquad (15.41)$$

The solution to this equation can be found in standard integral tables (Gradshteyn and Ryzhik, 1980) and involves the hyperbolic tangent (tanh):

$$V = r + \frac{h-r}{2}\left\{1 + \tanh\left(\frac{A(h-r)(x+vt)}{2}\right)\right\} \qquad (15.42)$$

where A and v are given by (15.40), and $x + vt$ has been substituted for z. This is an analytical solution for the leading edge and velocity of the action potential produced by (15.33). Depending upon whether A and v are positive or negative, the wave moves to the left or right. In front of the wave, $V = r$, the resting potential, while after the wave passes $V = h$, the maximum wave height (see Fig. 15.6).

To summarize, we have used Theorem 17 to prove that the Hodgkin–Huxley approximation in (15.33) will produce spikes that travel with constant velocity and shape along an axon. The leading edge of these propagating spikes is a traveling wave solution to the diffusion–reaction equation. As a check on the accuracy of the solution in (15.42), it is interesting to compare it with a simulation of the full diffusion–reaction system in (15.33). Before doing so, however, a brief digression on numerical methods is in order.

Simulations of diffusion–reaction equations depend on numerical methods derived from Taylor series approximations in ways analogous to those developed in Chapter 5. Expanding the diffusion–reaction equation (15.28) to lowest order in both partial derivatives, as in Euler's method, produces the formula:

$$V(x, t + \Delta t) = V(x, t) + \frac{\Delta t}{\tau}\left\{D^2\frac{V(x + \Delta x, t) - 2V(x, t) + V(x - \Delta x, t)}{(\Delta x)^2} - F(V(x,t))\right\}$$

$$(15.43)$$

where Δt and Δx are the temporal and spatial intervals between adjacent points. The expression on the right divided by $(\Delta x)^2$ is just the discrete approximation to the spatial second partial derivative. As discussed by Press *et al.* (1986), it is necessary (although not always sufficient in nonlinear problems) that Δt and Δx be chosen to satisfy the following inequality in order for the simulations to converge to meaningful results:

$$\frac{2\Delta t\, D^2}{\tau(\Delta x)^2} < 1 \tag{15.44}$$

Although it is easy in principle to choose values of Δx and Δt satisfying (15.44), in practice this may result in an extremely large number of very small time steps and a very long time for computation. More complex approximation methods that partially alleviate this problem are developed by Press *et al.* (1986), and the interested reader is referred to that treatment.

The diffusion–reaction equation for action potential propagation (15.33) has been simulated in MatLab script **Spike_Propagate.m** using the approximation method in (15.43). Parameter values were $\tau = 0.8$ ms from (15.33), $D = 0.25$ mm as in the analytical solution above, $\Delta x = 0.08$ mm, and $\Delta t = 0.01$ ms. Zero flux boundary conditions were used in the simulation. Running the script will produce an animated solution of a spike that propagates from left to right as a result of initial depolarization at the left end of the axon. By changing the site of the initial depolarization it is possible to explore effects of initial depolarization at the axon center or at both ends at once (the spikes annihilate when they meet). The simulated action potential is compared with the analytical form for the leading edge in Fig. 15.6. The analytical solution is extremely accurate almost to the peak of the spike, where R (i.e. K^+ conductance) truncates the simulated spike and leads to the recovery phase. Spike velocity in the simulation is 1.33 m/s, which is about 10% slower than the velocity calculated from (15.40). This is a consequence of the higher peak of the propagating wave in the analytical approximation, which is $h = 0.484$ as compared to 0.36 in the simulation. If v is calculated from (15.40) with $h = 0.36$ from the simulation, the result is $v = 1.31$ m/s, indicating that the overestimate of velocity indeed results from the greater maximum of V in the analytical solution.

In addition to producing traveling waves, such as action potential propagation, diffusion–reaction equations can spontaneously generate complex one- and two-dimensional spatial patterns. The mathematical basis for this is similar to that underlying

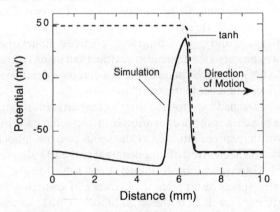

Fig. 15.6 Simulation (15.33) for spike propagation compared with analytical solution (tanh) in (15.42). Both are moving to the right.

visual hallucinations (see Chapter 7). In diffusion–reaction systems, different diffusion length constants for different chemicals play roles analogous to different excitatory and inhibitory spread functions in equations such as (7.22). Many dynamical models of pattern generation during morphological development are based on diffusion–reaction dynamics, and these models explain how leopards get their spots, tigers their stripes, and butterflies their wing patterns. These aspects of diffusion–reaction equation theory are elegantly developed by Murray (1989).

15.7 Compartmental models and dendritic spikes

Our analysis of action potential propagation shows that nonlinear diffusion reaction problems are sometimes amenable to analytical treatment. More generally, however, nonlinear diffusion problems must be solved by numerical approximation. Complex dendritic trees contain not just synaptic currents but also a wide range of active, nonlinear Na^+ and Ca^{2+} currents. To deal with such complex dendritic trees, Rall (1962, 1967) introduced what are known as compartmental models. **Compartmental models** are derived directly from (15.1) and (15.2) without passing to the limit $\Delta x \to 0$ that resulted in the diffusion equation. The result is a series of coupled ordinary differential equations in which time varies continuously but the spatial extent of dendrites is chopped into a finite number of discrete compartments of finite length, each of which interacts with the compartments adjacent to it. It is assumed that each compartment is sufficiently small so as to be approximately isopotential. Thus, the kth compartment of a dendritic tree would be described by an equation of the form:

$$C \frac{dV_k}{dt} = -\sum_{j=1}^{N} I_j + g_{k-1,k}(V_{k-1} - V_k) + g_{k,k+1}(V_{k+1} - V_k) \tag{15.45}$$

where I_j are the various ionic currents (synaptic, leakage, voltage gated, injected, etc.) passing through the membrane of the compartment. The two final terms in (15.45) represent the current flow or flux (see 15.1) into the compartment from the adjacent compartments, with $g_{k-1,k}$ and $g_{k,k+1}$ being the respective conductances. Note that in this case the conductance can vary along a dendrite rather than being constant as was assumed above. Segev *et al.* (1989) describe in detail how a complex dendritic tree with multiple branches can be reduced to a compartmental model.

To see what can be learned by implementing a compartmental dendritic model, let us consider the nature of active spike propagation from the soma back into dendrites. It has recently been shown that spikes initiated at the soma or axon hillock are actively propagated back up the apical dendrite of hippocampal and cortical pyramidal cells (Stuart and Sakmann, 1994; Magee and Johnston, 1995, 1997; Johnston *et al.*, 1996; Stuart *et al.*, 1997). These dendritic spikes result from the presence of voltage-gated Na^+ channels in the dendritic membrane as well as in the axon hillock. Unlike axonal spikes, however, dendritic spikes decrease in amplitude as they propagate back up the dendrite away from the soma. In addition, dendritic spikes can apparently only be initiated at the axon hillock but not in the dendrite itself (Stuart and Sakmann, 1994). Mainen *et al.* (1995) developed a compartmental model to study the nature of dendritic spike propagation, and the model developed below reproduces the key features of their model in simplified form.

In developing a compartmental model, the first question must always be: how many compartments are necessary? Mainen *et al.* (1995) used approximately 275 compartments to simulate the dendritic tree of a particular rat layer 5 pyramidal cell. As this resulted in a model with approximately 1000 coupled nonlinear differential equations, it will be necessary to use a greatly reduced number of compartments here. Recently, Bush and Sejnowski (1993, 1995) have developed a method permitting as many as 400 compartments to be reduced to eight or nine, and in another context Destexhe *et al.* (1996) have shown that models with as few as three compartments can reproduce many results obtained with 230 compartments. Accordingly, let us develop four-compartment model to study dendritic action potentials. The model consists of a soma compartment (assumed to include the axon hillock) and a chain of three dendritic compartments as illustrated in the top inset in Fig. 15.7. Each compartment will be assumed to contain both Na^+ and K^+ currents as described the cortical neuron equation (9.10). To further simplify, the conductances g between compartments in (15.45) will be assumed to all be identical; $g = 4$ provides reasonably accurate results. Thus, the equations for the soma will be:

$$\frac{dV_1}{dt} = -\{17.81 + 47.58V_1 + 33.8V_1^2\}(V_1 - 0.48)$$
$$- 26R_1(V_1 + 0.95) + I + 4(V_2 - V_1) \tag{15.46}$$

$$\frac{dR_1}{dt} = \frac{1}{5.6}\left(-R_1 + 1.29V_1 + 0.79 + 3.3(V_1 + 0.38)^2\right)$$

The soma (compartment 1) is only coupled to one dendritic compartment, so only one coupling term appears in the first equation. Note that coupling only occurs through current flux between compartments, so only the V_1 equation is coupled to the dendritic potential V_2.

In writing equations for the dendritic compartments, we must take account of data indicating that dendritic Na^+ channel densities are much lower than those at the axon hillock (lumped with the soma in the current model) (Mainen *et al.*, 1995; Stuart *et al.*, 1997). Were it not for this important difference, the dendrites would be essentially identical to the axon hillock. Making the simplifying assumption that both Na^+ and K^+ conductances in the dendrites are 0.05 times those in the soma, the equations for the dendritic compartments become:

$$\frac{dV_k}{dt} = -0.05\{17.81 + 47.58V_k + 33.8V_k^2\}(V_k - 0.48) - 1.3R_k(V_k + 0.95) + I$$
$$+ 4(V_{k-1} - V_k) + 4(V_{k+1} - V_k) \tag{15.47}$$

$$\frac{dR_k}{dt} = \frac{1}{5.6}(-R_k + 1.29V_k + 0.79 + 3.3(V_k + 0.38)^2)$$

for $k = 2, 3, 4$ (the second coupling term being dropped when $k = 4$).

Spikes, decisions, and actions

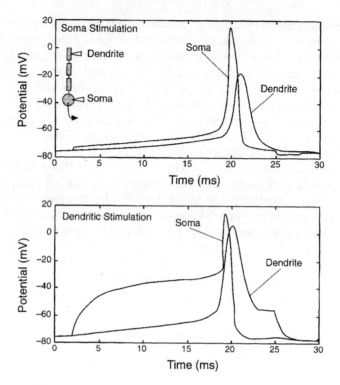

Fig. 15.7 Responses of four-compartment neuron (insert) described by (15.46) and (15.47). In the top panel stimulating current was delivered to the soma, while in the bottom the most distal dendritic compartment was stimulated. Compare with data in Fig. 15.8.

The four-compartment model described by (15.46) and (15.47) comprises just eight differential equations and has been implemented in MatLab script **ActiveDendrite.m**. As shown in the inset in Fig. 15.7, the model may be stimulated via current injection either into the soma or into the most distal dendritic compartment. In both cases the current strength is $I = 0.45$ nA, and the current step lasts 23 ms. Results following stimulation of the soma are plotted in the top panel of Fig. 15.7, where the two curves show potential at the soma (V_1) and in the distal dendritic compartment (V_4). In this case an action potential is first elicited in the soma, and it subsequently propagates back up the dendrite, reaching the distal compartment 1.16 ms later. In agreement with experimental data reproduced in Fig. 15.8 (Stuart and Sakmann, 1994), the dendritic spike has dropped in amplitude to about 64% of that triggered in the soma. These figures for attenuation and latency are typically found at a distance of about 225 μm from the soma (Stuart and Sakmann, 1994), so setting the length of each model dendritic compartment to about 75 μm produces good agreement with the data.

A more surprising result occurs when the distal dendrite rather than the soma is stimulated. In this case, depicted in the bottom panel of Fig. 15.7, the distal dendritic compartment depolarizes first, but it reaches a plateau, and the action potential is again generated first at the soma. This somatic spike then propagates back along the dendrite

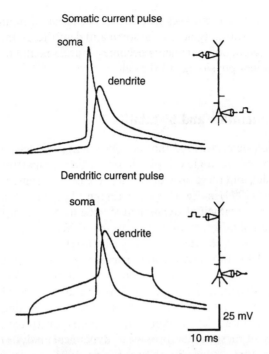

Fig. 15.8 Responses of a neocortical neuron to stimulation of the soma (top) or distal dendrite (bottom). Data reproduced with permission from Stuart and Sakman (1994), copyright Macmillan Magazines Ltd. Compare with simulation results in the previous figure.

producing a subsequent dendritic spike. Thus, spikes are always generated at the soma (or axon hillock) regardless of where the neuron receives a depolarizing input. As shown in the bottom panel of Fig. 15.8, this is also true for neocortical neurons (Stuart and Sakmann, 1994). This counter-intuitive behavior results from the much lower Na^+ and K^+ conductances in the dendrites relative to the soma. When current is injected into the dendritic compartment, conduction along the dendrite occurs too rapidly for the low Na^+ conductance to generate sufficient additional current to trigger a spike. Stimulation of the soma, however, produces a large Na^+ conductance increase sufficient to trigger a spike, and this in turn triggers spikes in the successive dendritic compartments, albeit with attenuated amplitudes.

To demonstrate the importance of dendritic Na^+ channels in boosting the amplitude of spikes propagated back from the soma, it is possible to simulate effects of TTX application to the dendritic compartments as was done in Chapter 10. This results in a decrease in the amplitude of the dendritic V_4 potential propagated from the soma by about 33%, which is similar to the decrease observed with experimental TTX application by Stuart and Sakmann (1994).

In addition to elucidating the nature of dendritic spike propagation, this example demonstrates the theoretical importance of compartmental models. A single compartment or equipotential neuron obviously could not provide any insight into amplitude attenuation of dendritic spikes. The fundamental conceptual importance of compartmental models thus emerges when there are different nonlinear ion currents in different parts of the neuron, typically the soma and dendrite. In this case the separation of channels into different compartments becomes a fundamental factor in the nonlinear dynamics and emergent physiology of the cell.

15.8   Plateau potentials and bistability

The previous section demonstrated that compartmental models are crucial for understanding certain aspects of single-cell physiology. Four compartments were used in the active dendrite model, and these provide results in reasonable agreement with the data of Stuart and Sakmann (1994) in Fig. 15.8 and with the 275-compartment model of Mainen *et al.* (1995). If the fundamental dynamical significance of compartmental models is indeed the separation of different ionic currents into different parts of the neuron, then it might be possible to study the dynamics of such current separation in a truly minimal compartmental model with just two compartments. This reasoning led Rinzel and colleagues to develop a series of elegant two-compartment models for a variety of complex neural phenomena (Pinsky and Rinzel, 1994; Booth and Rinzel, 1995; Li *et al.*, 1996; Booth *et al.*, 1997). The basic anatomy of these models, consisting of a soma plus one dendritic compartment coupled by conductance g_C, is illustrated in Fig. 15.9. One of the most fascinating of these models provides a dynamical analysis of plateau potentials and bistability in motorneurons (Booth and Rinzel, 1995; Booth *et al.*, 1997), so let us develop it here.

For years it was believed that motorneurons did little more than receive synaptic input and send axons to muscle fibers, where arriving spikes produced contraction, thus functioning as the 'final common pathway' controlling muscle activity. More recently,

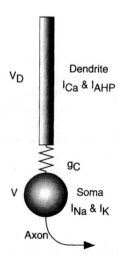

Fig. 15.9   A two-compartment neuron with ion channels distributed as indicated. The conductance between channels is g_C.

however, it has been discovered that individual motorneurons possess membrane properties that provide them with a form of short-term memory! The basis for this is a Ca^{2+} current present in the dendrites that can be switched between a low or 'off' state and a higher or 'on' state, the latter generating a plateau potential (Hounsgaard *et al.*, 1988; Elken and Kiehn, 1989; Kiehn, 1991). The Ca^{2+} plateau potential can be switched on by brief depolarization of the neuron, as is shown by the data from a cat spinal motorneuron in Fig. 15.10 (Hounsgaard *et al.*, 1988). Here brief depolarization sufficient to drive the neuron at a high rate results in continuing activity at a steady rate after the excitation has stopped. The ongoing activity continues until terminated by a subsequent hyperpolarizing pulse. Thus, brief excitation can switch on a plateau potential that maintains a roughly constant depolarization of the neuron until actively inhibited. Eken and Kiehn (1989) have suggested that plateau potentials may function as a local motor memory while an animal maintains a stable posture. Such bistable firing behavior suggests that the Ca^{2+} current dynamics of this neuron must generate a hysteresis loop of the sort studied in Chapter 6.

The two-compartment model for plateau potentials incorporates four ion currents: the Na^+ and K^+ currents involved in spike generation in the soma, and I_T-type Ca^{2+} plus slow I_{AHP} currents in the dendritic compartment (see Fig. 15.9). Coupling between the two compartments involves two factors: an intercompartmental conductance g_C, and a factor p representing the proportion of total neural membrane in the soma. The factor p is important, because the dendrites typically have a significantly larger membrane area than the soma with the result that current exchange between the two compartments can have unequal effects. The equations for this two-compartment model thus become:

$$C \frac{dV}{dt} = -I_{Na} - I_K + I + \frac{g_C}{p}(V_D - V)$$

$$C_D \frac{dV_D}{dt} = -I_{Ca} - I_{AHP} + \frac{g_C}{p}(V - V_D)$$

$$(15.48)$$

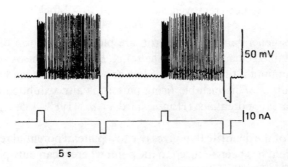

Fig. 15.10 Effect of plateau potentials in a motorneuron (reproduced with permission, Hounsgaard *et al.*, 1988). A brief depolarizing pulse switches the neuron on, and firing is maintained until a hyperpolarizing pulse switches the neuron off again. Once switched on, the neural activity is self-sustaining due to depolarization by a plateau potential.

As with all compartment models, it is necessary to define a new potential variable V_D for the dendritic compartment. The actual equations used in the plateau potential model are derived from (9.10) for the Na$^+$ and K$^+$ currents and adapted from (10.5) for the Ca^{2+} and I_{AHP} currents. The two equations for the soma are identical to (15.46) (except for the change in the coupling term to the form in 15.48), so only the equations for the dendritic compartment will be given:

$$2\frac{dV_D}{dt} = -(V_D + 0.754)(V_D + 0.7)(V_D - 1.0) - g_{AHP}C(V_D + 0.95)$$

$$+ \frac{g_C}{1 - p}(V - V_D) \tag{15.49}$$

$$\frac{dC}{dt} = \frac{1}{20}(-C + 0.5(V_D + 0.754))$$

Equation (15.49) has been constructed so that $V = V_D = -0.754$ at the resting equilibrium state, which is asymptotically stable. These equations were obtained from (10.5) by shortening the time constants and substituting the equilibrium value of X into the V_D equation. This convenient approximation reduces the number of equations for the dendritic compartment from three to two.

The script **Plateau.m** implements this two-compartment plateau potential model with parameter values $p = 0.37, g_C = 0.1$, and $g_{AHP} = 1.0$. The results of a 75 ms current pulse to the soma ($I = 1.0$ nA) followed 1.5 s later by a hyperpolarizing 75 ms pulse are plotted in Fig. 15.11. It is apparent that current flowing into the dendrite from the soma is sufficient to switch on the plateau potential, which then maintains a steady firing rate (12 spikes/s) until the hyperpolarizing off pulse. The key to this dynamical behavior is revealed by an examination of the dendritic (V_D, C) phase plane with isoclines from (15.49) (plotted for $V = -0.754$, its equilibrium value), which is plotted by **Plateau.m**. The parameters in (15.49) guarantee that this plane will have three steady states: two asymptotically stable and one unstable. You might therefore expect that this neural model would exhibit hysteresis, and this is so as shown in Fig. 15.12. The dendritic hysteresis loop at the bottom shows the three steady states between soma currents A and B (spikes in the soma have been removed by simulated TTX blocking of Na$^+$ channels as in Chapter 10).

The spike rates generated by the neuron are plotted in the top of Fig. 15.12 under conditions of sustained current input to the soma. Between points A and B the neuron is bistable and will maintain firing at either a low or a high rate, depending on the previous history of stimulation. This bistable firing pattern is also exhibited experimentally by neurons with plateau potentials (Hounsgaard *et al.*, 1988; Eken and Kiehn, 1989; Kiehn, 1991).

The bistability of the dendrite that gives rise to a plateau potential results from the fact that the I_{AHP} current has been reduced to the point where it can only partially counteract the inward Ca^{2+} current. If the conductance is increased to $g_{APH} = 1.5$, rerunning **Plateau.m** shows that the dendrite no longer has three steady states, and the plateau potential vanishes. There is evidence that plateau potentials require the presence of serotonin (Kiehn, 1991), and it was indicated in Chapter 13 that serotonin plays a role in

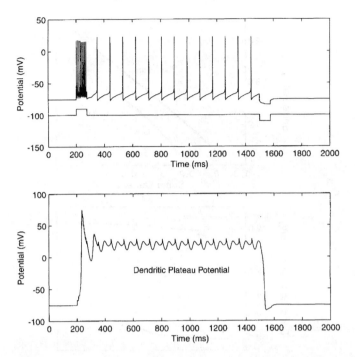

Fig. 15.11 Simulation of plateau potentials using the two-compartment neuron model in (15.48) and (15.49). Brief depolarization at On triggers a plateau potential in the dendrite that sustains firing by the soma until a brief hyperpolarization is delivered at Off. Compare with data in the previous figure.

modulating I_{AHP} currents in the lamprey. Thus, motorneurons can apparently be parametrically switched to and from a plateau potential mode via serotonin modulation.

The dynamical explanation of plateau potentials as a result of dendritic hysteresis may now seem obvious, but one might question whether the use of a two-compartment model is necessary in this dynamical example. Booth and Rinzel (1995) showed that a two-compartment model was indeed essential by a clever parametric manipulation. When $g_C = 0$ in (15.48) the soma and dendritic compartments are entirely independent, so hysteresis in the dendrite obviously cannot affect firing in the soma. In the opposite case when the coupling conductance is large, such as $g_C = 1$, any differences between V and V_D equilibrate so rapidly that the two compartments effectively become one electrically. Running **Plateau.m** with $g_C = 1$ will show that both compartments produce spikes in this case, and there is no plateau potential. Furthermore, the dendritic phase plane now has only one steady state. This will be more obvious if the program is run with $g_C = 0.20$. Coupling with the soma, approximated as being at equilibrium: $V = -0.754$, causes a bifurcation at which the plateau and unstable steady states of the dendrite vanish as g_C increases. Thus, only intermediate values of the coupling conductance g_C are capable of producing plateau potentials capable of driving spike generation, so separation of ionic currents into somatic and dendritic compartments is indeed essential to the dynamics of plateau-generating motorneurons.

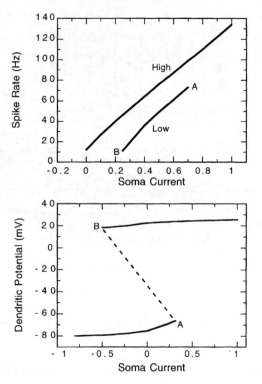

Fig. 15.12   Effects of plateau potential in two-compartment motorneuron model. The top graph shows high
and low spike rates generated between A and B depending on activation of the plateau potential. The bottom
graph shows the hysteresis loop in the dendrite under simulated TTX to block spikes in the soma. Between A
and B two asymptotically stable states exist separated by an unstable state (dashed line)

15.9   How many compartments suffice?

The elegant, two-compartment models of Rinzel and colleagues demonstrate that certain
complex dynamical behaviors require partial electrotonic separation between the soma
and dendrite. One need only define a second voltage variable and couple the compartments with conductance g_C as in (15.48). So, two compartments are computationally
efficient and sometimes necessary to explain neural dynamics, but are they sufficient? If
not, how many compartments should one use? Compartmental models of single neurons
range in size from two-compartments described by as few as four differential equations up
to at least 400-compartments described by several thousand differential equations.
Intermediate within this range is the CA3 hippocampal pyramidal cell model developed
by Traub *et al.* (1994) comprising 64 compartments.[1] The larger models provide much
more detailed descriptions of single neurons but at a price: many of the parameters are
unknown experimentally and can only be guessed or approximated, and the resulting

---

[1] The program NEURON by Michael Hines (1989, 1993) is freely available and recommended for detailed neural
simulations involving large numbers of compartments.

system of differential equations is generally intractable analytically. Extremely large multi-compartment neural simulations are perhaps best viewed as 'virtual neurons' (in the spirit of 'virtual reality') upon which to conduct virtual physiological experiments. On the other hand, two-compartment models represent a major simplification, but they are tractable and may capture the essence of neural dynamics. Indeed, Mainen and Sejnowski (1996) were able to adapt the Pinsky and Rinzel (1994) two-compartment model and show that the range of neocortical activity patterns could be generated simply by varying the parameter p (see 15.48). In general, it is a good strategy to use as few compartments as are necessary for understanding the problem at hand: small numbers for large network simulations, but larger numbers for detailed studies of one or a few neurons.

15.10 Exercises

1. Solve for the steady state of the cable equation (15.5) subject to the boundary conditions: $V(0) = -60\,\text{mV}$ and $V(2D) = -20\,\text{mV}$. Plot $V(x)$ for $0 \le x \le 2D$.

2. Assuming zero flux boundary conditions, generalize (15.24) to include sine wave terms. Write down the exact solution to the cable equation when the initial distribution is $V = 2\sin(\pi x/2) + \sin(5\pi x/2) - \sin(9\pi x/2)$. Produce an animation of this solution by modifying the script **Diffusion.m**.

3. The temporal order of synaptic events can have a dramatic effect on the postsynaptic potential at the soma. To study this in a passive dendrite, modify **Dendrite.m** by setting NumPSP $= 3$ so that you can incorporate three postsynaptic potentials. Plot the time course for 10 ms of the following distribution of PSP locations and times (note that the first event in the program always occurs at $t = 0$ ms):

(a) $x = 40\,\mu\text{m}, t = 0\,\text{ms}; x = 70\,\mu\text{m}, t = 2\,\text{ms}; x = 100\,\mu\text{m}, t = 4\,\text{ms}$.

(b) $x = 100\,\mu\text{m}, t = 0\,\text{ms}; x = 70\,\mu\text{m}, t = 2\,\text{ms}; x = 40\,\mu\text{m}, t = 4\,\text{ms}$.

(c) $x = 100\,\mu\text{m}, t = 0\,\text{ms}; x = 100\,\mu\text{m}, t = 2\,\text{ms}; x = 40\,\mu\text{m}, t = 4\,\text{ms}$.

(d) $x = 70\,\mu\text{m}, t = 0\,\text{ms}; x = 70\,\mu\text{m}, t = 2\,\text{ms}; x = 70\,\mu\text{m}, t = 4\,\text{ms}$

Which sequence of synaptic events produces the maximum peak PSP? Which sequence produces the greatest mean PSP? How would you expect the spike rate of this neuron to differ in these two cases (assume the threshold is a PSP of 0.45)?

4. The analytical solution for action potential propagation in (15.40) and (15.42) predicts that spike velocity will be directly proportional to the diffusion length constant D. Using the script **Spike_Propagate.m** test this prediction for $D = 0.4, 0.2$, and $0.1\,\text{mm}$. Plot simulated velocity as a function of D to determine whether the relationship is linear.

5. The formula for v in (15.40) also makes an implicit prediction of the dependence of spike velocity on the Na^+ equilibrium potential, E_{Na}. This is because the height of the action potential, h, is mainly determined by E_{Na} (see Fig. 9.2). Assuming that $h = 0.88E_{\text{Na}}$, which is the relationship in our analysis of (15.33), calculate v using (15.40) for $E_{\text{Na}} = 0.60, 0.40$, and 0.25 (i.e. 60, 40, and 25 mV). Let $D = 0.25\,\text{mm}$ in each case. Plot your calculated velocities and compare with the results of simulations using **Spike_Propagate.m** for these same values of E_{Na} ($E_{\text{Na}} = 0.55$ in the script).

6. Cortical spreading depression is a phenomenon in which a wave of unusually high extracellular K^+ concentration spreads across the cortex at a rate of several mm per minute. Grafstein (1963) proposed a quantitative biophysical model of this K^+ wave that is described by the equation:

$$\tau \frac{\partial K}{\partial t} = D^2 \frac{\partial^2 K}{\partial x^2} + 2K(2.3 - K)(9.5 - K)$$

where K is the excess extracellular K^+ relative to the resting state. The diffusion length constant is $D = 0.02$ mm, and $\tau = 0.5$ s. Solve this equation for the traveling wavefront, graph it, and calculate the velocity of propagation. Convert the velocity to mm/min. Cortical spreading depression is thought to be the cause of visual auras in migraine attacks, and these auras spread across the visual field at a rate equivalent to about 3 mm/min on the surface of the visual cortex. How does your computed velocity compare with this?

7. Reduce the four-compartment model for a neuron with active dendritic ion channels Na^+ in (15.46) and (15.47) to produce a two-compartment Rinzel-type model with coupling between the soma and dendrite described by (15.48). Use $p = 0.33$ (proportion of soma membrane), reduce dendritic ion conductances by a factor of 0.05 as in (15.47), and stimulate with $I = 0.2$ nA. Find a value of the intercompartment conductance g_C for which the dendritic spike always occurs after the soma spike for either dendritic or soma current injection. Plot your results and compare with Fig. 15.7.

8. Consider the following two-compartment model for a motorneuron:

$$\frac{dV}{dt} = -4\left(V^2 - \frac{V}{10}\right)(V - 1) - R\left(V + \frac{1}{5}\right) + I + \frac{g_C}{1 - p}(V_D - V)$$

$$\frac{dR}{dt} = \frac{1}{5}(-R + 3V^2)$$

$$2\frac{dV_D}{dt} = -4\left(V_D^2 - \frac{V_D}{10}\right)(V_D - 1) - C\left(V_D + \frac{1}{5}\right) + \frac{g_C}{1 - p}(V - V_D)$$

$$\frac{dC}{dt} = \frac{1}{20}(-C + 0.9V_D)$$

These are based on canonical neocortical neuron equations discussed in the problems of Chapter 9. First show by simulation (you can adapt **CanonicalNeuron.m** from Chapter 9) that these equations exhibit plateau potential behavior for $g_C = 0.047, p = 0.48$ (stimulate with $I = 0.05$ for about 50 time units). Next, analyze the phase plane for the second two equations using the approximation that the soma remains at the resting potential $V = 0$. For what range of g_C and p values can the dendrite sustain a plateau potential in this approximation?

16   *Nonlinear dynamics and brain function*

We have now completed our exploration of the dynamical foundations of neuroscience. Perhaps the most striking message to be taken away from this experience is that virtually all important aspects of neural function are inherently nonlinear. Action potentials and neural bursting are highly nonlinear limit cycle phenomena that cannot occur in a linear system. Motor control requires rhythmic oscillations that can be phase locked even in the presence of noise. This again is only conceivable in terms of limit cycle oscillations. At a higher cognitive level, categorization, decision making, and long-term memory are all inherently nonlinear phenomena related to the dynamics of networks with multiple steady states.

Nonlinear neural networks with multiple steady states unavoidably exhibit bifurcations and hysteresis as inputs or parametric variations switch the system between states. The positive aspect of hysteresis is that once a bifurcation point is passed, neither noise nor a small reduction in relevant input can immediately switch the system back into the previous state. At the level of cognitive decisions, hysteresis permits the brain to ignore modest fluctuations in the evidence available to it. The alternative in complex circumstances is perseveration and crippling indecision. Unfortunately, hysteresis inherent in neural decisions and memory can also have deleterious side effects. Depending on the range over which hysteresis operates, it can lock in judgments and categorizations based on partial and misleading early data. In short, hysteresis and generalization in neural networks can unfortunately lead to permanent storage of prejudices and bigotry based on early experiences with small samples. Such side effects of hysteresis may form one basis for the persistence of inappropriate social stereotypes.

As implied by the relationship among hysteresis, bifurcation, and prejudice just suggested, the conceptual framework provided by nonlinear dynamics has a range of application that vastly transcends the neural topics to which this book has been devoted. Surely human interactions, controlled by nonlinear nervous systems, will produce large-scale mirrors of nonlinear phenomena in the sociological realm. Thus, the foundations of nonlinear dynamics developed here would be expected to have application to a wide range of other scientific domains. This is manifestly true, and mention of salient examples will follow shortly. First, however, let us review the basic concepts of nonlinear neurodynamics to extract some general principles. This will lead to several guidelines on the inclusion of nonlinear dynamics in one's own research. Following this is a very brief survey of links between the nonlinear dynamics relevant to neural problems and dynamical issues in other fields. Finally, I shall close by posing the ultimate nonlinear dynamical problem: how can a nonlinear dynamical brain discover nonlinear dynamics?

16.1   Essence of nonlinear neurodynamics

Throughout this book, certain themes in nonlinear dynamics have emerged repeatedly. To begin, let me remind you of the mathematical basis for our understanding of nonlinear phenomena. Looking back at the first chapter, two principles were emphasized: the exponential function and Taylor series. As promised there, the combination of these two principles may now be recognized as the basis of nonlinear dynamical analysis. The primary strategy throughout the book has been to first solve for the steady states and then linearize about each based on Taylor series expansion. The local stability of each steady state is then determined by the eigenvalues of the Jacobian, which are solutions of the characteristic equation. If the eigenvalues are imaginary, oscillatory sine and cosine solutions result. It is the properties of the exponential function which reduce differential equations to solutions of polynomials. These concepts, exponential functions with real or imaginary exponents and Taylor series, constitute the bedrock upon which much of nonlinear dynamics is founded.

In addition, a number of geometrical principles have emerged that extend linearized analysis in important ways. The first of these we encountered was the Poincaré–Bendixon theorem. Although limited to two dimensions, this elegant theorem can prove the existence of limit cycles throughout the phase plane using inherently geometric and global considerations. The Hopf bifurcation theorem is also predicated on geometric considerations: a multi-dimensional system must decay onto a two-dimensional subspace within which limit cycle analysis becomes possible.

Without doubt, the most intellectually sophisticated application of geometry to nonlinear dynamics results from the theorems of Lyapunov. This brilliant mathematician realized that nonlinear dynamics could be understood by defining a landscape of hills and valleys where solutions to nonlinear differential equations descended to the lowest points. This is clearly the most global approach to analyzing nonlinear systems. Although the Lyapunov function approach has inherent difficulties associated with the creation of appropriate functions, I shall argue below that the geometrical interpretation of dynamics revealed by Lyapunov even provides insight into the inception of mathematics in the brain.

Turning to those aspects of nonlinear dynamics of greatest relevance for neuroscience, several further principles have emerged. First, many neural problems have at least two widely different time scales associated with the dynamics. This has permitted us to analyze the fast variation with the slow assumed constant, thereby reducing the dimensionality and complexity of the problem. Neuronal bursting is a classical example of this, because the bursting is typically generated as the slower variable (e.g. Ca^{2+} current) sweeps the faster variables (e.g. Na^+ spiking) back and forth through a bifurcation.

Another very important observation is that neurodynamics almost always incorporates either sigmoid (e.g. Naka–Rushton or logistic functions) or cubic nonlinearities. (As the cube root is also sigmoidal, albeit without finite asymptotes, cubics are sigmoidal along the orthogonal axis.) Restriction to such a very small subset of all possible nonlinearities has enabled us to develop a comprehensive treatment of the dynamical foundations of neuroscience. Sigmoid functions and cubics share the important property that they can be intersected by a straight line in either one or three points, and this property underlies the existence of multiple steady states and hysteresis.

An interplay between positive and negative feedback constitutes the final key theme in neural dynamics. At the level of networks this takes the form of a balance between recurrent excitation and inhibition, as exemplified by the Wilson–Cowan (1972, 1973) equations. At the level of single neurons, positive feedback is exemplified by the voltage-dependent opening of ion channels (e.g. Na^+ or Ca^{2+}) that cause further depolarization of the cell, while negative feedback is manifested by voltage-dependent channels producing hyperpolarization. In both cases the underlying dynamical characteristics are remarkably similar. Both networks and isolated neurons can produce limit cycles, and both can also produce hysteresis switching as witnessed by short-term memory circuits in networks and plateau potentials in neurons. Thus one might epitomize the dynamics of neuroscience as the nonlinear sigmoid dynamics of interacting positive and negative feedback pathways.

16.2 Strategies for neural modeling

As in any active and creative area of science, there are a vast range of strategies for attacking neural modeling problems. However, certain considerations and compromises are faced again and again. The first and most important consideration is the level of description of the neural elements. The alternative possibilities can be rank ordered in terms of increasing complexity. The simplest description of a neuron is certainly in terms of its spike rate as described by a Naka–Rushton or other sigmoid function. A more complex description is the isopotential neuron with multiple ionic currents that combine to produce individual spike outputs. However, as demonstrated elegantly by Booth and Rinzel (1995), firing patterns such as those involving plateau potentials require a two-compartment model for their description. Two, of course, is the lower limit of compartmental models, so the most complex descriptive level is the multi-compartment neuron, where multi typically ranges from 10 to several hundred.

In choosing among these levels of neural description one is faced with a number of trade-offs. It is obviously possible to simulate 100 times as many neurons if the description of each involves just one differential equation rather than 100. In the opposite direction, increasing realism is certainly achieved in moving from a spike rate description to a two or more compartment model. This consideration might seem to push one towards the maximum number of compartments compatible with computer memory. However, neurons simulated with very large numbers of compartments suffer in two important respects. First, there are almost always far more parameters in these models than can be reliably estimated from the data available. Second, very high dimensional nonlinear systems become impossible to analyze mathematically, so the responses of huge compartmental models frequently fail to provide any insight into underlying dynamical causes. Thus, simulations of very large compartmental models are probably best regarded as empirical experiments on 'virtual neurons'. As with experiments on any real neuron, one is also confronted here with the issue of whether the results obtained are generic as opposed to idiosyncratic.

So, what is the best strategy for choosing a level of neural description? In part the answer is dictated by the problem: those wishing to simulate visual cortex or the hippocampus will of necessity employ large numbers of neurons with the simplest possible description of each (probably spike rates), while those interested in the detailed ionic contributions to intracellular recordings will choose multi-compartment models. My bias is to choose the simplest possible description consistent with the data one wishes to explain. The striking success of the two-compartment model (Pinsky and Rinzel, 1994; Booth and Rinzel, 1995; Booth *et al.*, 1997) in explaining not only plateau potentials but also the range of spiking patterns observed in neocortical neurons (Mainen and Sejnowski, 1996) suggests that it may seldom be necessary to move beyond this level of neural description.

In keeping with the strategy of simplifying neural models as much as possible, one final point deserves mention. When a group of similar neurons are interconnected, it is frequently possible to represent their activity by that of a single neuron. We have seen examples of this subsampling technique in the Wilson–Cowan equations (1972, 1973) and in the lamprey simulations of Wallén *et al.* (1992). Thus, quite large neural network problems may frequently be reduced to manageable proportions. It is worth remembering that the goal of modeling complex neural systems is insight into the underlying computational dynamics, and simplification is frequently crucial to insight.

16.3   Nonlinear dynamics in other fields

Spikes, decisions, and actions has focused exclusively on the nonlinear dynamical principles required to understand neural function. Despite this focus, virtually all of these mathematical principles find application in a wide range of other scientific fields. Foremost among these is certainly physics, the science with the longest history of mathematical sophistication. Two excellent books revealing the application of nonlinear dynamics to physics are *From Order to Chaos* by Kadanoff (1993) and *Order within Chaos* by Bergé *et al.* (1984).

Within biology, nonlinear dynamics has probably had its largest impact in neuroscience, due largely to the brilliant work culminating in the Hodgkin–Huxley (1952) equations. As a result, several other approaches to dynamics in neuroscience have also been developed. Books by Kelso (1995) and Haken (1996) both stress the importance of nonlinear dynamics in neuroscience, but both are slanted more toward approaches introduced in statistical physics, so they provide complementary treatments to *Spikes, decisions, and actions*.

In addition to neuroscience, dynamical approaches are now common in many areas of biology including ecology, cardiac rhythms, and developmental pattern formation. Excellent surveys of dynamical approaches in these areas are provided in books by Edelstein-Keshet (1988), Glass and Mackey (1988), and Murray (1989). To cite one example, Murray (1989) provides an elegant treatment of the ways in which nonlinear diffusion–reaction equations can generate the patterning of spots on leopards or the designs of butterfly wings. These turn out to be examples of self-organizing systems with similar dynamical principles to those underlying visual hallucinations in excitatory–inhibitory networks (Ermentrout and Cowan, 1979).

Dynamical modeling has even been fruitfully employed in sociology. Kadanoff (1993) studied nonlinear interactions of factors mediating the growth and decay of urban centers. There is currently also evidence that political and economic decision making can exhibit chaotic dynamics (Richards, 1990). It is also likely that dynamical models of species interactions in ecology will be relevant to understanding sociological interactions.

Finally, I think it is now becoming clear that nonlinear dynamical concepts are central to issues in both psychology and philosophy. As we have seen, perceptual categorization and decision making receive a natural explanation in terms of bifurcations in competitive networks. Furthermore, the existence of neural chaos is clearly relevant to discussions of free will and determinism, as chaotic systems are deterministic without being predictable. Recent evidence for chaos in perception (Richards *et al.*, 1994) is certainly germane to this issue. Several thoughtful evaluations of the significance of neural network models for philosophy are available in books by Churchland (1989), Bechtel and Abrahamsen (1991), and Clark (1989). This should emphasize that nonlinear concepts provide an indispensable qualitative framework for exploring a vast range of phenomena.

16.4   Mathematics in mind

Exploration of the dynamical principles underlying brain function leads finally to the ultimate question in nonlinear dynamics: how is it possible for the brain to create and understand mathematics? Otherwise stated, what sort of nonlinear neurodynamical system or 'mind' is required to instantiate mathematics? This may seem an impossibly complex question, and neuroscientists are certainly only beginning to nibble at its edges. Nevertheless, some promising directions for conceptualizing the broad outlines of an answer are beginning to emerge.

In his eminently readable book, *The Number Sense: How the Mind Creates Mathematics*, Dehaene (1997) provides a concise summary of research on the brain and mathematics. One fundamental observation is that all mammals show an ability to discriminate relative magnitudes that is similar to the average human who has not been trained in basic arithmetic. Mammals will consistently discriminate 2 from 3 food pellets or 10 from 15, although they will confuse 10 and 11. In short, mammals appear able to discriminate quantity on a scale where equal increments are scaled logarithmically. This is not surprising, as virtually all sensory systems convey information concerning the magnitude of relevant stimuli (e.g. brightness, loudness, pressure, etc.), but they do so in accord with Weber's law, indicating an approximately logarithmic neural code for stimulus intensity.

To explore the next steps in the evolution of mathematical capabilities it is necessary to make a very important evolutionary observation. The invention of mathematics by the human brain is certainly no older than the invention of writing, which occurred first in Mesopotamia around 3000 B.C. and perhaps independently in Egypt around the same time (Diamond, 1997). Furthermore, throughout world history until perhaps the past half-century, the vast majority of the world's population has been both illiterate and innumerate. Thus, there has neither been the time (only about 250 generations) nor the evolutionary pressure for an explicit mathematical or reading capability to have evolved in the brain (Donald, 1991). Strikingly, the earliest writing systems evolved from simple accounting systems, that is from records of numbers; prose writing arose much later

(Diamond, 1997). Given the obvious fact that I can write and you, the reader, can comprehend the material in this book, only one plausible possibility remains. Mathematical (and reading) capabilities must have arisen as a byproduct of evolutionary forces that optimized certain brain areas for other functions.

Current evidence suggests that mathematical abilities are associated with the evolution of eye–hand coordination and the manipulation of objects. There is much evidence, for example, that both counting and the language of numbers are closely related to enumeration of fingers and other body parts. Thus, all children begin counting by associating objects with their fingers, and in certain societies body parts including toes, elbows, wrists, and knees are used to count up to about 30 (Dehaene, 1997). The link with language is witnessed by the fact that in at least one language the word for 10 literally means 'two hands'.

A relationship between the visual manipulation of objects and the evolution of mathematical ability is suggestively conveyed by recent studies of brain activation during mathematical thinking. A PET (positron emission tomography) study has shown that parts of human parietal cortex are most active during mental division (Dehaene *et al.*, 1996). In a second study, direct recording from parietal neurons in humans via electrodes implanted for medical reasons showed major activation of parietal area 7 in mental addition and subtraction (Abdullaev and Melnichuk, 1996). This is consistent with evidence that damage to inferior parietal cortex can produce major deficits in numerical abilities, termed 'acalcula' (Dehaene, 1997). These same studies also demonstrate that mathematics is not a language in any ordinary sense: subjects with brain damage leading to acalcula can still retain completely normal language abilities. Far from being a language, mathematics represents a thoroughly independent and powerful mode of brain function! This conclusion is also supported by the dissociation between linguistic and mathematical-spatial abilities in the human population. Relative to their percentage in the population, left-handers are twice as highly represented as right-handers among the most creative individuals in fields requiring mathematical or spatial reasoning abilities (e.g. Newton, Einstein, DaVinci, Michelangelo, Picasso). However, lefties on average fall below their dexterous counterparts in linguistic abilities.

Neurophysiological studies of parietal cortex in monkeys indicate that it contains a number of distinct areas that are collectively involved in perception of motion and surface orientation, and the visual guidance of object manipulation (Milner and Goodale, 1995; Sakata *et al.*, 1997). In fact, there is both physiological and clinical evidence that separate subdivisions of the parietal cortex subserve vision of peripersonal space (i.e. space within reaching distance) and extrapersonal space (Sakata and Kusonoki, 1992; Cowey *et al.*, 1994). Neurons in these areas are involved in both the visual and tactile (somatosensory) perception of motion, position, and surface orientation in three dimensions. Clearly, any cortical area devoted to both vision and touch could only correlate such information within peripersonal space. To carry out their functions normally, parietal areas must perform multiple vector summation and subtraction operations in order to compute both direction of motion and the location of the hand given only joint angle information, and a neural network capable of carrying out these computations was analyzed in Chapter 7 (Wilson *et al.*, 1992; Wilson and Kim, 1994). Furthermore, many visually guided object manipulations are voluntary actions, so some aspects of parietal neural activity must be accessible to or even a component of consciousness. This is manifestly true for humans and is probably true for great apes as well.

The mathematician Hadamard (1945) argued that mathematical thought was primarily visual and kinematic in nature, and reflection on the contents of *Spikes, decisions, and actions* reveals the striking degree to which our understanding of nonlinear dynamics relies on spatial visual concepts. Thus, phase plane analysis clearly involves visualization of vector flow fields in two dimensions, and Lyapunov function theory shows that nonlinear differential equations are intimately associated with surfaces in multi-dimensional spaces. Even chaos requires a visual appreciation of the fact that three dimensions are required for a trajectory to wander about in a bounded region without ever coming to rest or intersecting itself. Finally, three major discoveries in the history of mathematics reveal the spatial aspects of abstract mathematics: the Pythagorean theorem, the discovery of conic sections by the Greeks and the invention of analytical geometry by Descartes. The Pythagorean theorem was believed by Herodotus (450 B.C.) to have been inspired by the practices of Egyptian surveyors using 3–4–5 rope triangles to create right angles. Analytical geometry showed that the solutions of simultaneous algebraic equations could be viewed as intersections of lines or surfaces in space. Regarding conic sections, I am reminded of an episode that occurred when I was an undergraduate at Wesleyan University. Rich Young, a blind classmate and mathematics student, asked a group of us one day why parabolas, ellipses, circles, and hyperbolas were called conic sections. Someone quickly made a cone from a sheet of paper and let Rich feel the different ways in which he could grasp the cone with his fingers. An instant smile of recognition spread across his face as he saw the concept with his hands! Recall that parietal cortex integrates both visual and tactile information.

Further confirmation of a correlation between visually guided manipulation and mathematics comes from studies of gesturing during conversation. McNeill (1992) videotaped conversations about abstract mathematics carried on by two professional mathematicians. The striking result was that spatially meaningful hand gestures were almost perfectly correlated with the verbal expression of certain abstract mathematical concepts. Even more striking, there were cases of verbal errors when the hand gesture was nevertheless mathematically correct! Thus, the hand gestures represented a more accurate reflection of mathematical thought than did words.

In sum, therefore, mathematical ability seems to rest on two fundamental aspects of brain function. The simpler is a general appreciation of relative quantity, which is ubiquitous in mammalian sensory perception. Our nascent understanding of deeper aspects of mathematics indicates that they are localized in parietal association cortex. These cortical areas, which combine visual and somatosensory information, are also involved in vector computations related to eye–hand coordination, the perception of surfaces, and perception of motion in depth. Such vector-based computations are both inherently spatial and incorporate addition and subtraction as special cases. As these networks also appear capable of categorical decision making (e.g. coherent versus transparent motion, see Chapter 7), these cortical areas can create precise perceptual dichotomies. It is not inconceivable that more abstract dichotomies, equal–unequal, true–false, might also arise in such neural networks. Granted evolving capabilities to reflect consciously on visually guided manipulation within peripersonal space, we can thus begin to envision (metaphor intended) the emergence of mathematical abilities in nonlinear dynamical networks of the parietal cortex.

Understanding of mathematical ability and a great many other aspects of brain function awaits decades of future research. This research will certainly be accompanied by the development of more sophisticated mathematical tools for studying neural networks. Nevertheless, I think it reasonable to suggest that the nonlinear dynamical principles which will form the foundations of that understanding are largely available today.

Appendix: MatLab™ and the MatLab scripts

This appendix contains a very short introduction on the use of MatLab™ to run the simulations accompanying this book. It is assumed that the reader is already familiar with MatLab basics, at least at the level of the tutorials provided in the MatLab manual. The disk is readable by either Macintosh or Windows computers and contains two folders of MatLab scripts: **Mac MatLab**, and **WinMatLab**. The scripts in the two folders are identical except for a few differences in text formatting required by the two different operating systems. The appropriate folder should be copied onto the reader's hard disk so that all examples can be worked using these copies. That way, should you make any changes that prevent the script from working properly, you can always make another copy of the original. The disk with this book has been locked to prevent modification of the original scripts. MatLab is also available to run under UNIX, but I have not tested the scripts in a UNIX environment. Most scripts should run under the student version of MatLab. However, several of the scripts later in the book may require vectors longer than the student version permits. This was unavoidable in order to demonstrate some points concerning more complex neural networks. Also, some of these scripts may take several minutes to run even on a Macintosh G3. This was again unavoidable due to the necessary size and length of certain simulations.

The MatLab scripts were created on a Macintosh and then translated using TransMac under Windows NT. TransMac under Windows correctly maintains file name case and length and properly translates end-of-line and end-of-file conventions. From Windows, correctly configured contemporary copying mechanisms to and from UNIX all implement these conventions correctly in both directions. These include Samba, modern ftp clients, and modern zipping utilities, including gzip on the UNIX side and Winzip 7.0 on the NT side. In translating the files for UNIX, it is crucial that file names and letter cases be retained for the scripts to work properly.

Users of the WinMatLab directory on either Windows or UNIX may see directories named Resource.frk and files named Finder.dat. These support the production of a single diskette which can be read either by Macintosh or Windows machines, and may be safely ignored in those environments. Note that many scripts produce several figures in different windows, and each must be moved to the front for viewing (see MatLab manuals). Please check the book web site http://spikes.bsd.uchicago.edu for current information concerning the MatLab scripts and to report any problems.

Each MatLab script is designed to be run by the reader to verify assertions at appropriate points in the book. Following this, it is expected (and suggested at points in the text) that the reader will modify the scripts to solve problems at the end of chapters and hopefully also to begin to explore neuroscience problems related to her/his own interests. As acquaintance rather than expertise with MatLab is all that is expected of the reader, the points where mathematically appropriate modifications should be made are generally contained in statements between two lines of asterisks. This makes it rather easy to find and make the suggested parameter changes.

Note that many of the scripts provided generate several windows, which MatLab may plot on top of one another. All windows can be viewed successively by using the Windows menu, however.

Among the MatLab scripts is a folder entitled 'Basic Scripts'. This contains a very simple plotting program that will be adequate for any of the earlier problems in the book. It also contains the basic programs for finding the zeros of a transcendental function as outlined below.

Three MatLab functions that will be of general utility are **fzero()**, **roots()**, and **eig()**. Their functions are most easily understood through the following examples.

The function **fzero()** finds the zero of a function that is closest to an initial guess provided by the user. Suppose, for example, that it was necessary to solve the following transcendental equation:

$$x = \frac{10}{1 + e^{-x+5}}$$

This type of problem is fairly common in certain neural network formulations. This can be converted into a form suitable for use with fzero by moving both terms to one side:

$$x - \frac{10}{1 + e^{-x+5}} = 0$$

The left side has been entered into the MatLab function script **TransZero.m**, which is in the Basic Scripts folder. The use of fzero to call and solve this function is implemented by the script **Zero_Finder.m**. If this script is run with a range of initial guesses $0 \le x \le 20$, three distinct roots will be found. To solve other transcendental equations, **TransZero.m** must have the appropriate function entered and must then be *saved in the same folder* as **Zero_Finder.m**.

The **roots()** function is very simple to use and may be run from the Command window. It finds the roots of a polynomial expressed as a coefficient vector beginning with the coefficient of the highest power of the unknown. Suppose you wished to know the roots of the equation:

$$5x^3 - 2x^2 + 7 = 0$$

The coefficient vector in this case would be written in MatLab syntax as $[5, -2, 0, 7]$. Note that every coefficient must be included; the 0 indicates that the coefficient of x is zero. Typing the command roots($[5, -2, 0, 7]$) and hitting the carriage return will produce a listing of the roots, which are -1 and a complex conjugate pair in this case (try it).

The **eig()** function returns the eigenvalues of a square matrix if they exist. In MatLab syntax, the following matrix:

$$\overset{\leftrightarrow}{A} = \begin{pmatrix} 0 & 1 & 2 \\ 3 & 2 & 1 \\ 1 & 0 & 5 \end{pmatrix}$$

would be written as: $A = [0, 1, 2; 3, 2, 1; 1, 0, 5]$. Note that commas separate entries within a row, while semicolons separate the rows. Now the command eig(A) will generate the three eigenvalues, the first being -1.2073.

MatLab simulations are described in the chapter where they are first used along with a brief indication of how they should be used. Frequently one of the chapter figures was generated by the script being discussed, which should provide a cue to the expected MatLab results. However, two of the more general and complex scripts deserve brief mention here.

LinearOrder2.m prints out analytical solutions to autonomous, second order homogeneous linear differential equations for any specified initial conditions. This program automates the solution of second order equations and will be useful throughout the book. The user is prompted for the coefficients of the Jacobian matrix and for the initial conditions. In the special and highly improbable case where both roots are identical (critical damping), the solution is printed for $x(t)$ only, on the assumption that the second variable and initial condition relate to dx/dt. This program *should not be modified* by the user. The program plots both the temporal solutions $x(t)$ and $y(t)$ and the x–y trajectory in the state space (except for critical damping). The small arrows in the state space plot indicate the direction of local trajectories throughout the space.

RungeKutta4.m implements a fourth order Runge–Kutta routine with constant step size for a system of any order. The reader must modify the line indicating the number of equations N, the time step DT, and the entries in the initial condition vector. Most importantly, the reader must type in each equation using as variables for the right-hand side XH(1), XH(2), XH(3), . . . , XH(N). For example, the FitzHugh–Nagumo equations:

$$\frac{dV}{dt} = 10\left(V - \frac{V^3}{3} - R + I_{input}\right)$$

$$\frac{dR}{dt} = -0.8R + V + 1.2$$

would be written as:

$$K(1, \text{ rk}) = 10^*DT^*(XH(1) - (XH(1)^{\wedge}3)/3 - XH(2) + \text{Input});$$
$$K(2, \text{ rk}) = DT^*(-0.8^*XH(2) + XH(1) + 1.2);$$

The variables on the left-hand side, K(1, rk), etc., are the intermediate Runge–Kutta variables in eqns (5.26), (5.29), and their generalizations. The second argument, rk, indexes which of the four intermediate values is being calculated by the loop. All simulations suggested in the problems can be performed by appropriate modification of **RungeKutta4.m**. As the book progresses, however, many simulations can be carried out more easily by modification of scripts introduced in each chapter. Virtually all neural simulations in the book are based on this fourth order Runge–Kutta routine.

References

Abbott, L. F. (1994) Decoding neuronal firing and modelling neural networks. *Qtr. Rev. Biophys.* **27**, 291–331.

Abdullaev, Y. G. and Melnichuk, K. V. (1996) Counting and arithmetic functions of neurons in the human parietal cortex. *Neuroimage* **3**, S216.

Adini, Y., Sagi, D., and Tsodyks, M. (1997) Excitatory–inhibitory network in the visual cortex: psychophysical evidence. *Proc. Natl. Acad. Sci.* **94**, 10426–10431.

Agmon, A. and Connors, B. W. (1989) Repetitive burst-firing neurons in the deep layers of mouse somatosensory cortex. *Neuroscience Letters* **99**, 137–141.

Albrecht, D. G. and Hamilton, D. B. (1982) Striate cortex of monkey and cat: contrast response function. *J. Neurophys.* **48**, 217–237.

Albright, T. D. (1992) Form-cue invariant motion processing in primate visual cortex. *Science* **255**, 1141–1143.

Aram, J. A., Michelson, H. B., and Wong, R. K. S. (1991) Synchronized GABAergic IPSPs recorded in the neocortex after blockade of synaptic transmission mediated by excitatory amino acids. *J. Neurophysiol.* **65**, 1034–1041.

Avoli, M. and Williamson, A. (1996) Functional and pharmacological properties of human neocortical neurons maintained *in vitro*. *Prog. Neurobiol.* **48**, 519–554.

Avoli, M., Hwa, G. G. C., Lacaille, J.-C., Olivier, A., and Villemure, J.-G. (1994) Electrophysiological and repetitive firing properties of neurons in the superficial/middle layers of the human neocortex maintained *in vitro*. *Exp. Brain Res.* **98**, 135–144.

Bal, T. and McCormick, D. A. (1993) Mechanisms of oscillatory activity in guinea pig nucleus reticularis thalami *in vitro*: a mammalian pacemaker. *J. Physiol.* **468**, 669–691.

Barenghi, C. F. and Lakshminarayanan, V. (1992) Spontaneous symmetry breaking and the onset of chaos in a muscle model. *Neurol. Res.* **14**, 228–232.

Bechtel, W. and Abrahamsen, A. (1991) *Connectionism and the Mind*. Blackwell, Cambridge, MA.

Bergé, P., Pomeau, Y., and Vidal, C. (1984) *Order within Chaos*. Wiley, New York.

Bergen, J. R. and Julesz, B. (1983) Parallel versus serial processing in rapid pattern discrimination. *Nature* **303**, 696–698.

Bertram, R., Butte, M. J., Kiemel, T., and Sherman, A. (1995) Topological and phenomenological classification of bursting oscillations. *Bull. Math. Biol.* **57**, 413–439.

Best, E. N. (1979) Null space in the Hodgkin–Huxley equations. *Biophys. J.* **27**, 87–104.

Beuter, A., Bélair, J., and Labrie, C. (1993) Feedback and delays in neurological diseases: a modeling study using dynamical systems. *Bull. Math. Biol.* **55**, 525–541.

Blake, R. (1989) A neural theory of binocular rivalry. *Psych. Rev.* **96**, 145–167.

Bliss, T. V. P. and Collingridge, G. L. (1993) A synaptic model of memory: long-term potentiation in the hippocampus. *Nature* **361**, 31–39.

Bloomfield, S. A., Hamos, J. E., and Sherman, S. M. (1987) Passive cable properties and morphological correlates of neurones in the lateral geniculate nucleus of the cat. *J. Physiol.* **383**, 653–692.

Bonds, A. B. (1991) Temporal dynamics of contrast gain in single cells of the cat striate cortex. *Visual Neurosci.* **6**, 239–255.

Booth, V. and Rinzel, J. (1995) A minimal, compartmental model for a dendritic origin of bistability in motoneuron firing patterns. *J. Comput. Neurosci.* **2**, 299–312.

Booth, V., Rinzel, J., and Kiehn, O. (1997) Compartmental model of vertebrate motoneurons for Ca^{2+} dependent spiking and plateau potentials under pharmacological treatment. *J. Neurophysiol.* **78**, 3371–3385.

Buchanan, J. T. (1992) Neural network simulations of coupled locomotor oscillators in the lamprey spinal cord. *Biol. Cybern.* **66**, 367–374.

Burbeck, C. A. (1986) Negative afterimages and photopic luminance adaptation in human vision. *J. Opt. Soc. Am. A* **3**, 1159–1165.

Burkhardt, D. A. (1993) Synaptic feedback, depolarization, and color opponency in cone photoreceptors. *Visual Neurosci.* **10**, 981–989.

Burkhardt, D. A. (1995) The influence of center-surround antagonism on light adaptation in cones in the retina of the turtle. *Visual Neurosci.* **12**, 877–885.

Bush, P. C. and Sejnowski, T. J. (1993) Reduced compartmental models of neocortical pyramidal cells. *J. Neurosci. Meth.* **46**, 159–166.

Bush, P. C. and Sejnowski, T. J. (1995) Models of cortical networks. In *The Cortical Neuron*, ed. M. J. Gutnick and I. Mody. Oxford University Press, New York, pp. 174–189.

Cajal, S. (1911) *Histologie du Système Nerveux de l'Homme et des Vertébrés*, Vol. II, Maloine, Paris.

Carandini, M. and Heeger, D. J. (1994) Summation and division by neurons in primate visual cortex. *Science* **264**, 1333–1336.

Carr, C. E. (1993) Processing of temporal information in the brain. *Ann. Rev. Neurosci.* **16**, 223–43.

Chay, T. R. and Keizer, J. (1983) Minimal model for membrane oscillations in the pancreatic beta-cell. *Biophys. J.* **42**, 181–189.

Churchland, P. M. (1989) *A Neurocomputational Perspective.* MIT Press, Cambridge, MA.

Clark, A. (1989) *Microcognition: Philosophy, Cognitive Science, and Parallel Distributed Processing.* MIT Press, Cambridge, MA.

Clark, M. R. and Stark, L. (1974) Control of human eye movements: III. Dynamic characteristics of the eye tracking mechanism. *Math. Biosci.* **20**, 239–265.

Cohen, M. I. (1968) Discharge patterns of brainstem respiratory neurons in relation to carbon dioxide tension. *J. Neurophysiol.* **31**, 142–165.

Cohen, A. H. and Kiemel, T. (1993) Intersegmental coordination: lessons from modeling systems of coupled non-linear oscillators. *Amer. Zool.* **33**, 54–65.

Cohen, A. H., Holmes, P. J., and Rand, R. H. (1982) The nature of the coupling between segmental oscillators of the lamprey spinal generator for locomotion: a mathematical model. *J. Math. Biology* **13**, 345–369.

Cohen, A. H., Rossignol, S., and Grillner, S. (1988) *Neural Control of Rhythmic Movements in Vertebrates.* Wiley, New York.

Cohen, A. H., Ermentrout, G. B., Kiemel, T., Kopell, N., Sigvardt, K. A., and Williams, T. L. (1992) Modelling of intersegmental coordination in the lamprey central pattern generator for locomotion. *TINS* **15**, 434–438.

Cohen, M. A. and Grossberg, S. (1983) Absolute stability of global pattern formation and parallel memory storage by competitive neural networks. *IEEE Trans. Sys. Man Cyber.* **13**, 815–826.

Colonnier, M. (1968) Synaptic patterns on different cell types in the different laminae of the cat visual coretex: an electron microscope study. *Brain Res.* **9**, 268–287.

Connor, J. A., Walter, D., and McKown, R. (1977) Neural repetitive firing: modifications of the Hodgkin–Huxley axon suggested by experimental results from crustacean axons. *Biophys. J.* **18**, 81–102.

Connors, B. W. and Gutnick, M. J. (1990) Intrinsic firing patterns of diverse neocortical neurons. *TINS* **13**, 99–104.

Cooley, J., Dodge, F., and Cohen, H. (1965) Digital computer solutions for excitable membrane models. *J. Cell. Comp. Physiol.* **66**/supp. 2, 99–1.

Cowan, J. D. (1970) A statistical mechanics of nervous activity. In *Some Mathematical Questions in Biology*, 2, Am. Math. Soc. Providence, RI, pp. 1–57.

Cowey, A., Small, M., and Ellis, S. (1994) Left visuo-spatial neglect can be worse in far than in near space. *Neuropsychologia* **32**, 1059–1066.

Cronin, J. (1987) *Mathematical Aspects of Hodgkin–Huxley Neural Theory*. Cambridge University Press, Cambridge.

Crunelli, V. and Leresche, N. (1991) A role for GABAb receptors in excitation and inhibition of thalamocortical cells. *TINS* **14**, 16–21.

Davis, L. and Lorente de Nóó, R. (1947) Contribution to the mathematical theory of the electrotonus. *Studies from the Rockefeller Institute for Medical Research*, **131**, 442–496.

Debellis, R. J., Schiff, N. D., Plum, F., and Victor, J. D. (1998) Nonlinear dynamics in Wilson–Cowan type neural models resemble dynamical features of ictal EEG. Society for Neuroscience Abstracts, **24**, 1213.

Degn, H., Holden, A. V., and Losen, L. F. (1987) *Chaos in Biological Systems*. Plenum, New York.

Dehaene, S. (1997) *The Number Sense*. Oxford University Press, Oxford.

Dehaene, S., Tzourio, N., Frak, V., Raynaud, L., Cohen, L., Mehler, J., and Mazoyer, B. (1996) Cerebral activations during number multiplication and comparison: a PET study. *Neuropsychologia* **34**, 1097–1106.

Delcomyn, F. (1998) *Foundations of Neurobiology*. W. H. Freeman, New York.

Destexhe, A., McCormick, D. A., and Sejnowski, T. J. (1993) A model for 8–10 Hz spindling in interconnected thalamic relay and reticularis neurons. *Biophys. J.* **65**, 2473–2477.

Destexhe, A., Contreras, D., Sejnowski, T. J., and Steriade, M. (1994) A model of spindle rhythmicity in the isolated thalamic reticular nucleus. *J. Neurophysiol.* **72**, 803–818.

Destexhe, A., Contreras, D., Steriade, M., Sejnowski, T. J., and Huguenard, J. R. (1996) *In vivo*, *in vitro*, and computational analysis of dendritic calcium currents in thalamic reticular neurons. *J. Neurosci.* **16**, 169–185.

Diamond, J. (1997) *Guns, Germs, and Steel*. W. W. Norton, New York.

Dirac, P. A. M. (1958) *The Principles of Quantum Mechanics*. Oxford University Press, London.

Donald, M. (1991) *Origins of the Modern Mind*. Harvard University Press, Cambridge.

Douglas, J. K., Wilkens, L., Pantazelou, E., and Moss, F. (1993) Noise enhancement of information transfer in crayfish mechanoreceptors by stochastic resonance. *Nature* **365**, 337–340.

Douglas, R. J. and Martin, K. A. C. (1991) A functional microcircuit for cat visual cortex. *J. Physiol.* **440**, 735–769.

Douglas, R. and Martin, K. (1998) Neocortex. In *The Synaptic Organization of the Brain*, ed. G. M. Shepherd. Oxford University Press, New York, pp. 459–509.

Douglas, R. J., Koch, C., Mahowald, M., Martin, K. A. C., and Suarez, H. H. (1995) Recurrent excitation in neocortical circuits. *Science* **269**, 981–985.

Dowling, J. E. (1987) *The Retina: an Approachable Part of the Brain*. Harvard University Press, Cambridge.

Dowling, J. E. (1991) Retinal neuromodulation: the role of dopamine. *Visual Neurosci.* **7**, 87–97.

Dowling, J. E. (1992) *Neurons and Networks*. Harvard University Press, Cambridge, Mass.

Edelstein-Keshet, L. (1988) *Mathematical Models in Biology*. Random House, New York.

Ekeberg, Ö. (1993) A combined neuronal and mechanical model of fish swimming. *Biol.Cybern.* **69**, 363–374.

Eken, T. and Kiehn, O. (1989) Bistable firing properties of soleus motor units in unrestrained rats. *Acta Physiol. Scand.* **136**, 383–394.

Elman, J. L., Bates, E. A., Johnson, M. H., Karmiloff-Smith, A., Parisi, D., and Plunkett, K. (1996) *Rethinking Innateness: a Connectionist Perspective on Development*. MIT Press, Cambridge.

Ermentrout, B. (1994) An introduction to neural oscillators. In *Neural Modeling and Neural Networks*, ed. F. Ventriglia. Pergamon, Oxford, pp. 79–110.

Ermentrout, B. (1998) Neural networks as spatio-temporal pattern-forming systems. *Rep. Prog. Phys.* **61**, 353–430.

Ermentrout, G. B. and Cowan, J. D. (1979) A mathematical theory of visual hallucination patterns. *Biol. Cybernetics* **34**, 137–150.

Ermentrout, G. B. and Kopell, N. (1991) Multiple pulse interactions and averaging in systems of coupled neural oscillators. *J. Math. Biol.* **29**, 195–217.

Ermentrout, B. and Kopell, N. (1994) Inhibition-produced patterning in chains of coupled nonlinear oscillators. *SIAM J. Appl. Math.* **54**, 478–507.

Everson, R. M. (1987) Quantification of chaos from periodically forced squid axons. In *Chaos in Biological Systems*, ed. H. Degn, A. V. Holden and L. F. Losen, Plenum, New York, pp. 133–142.

Feller, M. B., Wellis, D. P., Stellwagen, D., Werblin, F. S., and Shatz, C. J. (1996) Requirement for cholinergic synaptic transmission in the propagation of spontaneous retinal waves. *Science* **272**, 1182–1187.

Feller, M. B., Butts, D. A., Aaron, H. L., Rokhsar, D. S., and Shatz, C. J. (1997) Dynamic processes shape spatiotemporal properties of retinal waves. *Neuron* **19**, 293–306.

Finkelstein, M. A., Harrison, M., and Hood, D. C. (1990) Sites of sensitivity control within a long wavelength cone pathway. *Vision Res.* **30**, 1145–1158.

Fischer, B. (1973) Overlap of receptive field centers and representation of the visual field in the cat's optic tract. *Vision Res.* **13**, 2113–2120.

FitzHugh, R. (1969) Mathematical models of excitation and propagation in nerve. In *Biological Engineering*, ed. H. P. Schwan. McGraw-Hill, New York, pp. 1–85.

Fitzhugh, R. (1961) Impulses and physiological states in models of nerve membrane. *Biophys. J.* **1**, 445–466.

Foehring, R. C. and Wyler, A. R. (1990) Two patterns of firing in human neocortical neurons. *Neurosci. Letters* **110**, 279–285.

Foehring, R. C., Lorenzon, N. M., Herron, P., and Wilson, C. J. (1991) Correlation of physiologically and morphologically identified neuronal types in human association cortex in vitro. *J. Neurophysiol.* **66**, 1825–1837.

Fox, R. and Herrmann, J. (1967) Stochastic properties of binocular rivalry alternations. *Percept. Psychophys.* **2**, 432–436.

Frost, W. N. and Katz, P.S. (1996) Single neuron control over a complex motor program. *Proc. Natl. Acad. Sci.* **93**, 422–426.

Frost, W. N., Lieb, J. R., Tunstall, M. J., Mensh, B. D., and Katz, P. S. (1997) Integrate-and-fire simulations of two molluscan neural circuits. In *Neurons, Networks, and Motor Behavior*, ed. P. Stein *et al.* MIT Press, Cambridge, MA., pp. 173–179.

Fuster, J. M. (1995) *Memory in the Cerebral Cortex*. MIT Press, Cambridge, MA.

Gaskill, J. D. (1978) *Linear Systems, Fourier Transforms, and Optics*. Wiley, New York.

Georgopoulos, A. P., Schwartz, A. B., and Kettner, R. E. (1986) Neuronal population coding of movement direction. *Science* **233**, 1416–1419.

Georgopoulos, A. P., Taira, M., and Lukashin, A. (1993) Cognitive neurophysiology of the motor cortex. *Science* **260**, 47–52.

Getting, P. A. (1989a) Emerging principles governing the operation of neural networks. *Ann. Rev. Neurosci.* **12**, 185–204.

Getting, P. A. (1989b) Reconstruction of small neural networks. In *Methods in Neuronal Modeling*, ed. C. Koch and I. Segev. MIT Press, Cambridge, MA, pp. 171–194.

Getting, P. A. and Dekin, M. S. (1985) Tritonia swimming: a model system for integration within rhythmic motor systems. In *Model Neural Networks and Behavior*, ed. A. I. Selverston. Plenum, New York, pp. 3–20.

Glass, L. and Mackey, M. C. (1988) *From Clocks to Chaos: the Rhythms of Life*. Princeton University Press, Princeton.

Gloor, P. and Fariello, R. G. (1988) Generalized epilepsy: some of its cellular mechanisms differ from those of focal epilepsy. *TINS* **11**, 63–68.

Goodwin, B. C. (1963) *Temporal Organization in Cells*. Academic Press, New York.

Gradshteyn, I.S. and Ryzhik, I. M. (1980) *Table on Integrals, Series, and Products*. Academic Press, Orlando.

Grafstein, B. (1963) Neuronal release of potassium during spreading depression. In *Brain Function: Cortical Excitability and Steady Potentials*, ed. M. A. Brazier. University of California Press, LA, pp. 87–124.

Gray, J. (1968) *Animal Locomotion*. Weidenfield and Nicolson, London.

Gray, C. M. and McCormick, D. A. (1996) Chattering cells: superficial pyramidal neurons contributing to the generation of synchronous oscillations in the visual cortex. *Science* **274**, 109–113.

Green, D. M. (1976) *An Introduction to Hearing*. Lawrence Erlbaum, Hillsdale, NJ.

Grillner, S. (1996) Neural networks for vertebrate locomotion. *Sci. Am.* **274**, 64–69.

Grillner, S., Buchanan, J. T., and Lansner, A. (1988) Simulation of the segmental burst generating network for locomotion in lamprey. *Neurosci. Letters* **89**, 31–35.

Grillner, S., Deliagina, T., Ekeberg, Ö., El Manira, A., Hill, R. H. *et al.* (1995) Neural networks that coordinate locomotion in lamprey. *TINS* **18**, 270–279.

Grossberg, S. (1987) *The Adaptive Brain*. Elsevier, Amsterdam.

Gutnick, M. J. and Crill, W. E. (1995) The cortical neuron as an electrophysiological unit. In *The Cortical Neuron*, ed. M. J. Gutnick and I. Moody. Oxford University Press, New York, pp. 33–51.

Gutnick, M. J. and Moody, I. (1995) *The Cortical Neuron*. Oxford University Press, New York.

Guttman, R., Lewis, S., and Rinzel, J. (1980) Control or repetitive firing in squid axon membrane as a model for a neuroneoscillator. *J. Physiol.* **305**, 377–395.

Hadamard, J. (1945) *The Psychology of Invention in the Mathematical Field*. Princeton University Press, Princeton.

Haken, H. (1996) *Principles of Brain Functioning*. Springer, Berlin.

Harris-Warrick, R. M. and Cohen, A. H. (1985) Serotonin modulates the central pattern generator for locomotion in the isolated lamprey spinal cord. *J. Exp. Biol.* **116**, 27–46.

Hebb, D. O. (1949) *The Organization of Behavior*. Wiley, New York.

Heeger, D. J. (1992) Normalization of cell responses in cat striate cortex. *Visual Neurosci.* **9**, 181–197.

Hellgren, J., Grillner, S., and Lansner, A. (1992) Computer simulation of the segmental neural network generating locomotion in lamprey by using populations of network interneurons. *Biol. Cybern.* **68**, 1–13.

Helmholtz, H. von (1909) *Physiological Optics*. Dover, New York. (1962 English translation by J. P. C. Southall from the 3rd German edition of Handbuch der Physiologischen Optik, Vos, Hamburg.)

Hertz, A., Sulzer, B., Küühn, R., and van Hemmen, J. L. (1989) Hebbian learning reconsidered: representation of static and dynamic objects in associative neural nets. *Biol. Cybern.* **60**, 457–467.

Hertz, J., Krogh, A., and Palmer, R. G. (1991) *Introduction to the Theory of Neural Computation*. Addison-Wesley, New York.

Hille, B. (1992) *Ionic Channels of Excitable Membranes*. Sinauer, Sunderland, MA.

Hindmarsh, J. L. and Rose, R. M. (1982) A model of the nerve impulse using two first-order differential equations. *Nature* **296**, 162–164.

Hindmarsh, J. L. and Rose, R. M. (1984) A model of neuronal bursting using three coupled first order differential equations. *Proc. R. Soc. Lond. B* **221**, 87–102.

Hines, M. (1989) A program for simulation of nerve equations with branching geometries. *Int. J. Biomed. Computing* **24**, 55–68.

Hines, M. (1993) NEURON—a program for simulation of nerve equations. In *Neural Systems: Analysis and Modeling*, ed. F. H. Eeckman. Kluwer Academic, Boston, pp. 127–136.

Hodgkin, A. L. and Huxley, A. F. (1952) A quantitative description of membrane current and its application to conduction and excitation in nerve. *J. Physiol.* **117**, 500–544.

Hodgkin, A. L. and Rushton, W. A. H. (1946) The electrical constants of a crustacean nerve fibre. *Proc. R. Soc. Lond. B* **133**, 444–479.

Hood, D. C. (1998) Lower level visual processing and models of light adaptation. *Ann. Rev. Psychol.* **49**, 503–535.

Hood, D. C. and Birch, D. G. (1990) A quatitative measure of the electrical activity of human rod photoreceptors using electroretinography. *Vis. Neurosci.* **5**, 379–387.

Hopfield, J. J. (1982) Neural networks and physical systems with emergent collective computational abilities. *Proc. Natl. Acad. Sci. USA* **79**, 2554–2558.

Hopfield, J. J. (1984) Neurons with graded response have collective computational properties like those of two-state neurons. *Proc. Natl. Acad. Sci. USA* **81**, 3088–3092.

Hoppensteadt, F. C. and Izhikevich, E. M. (1997) *Weakly Connected Neural Networks*. Springer, New York.

Horton, J. C. and Hoyt, W. F. (1991) The representation of the visual field in human striate cortex: a revision of the classic Holmes map. *Arch. Ophthalmol.* **109**, 816–824.

Hounsgaard, J., Hultborn, H., Jespersen, B., and Kiehn, O. (1988) Bistability of alpha-motoneurones in the decerebrate cat and in the acute spinal cat after intravenous 5-hydroxytryptophan. *J. Physiol.* **405**, 345–367.

Huettner, J. E. and Baughman, R. W. (1988) The pharmacology of synapses formed by identified corticocollicular neurons in primary cultures of rat visual cortex. *J. Neurosci.* **8**, 160–175.

Humphries, D. A. and Driver, P. M. (1970) Protean defense by prey animals. *Oecologia* **5**, 285–302.

Jirsa, V. K. and Haken, H. (1997) A derivation of a macroscopic field theory of the brain from the quasi-microscopic neural dynamics. *Physica D* **99**, 503–516.

Johnston, D. and Wu, S. M. (1995) *Foundations of Cellular Neurophysiology*. MIT Press, Cambridge, MA.

Johnston, D., Magee, J. C., Colbert, C. M., and Christie, B. R. (1996) Active properties of neuronal dendrites. *Ann. Rev. Neurosci.* **19**, 165–186.

Jones, E. G. (1995) Overview: basic elements of the cortical network. In *The Cortical Neuron*, ed. M. J. Gutnick and I. Moody. Oxford University Press, New York, pp. 111–122.

Kadanoff, L. P. (1993) *From Order to Chaos*. World Scientific, New Jersey.

Kaplan, D. and Glass, L. (1995) *Understanding Nonlinear Dynamics*. Springer, New York.

Kelly, D. H. and Martinez-Uriegas, E. (1993) Measurements of chromatic and achromatic afterimages. *J. Opt. Soc. Am. A* **10**, 29–37.

Kelso, J. A. S. (1995) *Dynamic Patterns: the Self Organization of Brain and Behavior*. MIT Press, Cambridge, MA.

Kepler, T. B., Abbott, L. F., and Marder, E. (1992) Reduction of conductance-based neuron models. *Biol. Cybern.* **66**, 381–387.

Kiang, N., Watanabe, T., Thomas, E. C., and Clarke, L. F. (1965) *Discharge Patterns of Single Fibers in the Cat's Auditory Nerve*. MIT Press, Cambridge, MA.

Kiehn, O. (1991) Plateau potentials and active integration in the 'final common pathway' for motor behavior. *TINS* **14**, 68–73.

Kim, J. and Wilson, H. R. (1993) Dependence of plaid motion coherence on component grating directions. *Vision Res.* **33**, 2479–2489.

Kim, J. and Wilson, H. R. (1996) Direction repulsion between components in motion transparency. *Vision Res.* **36**, 1177–1187.

Kim, U., Bal, T., and McCormick, D. A. (1995) Spindle waves are propagating synchronized oscillations in the ferret LGNd *in vitro*. *J. Neurophysiol.* **74**, 1301–1323.

Kim, U., Sanchez-Vives, M. V., and McCormick, D. A. (1997) Functional dynamics of GAGAergic inhibition in the thalamus. *Science* **278**, 130–134.

Kleinfeld, D. (1986) Sequential state generation by model neural networks. *Proc. Natl. Acad. Sci. USA* **83**, 9469–9473.

Koch, C. and Segev, I. (1989) *Methods in Neuronal Modeling*. MIT Press, Cambridge, MA.

Kohonen, T. (1989) *Self Organization and Associative Memory*, 3rd edn. Springer Berlin.

Kopell, N. (1988) Toward a theory of modelling central pattern generators. In *Neural Control of Rhythmic Movements in Vertebrates*, ed. A. Cohen, S. Rossignol and S. Grillner. Wiley, New York, pp. 369–413.

Kopell, N. and Ermentrout, G. B (1988) Coupled oscillators and the design of central pattern generators. *Math. Biosci.* **90**, 87–109.

Lansner, A., Ekeberg, ÖÖ., and Grillner, S. (1997) Realistic modeling of burst generation and swimming in lamprey. In *Neurons, Networks, and Motor Behavior*, ed. P. Stein *et al*. MIT Press, Cambridge, MA, pp. 165–171.

La Salle, J. and Lefschetz, S. (1961) *Stability by Liapunov's Direct Method*. Academic Press, New York.

Lewis, J. E. and Kristan, W. B. (1998) A neuronal network for computing population vectors in the leech. *Nature* **391**, 76–79.

Li, Y. X., Bertram, B., and Rinzel, J. (1996) Modeling N-Methyl-D-aspartate-induced bursting in dopamine neurons. *Neurosci.* **71**, 397–410.

Lighthill, M. J. (1958) *Introduction to Fourier Analysis and Generalized Functions*. Cambridge University Press, Cambridge, UK.

Longtin, A. (1993) Stochastic resonance in neuron models. *J. Stat. Physics* **70**, 309–327.

Longtin, A. (1995) Synchronization of the stochastic Fitzhugh–Nagumo equations to periodic forcing. *Il Nuovo Cimento* **17**, 835–846.

Lorenz, E. N. (1963) Deterministic nonperiodic flow. *J. Atmospheric Sci.* **20**, 130–141.

Lorenz, E. (1993) *The Essence of Chaos*. University of Washington Press, Seattle.

Lorenzon, N. M. and Foehring, R. C. (1992) Relationship between repetitive firing and afterhyperpolarizations in human neocortical neurons. *J. Neurophysiol.* **67**, 350–363.

Lotka, A. J. (1924) *Elements of Mathematical Biology.* Williams and Wilkins.

Lukashin, A. V., Amirikian, B. R., Mozhaev, V. L., Wilcox, G. L., and Georgopoulos, A. P. (1996) Modeling motor cortical operations by an attractor network of stochastic neurons. *Biol. Cybern.* **74**, 255–261.

Lyapunov, M. A. (1892) Problème général de la stabilité du mouvement. Reproduced in 1947 as Annals of Mathematics Study 17, Princeton University Press, Princeton.

Lytton, W. W. and Sejnowski, T. J. (1991) Simulations of cortical pyramidal neurons synchronized by inhibitory interneurons. *J. Neurophysiol.* **66**, 1059–1079.

MacDonald, N. (1989) *Biological Delay Systems: Linear Stability Theory.* Cambridge University Press, Cambridge.

MacKey, M. C. and Milton, J. G. (1987) Dynamical diseases. *Ann. New York Acad. Sci.* **504**, 16–32.

Magee, J. C. and Johnston, D. (1995) Synaptic activation of voltage-gated channels in the dendrites of hippocampal pyramidal neurons. *Science* **268**, 301–304.

Magee, J. C. and Johnston, D. (1997) A synaptically controlled associative signal for Hebbian plasticity in hippocampal neurons. *Science* **275**, 209–213.

Mainen, Z. F. and Sejnowski, T. J. (1996) Influence of dendritic structure on firing pattern in model neocortical neurons. *Nature* **382**, 363–366.

Mainen, Z. F., Joerges, J., Huguenard, J. R., and Sejnowski, T. J. (1995) A model for spike initiation in neocortical pyramidal neurons. *Neuron* **15**, 1427–1439.

Mareschal, I. and Baker, C. L. (1998) A cortical locus for the processing of contrast-defined contours. *Nature Neurosci.* **1**, 150–154.

Marr, D. (1971) A theory for archicortex. *Phil. Trans. R. Soc. Lond. B* **262**, 23–81.

Marshak, W. and Sekuler, R. (1979) Mutual repulsion between moving visual targets. *Science* **205**, 1399–1401.

Mather, G. and Moulden, B. (1980) A simultaneous shift in apparent direction: further evidence for a 'distribution shift' model of direction coding. *Qrtr. J. Exp. Psych.* **32**, 325–333.

Mathieu, P. A. and Roberge, F. A. (1971) Characteristics of pacemaker oscillations in aplysia neurons. *Can. J. Physiol. Pharmacol.* **49**, 787–795.

McCormick, D. A. (1998) Membrane properties and neurotransmitter actions. In *The Synaptic Organization of the Brain*, ed. G. M. Shepherd. Oxford University Press, New York, pp. 37–75.

McCormick, D. A. (1989) GABA as an inhibitory neurotransmitter in human cerebral cortex. *J. Neurophysiol.* **62**, 1018–1027.

McCormick, D. A. and Williamson, A. (1989) Convergence and divergence of neurotransmitter action in human cerebral cortex. *Proc. Natl. Acad. Sci. USA* **86**, 8098–8102.

McCormick, D. A., Connors, B. W., Lighthall, J. W., and Prince, D. A. (1985) Comparative electrophysiology of pyramidal and sparsely spiny stellate neurons of the neocortex. *J. Neurophysiol.* **54**, 782–806.

McCulloch, W. S. and Pitts, W. (1943) A logical calculus of ideas immanent in nervous activity. *Bull. Math. Biophys.* **5**, 115–133.

McNaughton, B. L., Barnes, C. A., Gerrard, J. L., Gothard, K., Jung, M. W. *et al.* (1996) Deciphering the hippocampal polyglot: the hippocampus as a path integration system. *J. Exp. Biol.* **199**, 173–185.

McNeill, D. (1992) *Hand and Mind: What Gestures Reveal about Thought.* University of Chicago Press, Chicago.

Miller, G. F. (1997) Protean primates: the evolution of adaptive unpredictability in competition and courtship. In *Machiavellian Intelligence II: Extensions and Evaluations* (ed. A. Whiten and R. W. Byrne), Cambridge University Press, Cambridge, 312–340.

Milner, A. D. and Goodale, M. A. (1995) *The Visual Brain in Action.* Oxford University Press, Oxford.

Milton, J. (1996) *Dynamics of Small Neural Populations.* American Mathematical Society, Providence, Rhode Island.

Morris, C. and Lecar, H. (1981) Voltage oscillations in the barnacle giant muscle fiber. *Biophys. J.* **35**, 193–213.

Movshon, J. A., Thompson, I. D., and Tolhurst, D. J. (1978) Spatial summation in the receptive fields of simple cells in the cat's striate cortex. *J. Physiol.* **283**, 53–77.

Murray, J. D. (1989) *Mathematical Biology.* Springer, Heidelberg.

Nagumo, J. S., Arimoto, S., and Yoshizawa, S. (1962) An active pulse transmission line simulating a nerve axon. *Proc. IRE* **50**, 2061–2070.

Naka, K. I. and Rushton, W. A. (1966) S-potentials from colour units in the retina of fish. *J. Physiol.* **185**, 584–599.

O'Keefe, J. and Nadel, L. (1978) *The Hippocampus as a Cognitive Map.* Clarendon Press, Oxford.

Partridge, L. D. and Swandulla, D. (1988) Calcium activated non-specific cation channels. *TINS* **11**, 69–72.

Pearson, K. G. (1993) Common principles of motor control in vertebrates and invertebrates. *Ann. Rev. Neurosci.* **16**, 265–297.

Perkel, D. H. and Mulloney, B. (1974) Motor pattern production in reciprocally inhibitory neurons exhibiting postinhibitory rebound. *Science* **185**, 181–183.

Pinsky, P. F. and Rinzel, J. (1994) Intrinsic and network rhythmogenesis in a reduced Traub model for CA3 neurons. *J. Comput. Neurosci.* **1**, 39–60.

Plant, R. E. (1981) Bifurcation and resonance in a model for bursting nerve cells. *J. Math. Bio.* **11**, 15–32.

Press, W. H., Flannery, B. P., Teukolsky, S. A., and Vetterling, W. T. (1986) *Numerical Recipes: The Art of Scientific Computing.* Cambridge University Press, Cambridge.

Rack, P. H. M. and Westbury, D. R. (1969) The effect of length and stimulus rate on the tension in the isometric cat soleus muscle. *J. Physiol.* **204**, 443–460.

Rall, W. (1962) Theory of physiological properties of dendrites. *Ann. NY. Acad. Sci.* **96**, 1071–1092.

Rall, W. (1967) Distinguishing theoretical synaptic potentials computed for different soma-dendritic distributions of synaptic inputs. *J. Neurophys.* **30**, 1138–1168.

Rall, W. (1989) Cable theory for dendritic neurons. In *Methods in Neuronal Modeling: from Synapses to Networks,* ed. C. Koch and I. Segev. MIT Press, Cambridge, pp. 9–62.

Rand, R. H., Cohen, A. H., and Holmes, P. J. (1988) Systems of coupled oscillators as models of central pattern generators. In *Neural Control of Rhythmic Movements in Vertebrates,* ed. A. Cohen, S. Rossignol and S. Grillner. Wiley, New York, pp. 333–367.

Rapp, P. E., Zimmerman, I. D., Albano, A. M., Deguzman, G. C., and Greenbaun, N. N. (1985) Dynamics of spontaneous neural activity in the simian motor cortex: the dimension of chaotic neurons. *Physics Letters* **110A**, 335–338.

Richards, D. (1990) Is strategic decision making chaotic? *Behavioral Sci.* **35**, 219–232.

Richards, W., Wilson, H. R., and Sommer, M. A. (1994) Chaos in percepts? *Biol. Cybern.* **70**, 345–349.

Rinzel, J. (1978) On repetitive activity in nerve. *Fed. Proc.* **37**, 2793–2802.

Rinzel, J. (1985) Excitation dynamics: insights from simplified membrane models. *Fed. Proc.* **44**, 2944–2946.

Rinzel, J. (1987) A formal classification of bursting mechanisms in excitable systems. In *Mathematical Topics in Population Biology, Morphogenesis, and Neurosciences,* ed. E. Teramoto and M. Yamaguti. Springer, Berlin, pp. 267–281.

Rinzel, J. and Ermentrout, G. B. (1989) Analysis of neural excitability and oscillations. In *Methods in Neuronal Modeling: from Synapses to Networks*, ed. C. Koch and I. Segev. MIT Press, Cambridge, pp. 135–169.

Rinzel, J. and Lee, Y. S. (1987) Dissection of a model for neuronal parabolic bursting. *J. Math. Biol.* **25**, 653–675.

Rinzel, J., Terman, D., Wang, X. J., and Ermentrout, B. (1998) Propagating activity patterns in large scale inhibitory neuronal networks. *Science* **279**, 1351–1355.

Rodiek, R. W. (1998) *The First Steps in Seeing*. Sinauer Associates, Sunderland, MA.

Rogawski, M. A. (1985) The A-current: how ubiquitous a feature of excitable cells is it? *TINS* **8**, 214–219.

Rolls, E. T. and Treves, A. (1998) *Neural Networks and Brain Function*. Oxford University Press, Oxford.

Rose, J. E., Brugge, J. F., Anderson, D. J., and Hind, J. E. (1967) Phase-locked response to low-frequency tones in single auditory nerve fibers of the squirrel monkey. *J. Neurophys.* **30**, 769–793.

Rose, R. M. and Hindmarsh, J. L. (1989) The assembly of ionic currents in a thalamic neuron I. The three-dimensional model. *Proc. R. Soc. Lond. B* **237**, 267–288.

Rothwell, J. (1994) *Control of Human Voluntary Movement*. Chapman and Hall, London.

Sakata, H. and Kusonoki, M. (1992) Organization of space perception: neural representation of three-dimensional space in the posterior parietal coretex. *Curr. Opin. Neurobiol.* **2**, 170–174.

Sakata, H., Taira, M., Kusunoki, M., Murata, A., and Tanaka, Y. (1997) The parietal association cortex in depth perception and visual control of hand action. *TINS* **20**, 350–357.

Salzman, C. D. and Newsome, W. T. (1994) Neural mechanisms for forming a perceptual decision. *Science* **264**, 231–237.

Samsonovich, A. and McNaughton, B. L. (1997) Path integration and cognitive mapping in a continuous attractor neural network model. *J. Neurosci.* **17**, 5900–5920.

Satterlie, R. A. (1985) Reciprocal inhibition and postinhibitory rebound produce reverberation in a locomotor pattern generator. *Science* **229**, 402–404.

Scheibel, M. E. and Scheibel, A. B. (1970) Elementary processes in selected thalamic and cortical subsystems: the structural substrates. In *The Neurosciences: Second Study Program*, ed. F. O. Schmitt. Rockefeller University Press, New York, pp. 443–457.

Schnapf, J. L., Nunn, B. J., Meister, M., and Baylor, D. A. (1990) Visual transduction in cones of the monkey *macaca fascicularis. J. Physiol.* **427**, 681–713.

Schwartz, E. L. (1980) Computational anatomy and functional architecture of striate cortex: a spatial mapping approach to perceptual coding. *Vision Res.* **20**, 645–669.

Sclar, G., Maunsell, J. H. R., and Lennie, P. (1990) Coding of image contrast in central visual pathways of the macaque monkey. *Vision Res.* **30**, 1–10.

Segev, I., Fleshman, J. W., and Burke, R. E. (1989) Compartmental models of complex neurons. In *Methods in Neuronal Modeling: from Synapses to Networks*, ed. C. Koch and I. Segev. MIT Press, Cambridge, pp. 63–96.

Selverston, A. I. (1985) *Model Neural Networks and Behavior*. Plenum, New York, pp. 3–20.

Shapley, R. and Enroth-Cugell, C. (1984) Visual adaptation and retinal gain controls. *Progress in Retinal Research* **3**, 263–343.

Shepherd, G. M. (1994) *Neurobiology*. Oxford University Press, Oxford.

Shik, M. L., Severin, F. V., and Orlovsky, G. N. (1966) Control of walking and running by means of electrical stimulation of the midbrain. *BioPhysics* **11**, 756–765.

Sholl, D. A. (1956) *The Organization of the Cerebral Cortex*. Methuen, London.

Siegel, R. K. (1977) Hallucinations. *Sci. Am.* **237**, October, 132–137.

Silva, L. R., Amitai, Y., and Connors, B. W. (1991) Intrinsic oscillations on neocortex generated by layer 5 pyramidal neurons. *Science* **251**, 432–435.

Skinner, F. K., Kopell, N., and Marder, E. (1994) Mechanisms for oscillation and frequency control in reciprocally inhibitory nodel neural networks. *J. Comput. Neurosci.* **1**, 69–87.

Smith, A. T. (1994) The detection of second-order motion. In *Visual Detection of Motion*, ed. A. T. Smith and R. J. Snowden. Academic Press, London, pp. 145–176

Somers, D. C., Todorev, E. V., Siapas, A. G., Toth, L. J., Kim, D. S., and Sur, M. (1998) A local circuit approach to understanding integration of long range inputs in primary visual cortex. *Cereb. Cort.* **8**, 204–217.

Sompolinsky, H. and Kanter, I. (1986) Temporal association in asymmetric neural networks. *Phys. Rev. Letters* **57**, 2861–2864.

Squire, L. R. and Zola-Morgan, S. (1991) The medial temporal lobe memory system. *Science* **253**, 1380–1386.

Stein, P. S. G., Grillner, S., Selverston, A. I., and Stuart, D. G. (1997) *Neurons, Networks, and Motor Behavior*. MIT Press, Cambridge, MA.

Steriade, M., McCormick, D. A., and Sejnowski, T. J. (1993) Thalamocortical oscillations in the sleeping and aroused brain. *Science* **262**, 679–685.

Stuart, G. J. and Sakmann, B. (1994) Active propagation of somatic action potentials into neocortical pyramidal cell dendrites. *Nature* **367**, 69–72.

Stuart, G., Spruston, N., Sakmann, B., and Häusser, M. (1997) Action potential initiation and backpropagation in neurons of the mammalian CNS. *TINS* **20**, 125–131.

Szentágothai, J. (1967) The anatomy of complex integrative units in the nervous system. In *Recent Development of Neurobiology in Hungary*, Vol. 1, ed. K. Lissáák. Akadémiai Kiadóó, Budapest, pp. 9–45.

Theunissen, F. and Miller, J. P. (1991) Representation of sensory information in the cricket cercal sensory system. II. Information theoretic calculation of system accuracy and optimal tuning curve widths of four primary interneurons. *J. Neurophysiol.* **66**, 1690–1703.

Touretzky, D. S., Redish, A. D., and Wan, H. S. (1993) Neural representation of space using sinusoidal arrays. *Neural Comp.* **5**, 869–884.

Traub, R. D., Jefferys, J. G. R., Miles, R., Whittington, M. A., and Tóth, K. (1994) A branching dendritic model of a rodent CA3 pyramidal neurone. *J. Physiol.* **481**, 79–95.

Traub, R. D., Whittington, M. A., Stanford, I. M., and Jefferys, J. G. R. (1996) A mechanism for generation of long-range synchronous fast oscillations in the cortex. *Nature* **383**, 621–624.

Treisman, A. (1982) Perceptual grouping and attention in visual search for features and for objects. *J. Exp. Psychol. Hum. Percept. and Perform.* **8**, 194–214.

Treisman, A. and Gelade, G. (1980) A feature integration theory of attention. *Cognitive Psych.* **12**, 97–136.

van der Pol, B. (1926) On relaxation oscillations. *Phil. Mag.* **2**, 978–992.

Van Vreeswijk, C. and Sompolinsky, H. (1996) Chaos in neuronal networks with balanced excitatory and inhibitory activity. *Science* **274**, 1724–1726.

Van Vreeswijk, C., Abbott, L. M., and Ermentrout, G. B. (1994) When inhibition not excitation synchronizes neural firing. *J. Comput. Neurosci.* **1**, 313–321.

Verghese, P. and Nakayama, K. (1994) Stimulus discriminability in visual search. *Vision Res.* **34**, 2453–2467.

Wallén, P., Ekeberg, ÖÖ., Lansner, A., Brodin, L., Tråvén, H., and Grillner, S. (1992) A computer-based model for realistic simulations of neural networks. II. The segmental network generating locomotor rhythmicity in the lamprey. *J. Neurophys.* **68**, 1939–1950.

Wang, X. J. and Rinzel, J. (1992) Alternating and synchronous rhythms in reciprocally inhibitory model neurons. *Neural Comput.* **4**, 84–97.

Wang, X. J. and Rinzel, J. (1993) Spindle rhythmicity in the reticularis thalami nucleus: synchronization among mutually inhibitory neurons. *Neuroscience* **53**, 899–904.

Wäässle, H. and Boycott, B. B. (1991) Functional architecture of the mammalian retina. *Physiol. Rev.* **71**, 447–480.

Whishaw, I. Q., McKenna, J. E., and Maaswinkel, H. (1997) Hippocampal lesions and path integration. *Curr. Opin. Neurobiol.* **7**, 228–234.

Willems, J. L. (1970) *Stability Theory of Dynamical Systems*. Nelson, London.

Williams, T. L. (1992) Phase coupling by synaptic spread in chains of coupled neuronal oscillators. *Science* **258**, 662–665.

Willshaw, D. J., Buneman, O. P., and Longuet-Higgins, H. C. (1969) Non-holographic associative memory. *Nature* **222**, 960–996.

Wilson, H. R. (1994a) The role of second-order motion signals in coherence and transparency. In *Higher Order Processing in the Visual System*, ed. J. Goode. CIBA Foundation, Wiley, New York, pp. 227–244.

Wilson, H. R. (1994b) Models of two-dimensional motion perception. In *Visual Detection of Motion*, ed. A. T. Smith and R. J. Snowden. Academic Press, London, pp. 219–251.

Wilson, H. R. (1997) A neural model of foveal light adaptation and afterimage formation. *Visual Neurosci.* **14**, 403–423.

Wilson, H. R. (1999) Essential dynamics of human and mammalian neocortical neurons. Submitted for publication.

Wilson, H. R. and Cowan, J. D. (1972) Excitatory and inhibitory interactions in localized populations of model neurons. *Biophysical Journal* **12**, 1–24.

Wilson, H. R. and Cowan, J. D. (1973) A mathematical theory of the functional dynamics of cortical and thalamic nervous tissue. *Kybernetik* **13**, 55–80.

Wilson, H. R. and Humanski, R. (1993) Spatial frequency adaptation and contrast gain control. *Vision Res.* **33**, 1133–1149.

Wilson, H. R. and Kim, J. (1994) A model for motion coherence and transparency. *Visual Neurosci.* **11**, 1205–1220.

Wilson, H. R., Krupa, B. and Wilkinson, F. (1999) Dynamics of scintillation in form vision. Submitted for publication.

Wilson, H. R., Levi, D., Maffei, L., Rovamo, J., and DeValois, R. L. (1990) The perception of form: retina to striate cortex. In *The Neurophysiological Foundations of Visual Perception*, ed. L. Spillmann and J. S. Werner. Academic Press, New York, pp. 231–272.

Wilson, H. R., Ferrera, V.P., and Yo, C. (1992) Psychophysically motivated model for two-dimensional motion perception. *Visual Neurosci.* **9**, 79–97.

Witkovsky, P. and Schütte, M. (1991) The organization of dopaminergic neurons in vertebrate retinas. *Visual Neurosci.* **7**, 113–124.

Index

abstraction, levels of 3–5, 221, 281
action potential 10, 261, 263, 267, 268, 270
adaptation 81, 131–34
    spike frequency 156–59, 209, 218
afterdepolarization 163, 168;
        see also postinhibitory rebound
afterhyperpolarization 156–9, 208, 219
afterimages, negative 102–3
alpha function, synaptic 191–2, 197
amplification, cortical 110
anode break excitation 194
antiphase locking 190, 193, 210, 211, 220
asymptotic stability 37, 117, 227, 229, 246
auditory neurons 145
autoassociative network 239, 247, 248
autonomous system 11
axon 10, 237, 247

bifurcation 80–1, 144, 160, 181, 248, 283
    saddle-node to limit cycle 148, 149, 156,
        158, 169, 210
    see also Hopf bifurcation
binocular rivalry 133
boundary conditions 254, 257–8
bursting
    endogenous 164
    neural 161–9
    parabolic 165–6
    taxonomy of 170–1

CA3, *see* hippocampus
cable equation 251, 253, 259;
        see also diffusion equation
capacity, memory 240
cascade of equations 17–19
catastrophe and bifurcation 84
categorical decisions 97, 285
center 38–40, 235
central pattern generator 205, 220–1
chain rule 63
chaos 173–80, 183–4, 282, 283
characteristic equation 29, 32, 49, 50, 51
Clione 193–5, 220–1
Cohen–Grossberg theorem 244–6
command signal 205

compartmental models
    as 'virtual neurons' 275, 281
    of neurons and dendrites 268–71, 281
    two compartments 272–4, 276, 281
conductance 22
conic sections 285
conservation of energy 232, 235
conservative oscillation 235
conservative systems 232–6
constant of motion 234, 235
convolution 15, 43, 45
cortical network model 103–12
cosine weighting 94, 98
creativity 184
critical damping 40–3

decisions 85, 283
déja vu 239
delay, synaptic 188, 189
dendrite 9, 251, 254–6, 259–61
dendritic spikes 268–71
determinism 184, 283
development, retinal 112
diffusion equation 251–61
diffusion-reaction equation 262–3, 266–8, 282
Dirac δ function 58, 260
direction repulsion in motion 97
domain of attraction (or asymptotic
        stability) 227–9, 243, 245
dynamical diseases 57

eigenvalues 32–3, 49, 226, 288
eigenvectors 33–4
electroretinogram 18
epilepsy 112, 202
equilibrium point 22, 31, 73, 227, 263;
        see also steady state
equilibrium potential 22, 191
equivalent cylinder model of dendrites 254, 259
errors in simulations 65–7
Euler's formula 8
Euler's method 60–2, 65
evolution 283–4
excitation, recurrent, *see* positive feedback
exponential function 5–6, 13, 280
eye movements, saccadic 42–3